Mathematical Analysis for Management Decisions:

Introduction to Calculus and Linear Algebra

Mathematical Analysis for Management Decisions:

Introduction to Calculus and Linear Algebra

A. K. McAdams
Cornell University

with the assistance of
Christian T. Janssen
University of Alberta

THE MACMILLAN COMPANY
COLLIER-MACMILLAN LIMITED, LONDON

© Copyright, The Macmillan Company, 1970

All rights reserved. No part of this book may be reproduced or transmitted in any form or by any means, electronic or mechanical, including photocopying, recording or by any information storage and retrieval system, without permission in writing from the Publisher.

First Printing

Library of Congress catalog card number: 69–14173

THE MACMILLAN COMPANY
866 Third Avenue, New York, New York 10022
COLLIER-MACMILLAN CANADA, LTD., TORONTO, ONTARIO

Printed in the United States of America

PREFACE

The materials in this book have been used in draft form at the Graduate School of Business and Public Administration of Cornell University over the past several years. The book grew out of our need for materials designed for students who come to us from diverse backgrounds and whose interests are in practical applications of calculus and linear algebra, both in later course work and directly in the fields of business, public, and health care administration. Emphasis has been placed on intuitive appreciation of the various topics presented rather than on formal mathematical development, and coverage has been limited to functions that are considered most useful.

The book begins with basic concepts of set theory. These are used in developing the concepts of limits, continuity, and differentiation for selected functions of a single variable. The differential calculus is then used in the identification of maximum and minimum values of such functions. In Chapter 4 basic managerial and economic applications of the differential calculus are presented.

Selected functions of more than one variable are introduced in Chapter 5. The procedure for differentiating such functions—partial differentiation—is related back to that for functions of a single variable. Set theory again proves useful for a direct statement of this relationship. The use of partial differentiation for finding maximum and minimum values is explored and then the LaGrangian approach to finding maxima and minima subject to one or more constraints is emphasized. Considerable space is devoted to the interpretation of this process, especially the meaning of λ_i, the LaGrangian multipliers. Further useful applications are presented in the appendix to Chapter 5.

In Chapter 6 the concept of inverse functions is introduced as the basis for developing logarithmic functions from exponential functions, their more familiar inverses. In Chapter 7 the concept of inverse relationships is extended by the fundamental theorem of the calculus to introduce integration, the *process* that is inverse to differentiation. Basic concepts of integral calculus are developed and applied in this chapter.

Generally, concepts are first introduced, followed by examples to illustrate the mechanics of the various processes and then applications. This sequence of presentation is reversed in Chapters 8 and 9. In Chapter 8 an application of linear algebra—linear programming—is introduced along with the simplex algorithm, both on an informal basis in the context of a physical problem. The dual to the linear program and its economic interpretation are stressed in the tradition of

Baumol and others. The presentation is extended through sensitivity analysis and parametric programming, both useful extensions for the manager. In Chapter 9 the mechanics of the simplex procedure are shown to be matrix row and column operations and, as applied to the illustrative problem of Chapter 8, they lead to the development of the inverse to the matrix of coefficients for the problem. This reversal of the order of presentation is intended to motivate the study of linear algebra by establishing its usefulness and power in advance.

Many persons have contributed to the development of this book. Christian Janssen assisted greatly in the preparation of the manuscript, originally as research assistant, then as instructor, and finally as colleague. He has been extremely helpful in the development of problems, illustrations, and applications for the book. This is especially true for Chapters 5, 6, and 9. In the last chapter he also originated drafts of a significant portion of material. His contributions are warmly appreciated and are recognized on the title page.

Other persons who have contributed to the development of this volume include my colleagues Thomas Dyckman, Seymour Smidt, Joseph Thomas, and James Kinard. All have used the materials with classes and have made many useful suggestions. The set chapter was originally drafted by Thomas Dyckman and appears in substantially the same form in our volume, *Decision Making Under Uncertainty*. Cornell graduate students Larry Hastie, Roberto Saloman, and Ted Steele have made significant contributions to the development of the problem materials and solutions for the book.

Two colleagues from Carnegie-Mellon University, Barbara Schwartz and Morton Kamien, also offered useful suggestions. Professor Schwartz's contributions were made at the early stages of the project and Professor Kamien's during the later stages. Both were most helpful.

The arduous task of preparing the manuscript through its various drafts fell to Marge Snedden, assisted by Jean Tubbs, Judy Wickam, and Karen Hemecek, among others. Their patience and achievements were invaluable. I am deeply indebted to them and to the students who used and commented on the materials.

The errors that remain in the book are my responsibility.

The book is dedicated to Ann and our boys.

Ithaca, N. Y.   A. K. M.

CONTENTS

1. **Sets** 1

    1.1 Concept of a Set 2
    1.2 Special Sets 5
    1.3 Set Operations 9
    1.4 Counting 10
    1.5 Functions 15
    1.6 Point Sets 19
    1.7 Summary 23

    Appendix 1A: The X Element Technique 25

    Appendix 1B: Number Sets 26

    Appendix 1C: The Distinction Between a Function and an Equation 26

2. **Limits and Continuity** 35

    2.1 Limits 36
    2.2 Sequences 37
    2.3 The Limit of a Function 43
    2.4 Theorems on Limits 49
    2.5 Continuity 52
    2.6 Summary 54

3. **The Derivative, Maxima and Minima** 59

    3.1 General Discussion 60
    3.2 The Derivative 61
    3.3 Differentiation 66
    3.4 General Differentiation Rules 70
    3.5 Higher Order Derivatives 78
    3.6 Critical Points 79
    3.7 Maxima and Minima 82

3.8 Inflection Points   87
3.9 Summary   89

Appendix 3A: Proofs of Rules 3.4, 3.5, and 3.6 for Derivatives   92

4. Applications of the Derivative   99

4.1 Problems in Inventory Control: Optimal Reorder Quantity   100
4.2 Finding the Minimum Cost Inventory Policy   101
4.3 An Inventory Model   102
4.4 Problem of Profit Maximizing: Quantity to Order   105
4.5 Marginal Cost and Marginal Revenue   108
4.6 Summary   111

5. Partial Differentiation   115

5.1 Partial Differentiation   116
5.2 Maxima and Minima, Necessary Conditions   123
5.3 Sufficient Conditions   125
5.4 Implicit Differentiation   131
5.5 Constrained Maxima and Minima   133
5.6 More than One Constraint   139
5.7 Method of LaGrange Multipliers   144
5.8 The Meaning of λ 146
5.9 Summary   153

Appendix 5A: Applications to the Production Function and to Simple Linear Regression   155

6. Exponential and Logarithmic Functions   165

6.1 Inverse Functions   166
6.2 Exponential Functions   170
6.3 The Irrational Number e 173
6.4 Rules for the Derivative Functions of Exponential Functions   176
6.5 Logarithmic Functions   178

- 6.6 Rules for Derivatives of Logarithmic Functions 183
- 6.7 Applications to Business and Economics 190
- 6.8 Summary 201

    Appendix 6A: A Brief Review of Some Characteristics of Exponential and Logarithmic Functions 204

    Appendix 6B: Calculation of the Numerical Value of the Constant e 208

7. Integration 213

- 7.1 Integration, the Inverse Process to Differentiation 215
- 7.2 The Fundamental Theorem 221
- 7.3 Rules of Integration 229
- 7.4 Integration by Substitution and Integration by Parts 234
- 7.5 Integration by Numerical Approximation 242
- 7.6 Improper Integrals 244
- 7.7 Multiple Integration 247
- 7.8 Integration Rules 248
- 7.9 Summary 249

    Appendix 7A: Further Applications of Integration 251

    Appendix 7B: Table of Values of e^x 253

8. Linear Programming 259

- 8.1 The Problem 262
- 8.2 Graphic Solution 263
- 8.3 Algebraic Solution 267
- 8.4 The Simplex Method 270
- 8.5 The Dual Program 275
- 8.6 Sensitivity Analysis 281
- 8.7 Parametric Programming 286
- 8.8 Feasibility Variables 288
- 8.9 Summary 289

9. Vectors and Matrices 303

- 9.1 Vectors 304

9.2 Vector Spaces   309
9.3 Matrices   313
9.4 Special Matrices   316
9.5 Matrix Inversion   318
9.6 Row and Column Operations   321
9.7 Determinants   327
9.8 Cofactors and the Adjoint   335
9.9 Summary   339

Index   347

CHAPTER 1

Sets

The study of calculus and linear algebra begins with the concept of a set. This concept possesses an intuitive counterpart in everyday experience and, although disarmingly simple, it will be employed to develop most of the major concepts used in this book. We are *not* interested in sets as such, but their study will prove helpful in understanding many of the topics that follow.

1.1 The Concept of a Set

The following definition of a set is adopted.

> **Definition—Set:** *A set is a* well-defined *collection of* distinct *objects called elements.*

Examples include the real numbers, the positive integers less than ten, the members of a particular labor union, the stock certificates of a given company, and possibly, the assets of a particular firm. For the present, the convention of representing a set by a capital letter or by writing out its elements (members) and enclosing them within braces, that is, { }, is used. The elements of the set can be indicated either by listing them all; for example,

$$S = \{2, 4, 6, 8\}$$

or by using a defining property to identify them, in which case a lower-case letter will be used to represent the elements. For example:

$$S = \{x: x \text{ is a positive even integer less than } 10\}.$$

(Read this as: S is the set of all elements x, such that x is a positive even integer less than ten.) The letter x is called a variable over the set S since it can stand for any element in S.

It is somewhat more difficult to express the assets of a firm as a set. The list might begin as follows:

$S_1 = \{$Building at 406 Oak Street, Welding Machine serial number 60471, ...$\}$

Although the listing of each asset could be eventually completed (provided the term "asset" is well-defined), the process would take a long time for most firms. It is much easier simply to state that the set in question is the set of assets owned by Company X. If agreement could then be reached on what the words "asset" and "Company X" mean, a set would be defined.

The definition of a set includes both the characteristics *well defined* and *distinct*. Both of these terms deserve further comment. The term *well defined* means that, given any element x, one and only one of the following statements is true. Either (1) the element belongs to the set; that is, $x \in S$ (read $x \in S$ as x is an element of the set S), or (2) the element does not belong to the set; that is, $x \notin S$ (read $x \notin S$ as x is not an element of the set S).

The collection S, consisting of the positive, even integers less than 10 is well-defined because, for any element, it is possible to determine which of the above statements applies. For example, four is an element of this set, that is $4 \in S$, but five is *not* an element of this set, that is $5 \notin S$.

It is less clear whether the collection S_1, assets of Company X, is a set. Welding machine serial number 60471 is a member of this collection. The number four is not; that is, $4 \notin S_1$. But what about "an enthusiastic work force?" If S_1 is to be a set, the meaning of the noun "asset" must be established so that, for every possible element considered, its membership or nonmembership in S_1 is determined. If this cannot be done, then there will be one or more elements x for which it cannot be specified whether statement one or two applies, in which case the collection S_1 is not a set.

The importance of the characteristic of being *well defined* may be further illustrated by applying it to "qualities needed for success." There is probably little if any agreement upon those qualities that, if possessed, would lead to success. Indeed, different elements would be suggested by different individuals, and it would not be possible to say of any element that it does or does not belong to the set. Thus, although abstractly such a set of qualities may exist, it cannot be specified operationally.

The characteristic *distinct* means that no two elements in a set are identical. Thus, the Senior Senator from New York and Jacob D. Javits cannot be separate elements in the set known as the Eighty-Ninth Congress. However, they could represent two separate elements in the set that will be known as The Ninety-Fifth Congress.

Ordered Sets

Up to now, the definition of a set has not involved order. The set $\{2, 4, 6, 8\}$ is identical to the set $\{4, 6, 2, 8\}$. If order is relevant to a set, then the set must be denoted as an ordered set and the method of ordering must be given.

> ***Definition—Ordered Set:*** *The set A is said to be an* ordered set *if, for any two elements x and y in A, it is specified that either x precedes y or y precedes x.*

When the order of a set is relevant, parentheses are used in writing the set; for example, the set $S = (2, 4, 6, 8)$ is an ordered set.[1] Even when order is not necessary, the elements of a set are generally listed in a natural order where one exists. For example, the sets $A_1 = \{a, b, c, d\}$ and $A_2 = \{1, 2, 3\}$, which are listed in a natural order, for our purposes are not ordered sets. If it is intended that order be significant, then the sets must be written in parentheses; $A_1^* = (a, b, c, d)$ and $A_2^* = (1, 2, 3)$. Note that A_1^* is not the same set as A_1, since order is a property of A_1^* but not of A_1. For the sets A_1^* and A_2^*, the rule for ordering is clear although it is not stated. For the ordered set $A_3^* = $ (vice-president for finance, vice-president for marketing, and vice-president for production) the ordering rule is not obvious and, therefore, it would be helpful if it were specified.

Consider the set of nine consecutive measurements made of the diameter of washers produced by some process. Here a number of orderings is possible: by diameter, smallest first; by diameter, largest first; by diameter, in the order observed; and so on. Assume that the ordering desired is by diameter, in the order observed. This ordered set might be $S_7 = (2.5'', 2.2'', 2.4'', 2.1'', 2.4'', 2.8'', 2.9'', 3.1'', 3.0'')$. Two important observations should be made about this ordered set. First, the third and fifth observations are identical. The student might suspect that this violates the characteristic of distinctness in the definition of a set. It

---

[1] The concept of an ordered set is the same as that of an *n*-tuple, which is discussed in the section on pairs.

does *not*. The members of the ordered set are the diameters of separate and distinct washers. The first 2.4″ is the diameter of the third washer while the second 2.4″ is the diameter of the fifth washer. They are distinct items in this particular ordered set. This might be easier to see if the washer being observed were indicated. Thus S_7 could be written (diameter of washer one = 2.5″, diameter of washer two = 2.2″, diameter of washer three = 2.4″, ...). This would illustrate more plainly that the third and fifth elements of S_7 are distinct, but the word diameter is common to all elements and can therefore be omitted. Furthermore, the particular washer being observed is specified by its position in the ordered set. After these modifications, only the original specifications of the ordered set, that is, S_7 remains. If S_7 had *not* been specified as an ordered set, then the second 2.4″ would be redundant and would have to be deleted or ignored under the definition of a set, and the new set S_7 would be the set of eight *different* diameters observed in the ten observations.

The second observation to be made about this set is that the order itself may be important. Here, for instance, it may indicate a trend in the performance of the machine (toward larger diameters). Such a trend might provide information to the machine operator, which would serve as a signal for some specific action. If only unordered observations were available, information of this type would be lost.

Distinction Between an Element and a Set

It is important not to confuse the concept of an element—the fundamental unit of the system—with the concept of a set. An element belongs to a set, but it is not the same thing as the set. Consider the set S_8 which is The Chamber of Commerce in Ithaca, N.Y. Suppose that Ithaca's Chamber of Commerce consists of three members, Smidt, McAdams, and Dyckman. Then S_8 can be written {Smidt, McAdams, Dyckman}. The Ithaca Chamber of Commerce is a different notion from the elements that comprise it. This is more easily seen if a single-element set is considered. Therefore, suppose that in a dispute over a policy statement that they attempted jointly to author, the last two members resign. Then S_8 consists of a single member, Smidt. But the notion "Smidt" is not the same as the notion "Ithaca Chamber of Commerce." It may be true that, in speaking about the Chamber of Commerce, interest centers in some of the actions of Smidt, but the two notions are by no means identical. Smidt is a human being and in most respects has no connection with the Chamber of Commerce. The Chamber of Commerce in turn is inanimate and connotes the members, the properties, and the objectives of the organization, among other things. Furthermore, if there were only two Chambers of Commerce, Ithaca and Atlanta for example, and the latter were dissolved, then the set S_8^*, the set of all Chambers of Commerce, could be written as {{Smidt}} which, in turn, is neither identical to The Ithaca Chamber of Commerce, {Smidt}, nor to the element, Smidt.

As another example, consider the set of companies in which Mr. A owns stock. If this set contains only one element, say A.T.&T., then the set is a set with one element, A.T.&T. But the firm A.T.&T. is certainly a different notion from the set of companies in which Mr. A owns stock. Suppose, further, that Mr. B also owns stock. Then if B's portfolio is the set {G.E., G.M.}, the set of A's and B's portfolios can be written as {{A.T.&T.}, {G.E., G.M.}}. If Mr. B now sold his holdings, the *set of portfolios* could be written {{A.T.&T.}} which is a different notion than the *set of companies in which Mr. A owns stock*. The element a is thus not the same thing as the set whose single element is a, $\{a\}$. And the set whose single element is the set with a single element a, $\{\{a\}\}$, is different from both the element a and the set $\{a\}$.

The elements of a set then may themselves be sets. The set $L_2 = \{\{2, 4\}, \{6, 8\}\}$ is a set with two elements. These elements are the set $\{2, 4\}$ and the set $\{6, 8\}$. Thus, $\{2, 4\} \in L_2$; and $\{6, 8\} \in L_2$. Furthermore, the set L_2 is not equal to the set $S = \{2, 4, 6, 8\}$. The set S contains as elements the numbers 2 and 4 while the set L_2 contains as elements the *sets* $\{2, 4\}$ and $\{6, 8\}$. A set and an element are not the same notion, and therefore $L_2 \neq S$. Another example of a set of sets is the set $L_3 = \{\{2\}, \{2, 4\}, \{2, 4, 6\}, S\}$. In this case the sets which are elements of L_3 overlap, but each is a distinct element of L_3.

Set Equality

> **Definition—Set Equality:** *Two unordered sets are said to be* equal *if and only if they contain precisely the same elements.*

The set $S = \{2, 4, 6, 8\}$ is equal to the set $J = \{(-2)^2, 2, 6, 2^3\}$. On the other hand, the set S is not equal to the set $L = \{(-2)^2, 2, 6, (-2)^3\}$ since the element $(-2)^3$ is equal to -8. However, the set S is equal to the set $L_1 = \{(-2)^2, 2, 6, \sqrt{64}\}$ because the last element of L_1 is 8.

1.2 Special Sets

There are a few special sets that are used extensively. These are the null or empty set, the universe or population set, subsets, partitions, and power sets.

The Empty (Null) Set

> **Definition—Empty Set:** *A set with no elements is called an* empty *or* null *set and is written* ∅.

Suppose that Mr. C owns stock only in companies that experienced a loss last year. Then the set of companies that made a profit last year and in which Mr. C owns stock is an example of an empty set. The set of positive integers less than zero is another empty set. Although a set may be empty (that is, have no members), it is still a meaningful notion. Thus, the set of companies in which Mr. C owns stock is a meaningful notion different from the elements, if any, in the set. The set exists, although it may be empty.

For another illustration, consider again the Chamber of Commerce example. This set consisted, finally, of only one member, Smidt. Furthermore, Smidt and the Chamber of Commerce, the set S_8, were distinct notions, although Smidt is the sole member of the set. This distinctness is illustrated by the ability of Smidt to become disassociated from the set through his resignation. The set S_8 would then be empty; it would have no members. Nevertheless the set S_8 is still meaningful. It is meaningful to think about The Chamber of Commerce of Ithaca although for the moment it has no members.

Because the empty or null set is written as a "0" with a slash through it, $\emptyset$, it is easy to confuse the null set with the number zero. They are different notions; zero is not the null set, $0 \neq \emptyset$. Moreover 0 is not an element of $\emptyset$, $0 \notin \emptyset$; since the null set has no members. Finally the null set $\emptyset$ should be distinguished from $\{0\}$, which is a set with one member, 0. In the context of the previous discussion, $\{0\}$ is the set whose single member denotes *the number* of companies that made money last year, and in which Mr. C owns stock, or it is the set whose single member denotes the *number* of people belonging to the Ithaca Chamber of Commerce after Smidt resigns. The set $\emptyset$ would represent the set of Chambers of Commerce, the set S_8^*, if both the Atlanta and Ithaca chapters were dissolved.

The Universe or Population Set

The set of all logical possibilities in some discussion will often be of interest. Such a set constitutes the basis for a discussion and must be agreed upon, at least tacitly, before the discussion can be conducted unambiguously. For example, suppose that the set of incorporated business firms in Ohio which are in business today is under consideration. Firms incorporated in Indiana, among others, then are excluded and any statement made about an Indiana firm becomes irrelevant to the discussion.

> **Definition—Universe Set:** *The set of all elements relevant to a particular problem or discussion is known as the* universe set *or* universe.

The universe set for some discussion or investigation may be a very large set, such as the lengths to which a steel beam could be cut, or it could be a relatively small set, such as the customers who arrive at a particular store counter during a specified period. In either case, this concept of a universe set will mean the totality of things under consideration. Questions which may be asked in the latter situation, such as how many customers were male, will involve subsets of the universe set.

Subsets

> **Definition—Subset:** *Let A and B be sets. If every element in set A is an element of set B, then A is said to be a* subset *of B. This relationship is written* $A \subseteq B$.

Thus, if $S^* = \{2, 4\}$ and $S = \{2, 4, 6, 8\}$ then $S^* \subseteq S$. (Note that S^* is not a subset of $L_2 = \{\{2, 4\}, \{6, 8\}\}$ since the elements of L_2 are the *sets* $\{2, 4\}$ and $\{6, 8\}$ while the elements of S^* are the *numbers* 2 and 4. This is indicated by writing $S^* \nsubseteq L_2$.) If every element in S^* is also an element of S, but S^* does not contain *all* the elements of S (as is true in the present case), then S^* is said to be a *proper* subset of S.

> **Definition—Proper Subset:** *Let A and B be sets. Then if every element in a set A is an element in a set B, but the elements in A do not exhaust those in B, A is said to be a* proper subset *of B. This relationship is written $A \subset B$.*[2]

The null set is a subset of every set, therefore $\emptyset \subseteq S$. Intuitively the reasoning is as follows: Either $\emptyset \subseteq S$ or $\emptyset \nsubseteq S$. But if $\emptyset \nsubseteq S$, then $\emptyset$ must contain at least one element which is not in S. But $\emptyset$ contains no elements and therefore cannot contain an element not in S.

The set S is also a subset of itself, $S \subseteq S$, since the definition of a subset is satisfied. Every element in S is an element in S. Thus, if S is the universe set, any subset of S, including S itself, is a subset of the universe set. The relationships $S \subseteq S$ and $\emptyset \subseteq S$, are required in the development of the important concept of a power set discussed in the section following that on partitions.

Partitions

It is often convenient to divide a set into parts that do not overlap and that exhaust the elements of the original set. Such sets are said to be *mutually exclusive or disjoint*. For example, consider the single-element sets that can be made from the set S. They are $\{2\}, \{4\}, \{6\}, \{8\}$. Also the *elements* of the set L_2, $\{2, 4\}$ and $\{6, 8\}$, exhaust the elements of S and do not overlap. Another example might be the division of the employees of a firm into union and nonunion employees. A set may be divided in many ways and the particular division of interest depends upon the discussion or problem at hand. The employees of the firm in the example above could also have been divided by salary, time employed, supervisory level, or some other relevant characteristic.

When a set is divided into one or more sets which are *mutually exclusive or disjoint* (that is, they do not overlap) and which exhaust the original set, it is said that the original set is *partitioned*. This set of mutually exclusive and disjoint sets is called a *partition* of the original set. A partition then, is a set of sets. The division of the set of employees of a firm into union and nonunion employees is a partition of the original set. The division of S into single-element sets constitutes a partition of the set S; $\{\{2\}, \{4\}, \{6\}, \{8\}\}$, is a partition of S. A further example would be the partition of the set of customers of a salesman into those who placed orders on his last call and those who did not.

---

[2] Note the similarity between the symbol $\subseteq$ with $\leq$; and $\subset$ with $<$. However, one should be careful not to confuse these relations. The symbol $\subset$ stands for the subset relationship while the symbol $<$ means "less than."

Definition—Partition[3]: *Given a set A whose elements are the sets $A_i = \{b_1, b_2, \ldots, b_k\}$, then the set A is a partition of the set B if* (1) *for every element $A_i \in A$, $A_i \subseteq B$ and* (2) *for every element $b_i \in B$, b_i is an element in one and only one of the sets A_i.*

A partition of some set B, then, is a set of disjoint (nonoverlapping) subsets of B that exhaust B. Thus, in order to have a partition, every element of B must be included in one and only one of the subsets. For instance, the set $P_1 = \{\{2, 4, 6\}, \{8\}\}$ is a partition of S. Each of the two elements in P_1 is a subset of S, and every element in S is included in one and only one of these subsets. The set $P_2 = \{\{2\}, S\}$ is *not* a partition of S, but $P_3 = \{S\}$ is. (Why?)

The Power Set

Another set with particular characteristics that will be useful in later chapters is the set of all the possible subsets of a given set. This set is known as the *power set* of the given set.

Definition—Power Set: *Let S be a set. The set of all possible subsets of S is called the* power set *of S. The symbol for the power set of S is 2^S.*

The power set for the set, $S = \{2, 4, 6, 8\}$ can be used as an example. There are the one-element sets which are subsets of S

$$\{2\}, \{4\}, \{6\}, \{8\};$$

the two-element subsets of S

$$\{2, 4\}, \{2, 6\}, \{2, 8\},$$
$$\{4, 6\}, \{4, 8\},$$
$$\{6, 8\};$$

the three-element subsets

$$\{2, 4, 6\}, \{2, 4, 8\}, \{2, 6, 8\},$$
$$\{4, 6, 8\};$$

the four-element subset

$$\{2, 4, 6, 8\} = S;$$

and finally the null set $\emptyset$, which is by definition a subset of any set

$$\emptyset.$$

In all, there are 16 subsets in the power set of the set S, and 16 is 2^4. This is the significance of the symbol 2^S. Since S has four elements, there are 2^4 or 16 subsets of S, and therefore 16 elements in the power set for S. In general, for a set S with

---

[3] Using the set operations defined in Section 1,3, this definition can be written as $A_i \cap A_j = \emptyset$ for $i \neq j$ and $A_1 \cup A_2 \cup \ldots = B$; that is, the intersection of the elements of A is the null set, and the union of the elements of A is B. Also b_k is an element of one and only one A_i.

Sec. 1.3] SET OPERATIONS

n elements, there are 2^n elements in the power set 2^S. This result can be proved but it is not done here.[4]

1.3 Set Operations

It is useful to be able to combine sets in various ways. The purpose of this section is to consider the following operations defined for sets: union, intersection, difference, and complementation.

Set Union

> **Definition—Set Union:** The union *of two or more sets is the set of all elements belonging to at least one of the sets.*

Using set notation, the union of the sets M and K, written $M \cup K$, is:

$$M \cup K = \{x: x \in M \text{ or } x \in K\}.$$

(Read this as M union K is the set of all elements x such that x is an element of M or x is an element of K.) For example if $M = \{1, 2, 3\}$ and $K = \{2, 4, 6\}$, then $M \cup K = \{1, 2, 3, 4, 6\}$. Note that the element 2 that is a member of both M and K is included only once in $M \cup K$ since only distinct elements may belong to a set.

Set Intersection

> **Definition—Set Intersection:** The intersection *of two or more sets is the set of all elements common to all of the sets. Using set notation, the intersection of the sets M and K is written* $M \cap K = \{x: x \in M \text{ and } x \in K\}$.

How should this notation be read? For the present example, $M \cap K = \{2\}$. The element 2 is the only element common to both sets.

The set-union and set-intersection discussed so far can be expressed graphically using Venn diagrams.[5] A method for proving set relations called the *x*-element technique is given in Appendix 1A. The concepts of set union and set intersection are illustrated by the cross-hatched areas in Figure 1.1 for the two-set case. If two or more sets do not have any elements in common, that is, if they do not overlap, they are disjoint sets. If M and K are disjoint, then $M \cap K$ is the null set. Disjoint sets are also referred to as mutually exclusive sets.

---

[4] Intuitively, any subset, call it S_{**}, of another set, call it S_*, can be determined by selecting one of two alternatives for every element $x \in S_*$; either $x \in S_{**}$ or $x \notin S_{**}$. If there are n elements in S_*, the choice between the two alternatives must be made n times. Each different set of n choices leads to a different subset S_{**}. Since there are $2 \cdot 2 \cdots = 2^n$ possible choices for each subset, there are 2^n such subsets.

[5] Diagrams of the type shown in Figure 1.1 are named after the nineteenth-century English mathematician and logician John Venn, who first introduced them.

Figure 1.1

Set Union $M \cup K$ and Intersection $M \cap K$ Illustrated.

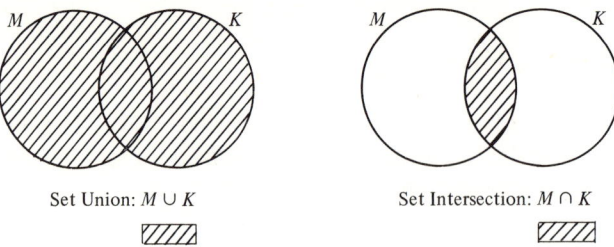

Set Difference and Complement

Definition—Set Difference: *The difference of two sets is the set of all elements in the first set which are not in the second.*

Thus, if M and K are sets, the set difference of M and K, written $M - K$, can be expressed in set notation as $M - K = \{x: x \in M \text{ and } x \notin K\}$, which is read the set of all elements x such that x is an element in M, and x is not an element in K. For the continuing example: $M - K = \{1, 3\}$. Note that the order of the sets is relevant. $M - K = \{1, 3\}$ is not the same set as $K - M = \{4, 6\}$ and, more generally, $A - B$ is not equal to $B - A$ unless A and B are the same set, in which case $A - B$ and $B - A$ are both the null set.

Definition—Set Complement: *Given two sets A and B where B is a subset of A, the set difference $A - B$ is known as the complement of B in A. (The symbol for the complement of B is B', which is read, not B in A.)*

Note that the complement of a set is a difference (is the reverse true?) and that the set within which the complement is defined should be specified to avoid ambiguity. (See Figure 1.2(c).) Thus we should say: the complement of B in A, written B' in A. B' alone means the set of all elements (in some set) which are *not* in B. If the set within which the complement is defined is not specified, the universe set is assumed. A more meaningful statement then would be: B' is the set of all elements in the universe under discussion that are not also in the set B. If the universe set under consideration is the set of all positive integers less than 10 and $B = \{2, 4, 6\}$, then $B' = \{1, 3, 5, 7, 8, 9\}$. The ideas of set difference and set complement are illustrated by the shaded areas of the Venn diagrams in Figure 1.2 for the two-set case.

1.4 Counting

The determination of the number of elements in a set is a common task. Suppose that a foreman is asked to determine the number of machines of a certain type in a plant. He might proceed as follows. Establish a starting point and some order

Sec. 1.4]   COUNTING   11

Figure 1.2

Set Difference A − B and Set Complements B′ and B′ in A Illustrated Using Venn Diagrams.

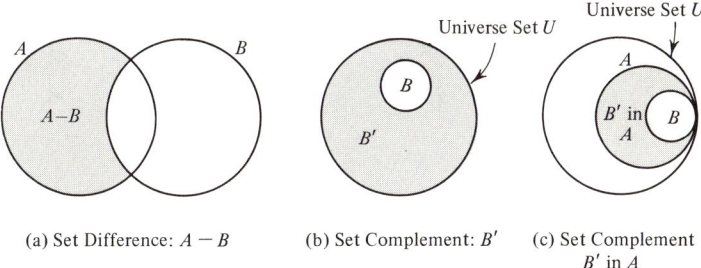

(a) Set Difference: A − B   (b) Set Complement: B′   (c) Set Complement B′ in A

of progression, then assign to the initial machine the number one, to the next machine the number two, and so on until the items to be counted have each been assigned a number. In set terms this can be described as taking two sets, the set of machines of the given type in the plant and the set of ordered positive integers (which will be denoted by I^+) and matching elements of the two sets until the elements of the first set are exhausted.[6] If there were 10 machines and the foreman had used a subset of I^+ consisting of the first 10 naturally ordered integers of I^+, then to each machine an integer would have been paired and vice versa. This intuitive discussion suggests the important concept of a one-to-one correspondence.

One-to-One Correspondence

Definition—One-to-One Correspondence: Two sets are said to be in one-to-one *correspondence if the elements in the sets are paired such that each element in the first set is paired with one and only one element in the second set and each element in the second set is paired with one and only one element in the first set.*

Definition—Equivalent Sets[7]*:* Two sets A and B are said to be equivalent sets *if they can be put into one-to-one correspondence. (This is written $A \leftrightarrow B$.)*

The set of corporations whose stock is traded on the New York Stock Exchange is equivalent to the set of symbols used to represent these stocks. The concept of a one-to-one correspondence is illustrated in Figure 1.3(a). Figures 1.3(b) and 1.3(c) illustrate sets that do not show one-to-one correspondence. (Why?)

When the machines in a plant or even the grains of sand in a thimble are counted, one would expect the pairing process to have an end. A set whose elements can

---

[6] The reader is referred to Appendix 1B to this chapter for a detailed schematic of the number system.

[7] The notion of equivalent sets should not be confused with the concept of equivalence which refers to the symmetric, reflexive, and transitive properties of relations such as the equality (=) relation.

Figure 1.3

Three Cases Illustrating the Conditions Required for a One-to-One Correspondence Using the Sets A and B, A and B*, and A and B**.

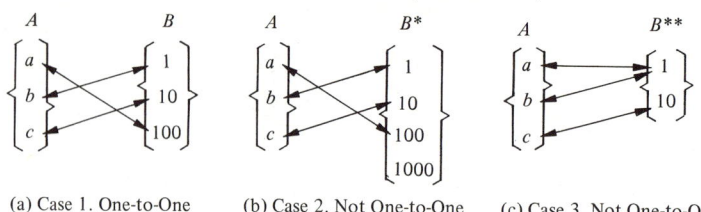

(a) Case 1. One-to-One (b) Case 2. Not One-to-One (c) Case 3. Not One-to-One

be counted by a pairing process that ends is termed *finite*. This notion may be stated more rigorously using set notation: where $I^+ = (1, 2, \ldots, n, \ldots)$, a *set A is finite* if $A \leftrightarrow I_n$ for some $n \in I^+$, where $I_n = \{x: x \in I; 1 \leq x \leq n\} = (1, 2, \ldots, n)$. An *infinite* set may now be defined as any set that is neither empty nor finite. The ordered set I^+ is an example of an infinite set; the pairing process has no end. An infinite set can be placed into a one-to-one correspondence with some proper subset of itself; a finite set cannot. For example, the set $I^+ \leftrightarrow I_e^+$ where I_e^+ is the set of even positive integers; that is, $I_e^+ = \{2, 4, \ldots, 2n, \ldots\}$. Each element in the set I^+ can be paired with an element in the set I_e^+, namely, twice itself.

Many infinite sets are *equivalent* to the set I^+. A set that is equivalent to the set I^+, is said to be *countably infinite* or countable. For example, consider the set of rational numbers. A rational number is a number formed by taking the ratio of two integers. The set of all rational numbers is a set that is equivalent to I^+. To prove this statement, a rule is needed that would establish a one-to-one correspondence between the rational numbers and the integers. This can be done although the proof is not given here. The subset of rational numbers between zero and one is also equivalent to the set I^+.

Now consider the set of irrational numbers. An irrational number is a number that cannot be expressed exactly as a ratio of two integers. The numbers π, $\sqrt{2}$, and e (the base of the natural logarithms) are common examples. A full discussion of the irrational numbers as well as the proofs that $\sqrt{2}$, π, and e are irrational would be beyond the scope of this book. The set of irrational numbers is neither empty nor finite, it is *infinite*. Furthermore, the set of irrational numbers cannot be put into a one-to-one correspondence with the set I^+; it cannot be counted. This statement can be proved, but again the proof is beyond the scope of this book. A set that is infinite but not countable, is said to be *noncountably infinite;* such a set is not equivalent to I^+.

The union of the set of rational numbers (which includes zero) and the set of irrational numbers is called the set of real numbers, R. The set of real numbers is therefore noncountably infinite. The set of real numbers can be put into a one-to-one correspondence with the points on an infinitely long straight line. This property is the basis for drawing graphs, which in effect translate statements in algebraic form into equivalent statements in geometric form, and vice versa.

The process of setting up a one-to-one correspondence is the basis for comparing the size (number of elements) of two sets. Two sets are said to be of the same size if they are equivalent to one another, or if each is equivalent to a common third set. To summarize, the basic size categories considered so far are the following:

> Finite sets: The elements in the set can be put into one-to-one correspondence with the first n positive integers.
>
> Countably infinite sets: The elements in the set can be placed into one-to-one correspondence with the elements in the set I^+.
>
> Noncountably infinite sets: The elements in the set can be placed into one-to-one correspondence with the elements in the set of real numbers.

In this book the largest sets that will be encountered will be sets that can be made equivalent to the set of real numbers; however, there are sets that contain so many elements that the elements cannot even be put into one-to-one correspondence with the real numbers.

Pairs

Definition—Pair: A pair *is a set consisting of two elements.*

Suppose that $S = \{a, b, c\}$ is a set. Then a pair of elements from S is a subset of S. For example, the pair of elements a, b make up a subset of S, and this set is written $\{a, b\}$, which is a set consisting of two elements, a and b. It will be useful at times to distinguish the location or order of selection of the elements in a pair. This can be accomplished by agreeing that the first member of the pair be the first element chosen. When a pair (remember, a pair is a set) is ordered, it is known as an ordered pair and denoted by (a, b). Note that $(a, b) \neq \{a, b\}$. Ordered pairs can also be formed by selecting one element from each of two sets. In this case the concept of an ordered pair applies even if the elements are identical.

Consider now two sets, called M and S, and all the ordered pairs that can be made by selecting the first element of the pair from the first set and the second element of the pair from the second set. The set of ordered pairs so formed, written $M \times S$, read "M cross S," is called the *Cartesian product* set of the two sets M and S. The set $R \times R$, where R denotes the set of real numbers, is the product set that yields the real plane of analytic geometry whose ordered pairs are called the Cartesian coordinate system. The set $R \times R$ suggests the origin of the name, Cartesian product set (see Figure 1.4).

> *Definition—Cartesian Product Set of Two Sets:* The Cartesian product set *of two sets, A and B, is the set of all ordered pairs (x, y) in which $x \in A$ and $y \in B$.*

A subset of a product set is called a *relation*. The first quadrant of the real plane is an example of a relation since it is a subset of $R \times R$.

For another example of a product set consider the following two sets: $S = \{2, 4, 6, 8\}$, and $M = \{P, Q, R\}$. Then consider this product set: $S \times M =$

Figure 1.4

The Real Plane of Analytic Geometry, R × R, with the First Quadrant Indicated.

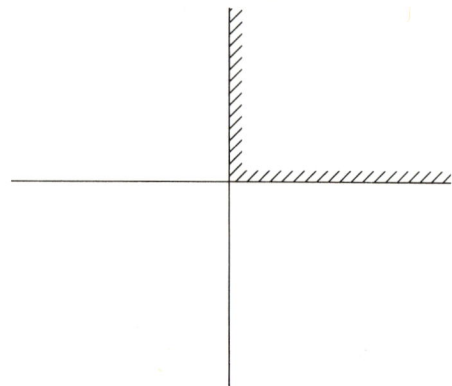

{(2,P), (2, Q), (2, R), (4, P), (4, Q), (4, R), (6, P), (6, Q), (6, R), (8, P), (8, Q), (8, R)}. Note that M × S is *not* the same set as S × M. (Why?) The set {(2, P), (2, Q), (2, R)} is an example of a relation formed from the product set S × M. (Can you write M × S? Is the set {(2, P), (4, P), (6, P), (8, P)} a relation that can be formed from the product set M × S?)

The concept of an ordered pair can be extended to an ordered triple, (a, b, c) for example, or in general, to an ordered n-tuple. An ordered n-tuple is simply an ordered set. This can be illustrated using an ordered 3-tuple formed by selecting the first two elements from S and the last element from M. One element of this Cartesian product set would be (2, 4, P). The set of all 3-tuples similar to the one given above would be S × S × M = {(2, 2, P), (2, 4,P), ..., (2, 2, Q), ..., (4, 2, P), ..., (6, 8, P), ..., (8, 8, P), ..., (8, 8, R)}. Because this set contains 48 elements, only a representative listing is given. Likewise, the notion of the product set can be extended to include any number of sets. Three-dimensional Euclidean space would be represented as R × R × R. (Suppose that a company wishes to test the effects of different types of advertising for different products in different media. If there are two types of ads, two products, and three different media, can you determine how many possibilities there are using the product-set concept? Try to develop a general rule for determining the number of possibilities.)

The symbol S^2 is used for the Cartesian product set S × S; S × S × S × S is represented by S^4 and S × S × M by S^2 × M. But one must be careful here; since the order of set multiplication is significant, the product set S × M × S cannot be represented as S^2 × M, but *must* be represented as shown initially: S × M × S. Further, S × M × S is not the same set as either (S × M) × S or S × (M × S). (Why?)

Consider the product set S × S for the illustrative set S = {2, 4, 6, 8}.

$$S \times S = \begin{Bmatrix} (2, 2), (2, 4), (2, 6), (2, 8), (4, 2), (4, 4), (4, 6), (4, 8), \\ (6, 2), (6, 4), (6, 6), (6, 8), (8, 2), (8, 4), (8, 6), (8, 8) \end{Bmatrix}$$

Figure 1.5

The Product Set S × S Graphed.

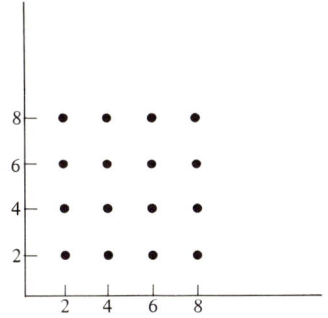

It has $4 \times 4 = 4^2 = 16$ elements. The reader should turn back now and compare the power set 2^S, which also has 16 elements with this Cartesian product set. (Are they the same set?) This product set is by definition the set of ordered pairs indicated graphically in Figure 1.5.

1.5 Functions

It is useful to begin the discussion of functions with a definition.

> **Definition—Function:** *Given two sets A and B and a rule that assigns to each and every element in A a uniquely specified element $y \in B$, then this rule specifies a set of ordered pairs, f, and this set is called a function from A into B. The function f is written $f = \{(x, y): $ for all $x \in A$ there is a unique $y \in B.\}$*

This definition requires comment. First, a function is a set. Second, a lower case letter, typically f, is used to represent this set. Hereafter sets that are functions may be represented by either lower case or capital letters (depending on whether the emphasis is on the function characteristics or the set characteristics). Third, the element y in the set B can also be written as $f(x)$ where x is an element in A. Thus, $f(x)$ and y are two names for the same thing. Finally, a function from A into B gives a set of ordered pairs of the form (x, y) or $(x, f(x))$, while a function from B into A would give a set of ordered pairs of the form (y, x) or $(y, f(y))$.

An example of a function is provided by a function from the set S into the set M given by $\{(2, P), (4, P), (6, Q), (8, Q)\}$. Here m, an element of M, takes on the value Q when s, an element of S, is 6. This can be given in more compact form by $m = f(s)$ generally, or $Q = f(6)$ specifically.

If f is a function from A into B, then the set A is known as the *domain* (or domain of definition) of the function f and the set B is known as the *range* of the function. The process of establishing the correspondence, the ordered pairs, is known as *mapping*. The set A is said to be mapped *into* set B; symbolically, $A \rightarrow B$. (If this mapping exhausts all of B, A is said to be a function *onto* B.) Figure 1.6 illustrates

Figure 1.6

Mapping of a Set A into a Set B.

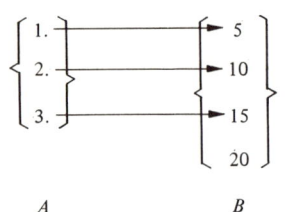

a mapping of *A* into *B*. The mapping here is indicated by the arrows and the function is {(1, 5), (2, 10), (3, 15)}. Note in particular that, although the mapping must involve each element of *A*, the entire set *B* need not be used.

Three important relationships can now be summarized moving from the more general to the more specific.

1. *A Cartesian product set* of two sets consists of all the ordered pairs that can be developed by selecting the first element from the first set listed in the product and the second element from the second set listed in the product. If there are n elements in S, and m elements in T, then the Cartesian product set $S \times T$ has $n \cdot m$ elements.

2. *A relation* is a subset of the Cartesian product set. It may or may not exhaust (involve all elements in) the first set of the product set, and each element from the first set may be paired with one or more elements in the second set.

3. *A function* is also a subset of a Cartesian product set. It is a relation that must exhaust the first set of the Cartesian product (which is called the domain) and maps each and every element from this first set to one and only one element in the second set of the product set (which is called the range). Thus it may not exhaust the range. However, different elements of the domain may be mapped into the same element in the range. The mapping is called an "onto mapping" if the range set is exhausted. Furthermore, an onto mapping in which the number of elements in the domain set is equal to the number of elements in the range set is called a one-to-one mapping and the function it determines is a one-to-one function. The word *mapping* implies that the set defined by the mapping is a *function*. The function given in Figure 1.6 could have been determined using the equation rule $y = 5x$ once the sets *A* and *B* had been specified.

It should be made explicit that while the pairing rule for a function can sometimes be expressed by an equation, the equation is *not* the function. The equation provides the element in the range set that is matched to a specified value in the domain. But it may also seem to imply matchings that are not part of the function. For example, the equation $y = 5x$ suggests that the pair (4, 20) belongs to the function from *A* into *B* just discussed, yet this pair is not an element in the function since 4 is not an element in the domain. If, however, the range and domain sets are known, the equation provides all the additional information about the function that would be given by the set of ordered pairs. Therefore, by convention, the pairing rule is often allowed to stand for the function when the domain and

Sec. 1.5] FUNCTIONS

range sets may be (or are implicitly) assumed to be the real numbers. (See Appendix 1C for a complete discussion of the problems that result if either assumption is not valid as well as for some other implications of the definition of a function.)

The several conventions frequently used when specifying functions can now be summarized.

1. The function is often specified by describing only its rule, often in the form of an equation.
2. The domain and the range of such a function are assumed to be the set of real numbers.
3. The symbol x is used to denote an element in the domain of definition.
4. The symbol $y = f(x)$ is used to denote an element in the range set.

These conventions are used here unless some other convention is specifically stated. A phrase such as "$y = 5x$ is a function" should be interpreted as shorthand for the more accurate statement: "The function in question is the function whose rule is $y = 5x$ and whose domain of definition and range are each the set of real numbers, where x is an element in the domain and $y = f(x)$ is an element in the range."

Although these conventions are useful, the reader must be careful not to fall into the habit of assuming that *every* equation can be used as the rule for a function. The equations (a) $y = 1/x$ and (b) $y^2 = 5x$ do not describe the rule for a function unless the range and the domain are both carefully specified: For $y = 1/x$, zero must be omitted from the domain set, while for $y^2 = 5x$, the rule defines a function if the domain and range sets are defined as the non-negative real numbers.

Finally, an equation such as $y = 5x$ is a common way but, as has already been shown, not the only way to state a rule for mapping the elements from the domain to the elements of the range. Other ways to establish a set of ordered pairs for a function include graphs, tables, verbal rules, and diagrams. For example, a diagram is used initially to establish the function $\{(1, 5), (2, 10), (3, 15)\}$ in Figure 1.6.

Set Functions

If x is an element in the domain of a function, the element $y = f(x)$ in the range that corresponds to x is known as the *image* of x under the mapping f. For example, 10 is the image of 2 under the mapping given by the rule $y = 5x$. When the term function is used and the notation $y = f(x)$ is employed, then x is known as the *argument* and y is the *value* of the function. Again, in the example, 10 is the value for the argument 2. Functions that have pairing rules given by equations are occasionally called point functions because the argument x can be considered as a point on the real line. If the elements in the domain are sets, the function is sometimes called a set function.

Consider the sets $A = \{0, 1\}$ and $B = I^+$, and suppose that $y = f(x) = (x + 2)$ is the pairing rule where $x \in A$ and $y \in B$. Then the function $f = \{(0, 2), (1, 3)\}$ is an example of a point function. Also since $f(0) = 2$, 2 is the image of 0 under the mapping f. Now consider the set $S = \{(a, a), (a, b), (b, a), (b, b)\}$. This set is a set of ordered pairs, and since an ordered pair is a set, S is a set of sets. Suppose that a mapping between the sets S and $B = I^+$ is accomplished by the rule

$y = f(s)$ = number of a's in the ordered pairs where $s \in S$ and $y \in B$. Then the function $f = \{((a, a), 2), ((a, b), 1), ((b, a), 1), ((b, b), 0)\}$ is formed. This is a function with four elements, each of which is an ordered pair. For example, $((a, a), 2) \in f$. Each of these ordered pairs in turn has two elements; the first is itself an ordered pair and the second is an integer. The function f in this case is a set function since its domain is a collection of sets. The argument of the function is, in general terms, s and the value is y. Thus, for the argument (a, a), the value of the function, 2, is given by the rule $y = f(s)$ = number of a's in the ordered pair.

Functions and Relations

An essential property of a function is that no two ordered pairs have the same first element (although two different ordered pairs may have identical second elements). Some texts prefer to call such functions single-valued functions. These texts broaden the concept of a function to include many-valued functions in which more than one element in the range is paired off with a given element in the domain. Here, many-valued functions are called relations. Thus, if $A = \{4\}$ is the domain set, $B = \{+2, -2\}$ is the range set, and $y^2 = x$ is the pairing rule where $x \in A$, and $y \in B$, a relation and not a function is defined since, by the rule, 4 is mapped into both plus and minus 2. If B were restricted to $\{+2\}$ or $\{-2\}$, a function would be defined.

Another example is given in Figure 1.7 where the set X is the set of real numbers. In this case, the graph defines a function from X into Y but *not* from Y into X. For each value in X, there is one and only one value in Y. On the other hand, unless Y is defined as the set of non-negative real numbers, there are elements in Y for which there is no corresponding element in X. Furthermore, even if Y is defined as the set of non-negative reals, there is no function from Y into X since more than one element in X is matched to each element in Y. In this latter case, however, there is a relation. Note finally that if the domain and range sets are both the non-negative reals, there exists a function from X into Y and also a function from Y into X. These functions are both one-to-one functions and are

Figure 1.7

Graph Illustrating a Function of X into Y but not *Y into X.*

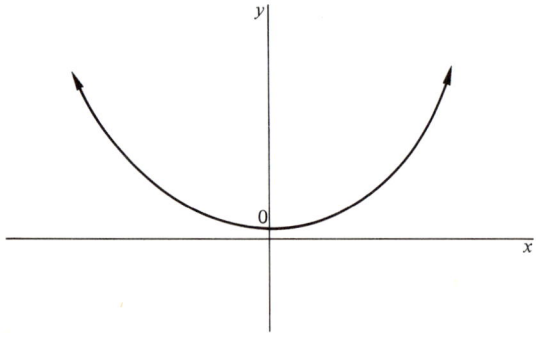

represented graphically by the portion of the curve in the top right quadrant in Figure 1.7.

1.6 Point Sets

The set of points or point set $H = \{(x, y): x, y \in R \text{ and } y = 2 - x\}$ is the set of ordered pairs that constitutes the points on the straight line graphed in Figure 1.8. Using Figure 1.8 and the rule for the function H, several other point sets can be illustrated in two space. For example, the set $K = \{(x, y): x, y \in R \text{ and } y < 2 - x\}$ includes all the points in the plane below and to the left of the set H. The set K is specified by a linear inequality and is called an open half space. The adjective "open" indicates that the boundary, $y = 2 - x$, is not contained in the set. That is, the set of points which constitutes the boundary to the set K is not a subset of K.

If the boundary as well as the open half space is intended to be included in a single set, the set is written $L = H \cup K = \{(x, y): x, y \in R \text{ and } y \leq 2 - x\}$. The set L is called a *closed* half space since it includes its boundary points.

A closed half space contains points of two types, *boundary* points and *interior* points that should be distinguished. If a circle with positive radius is constructed using a boundary point as its center, this circle, no matter how small, will include points outside the set. Such a circle is known as a *neighborhood* of the point at its center. Contrarily, a circle of positive radius can be constructed using an interior point as its center which, if small enough, will contain only points belonging to the set. In fact, the latter circle can be made small enough so that it includes only other interior points; the neighborhood of the point includes only interior points.

An *open point set* may now be described as a set made up only of interior points while a closed set contains all its boundary points. A closed half space is a closed set and an open half space is an open set. Observe that a given set may be

Figure 1.8

The Set $H = \{(x, y): x, y \in R \text{ and } y = 2 - x\}$.

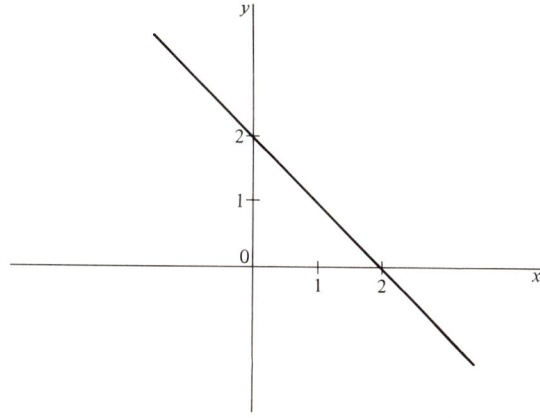

Figure 1.9

Intersections of Point Sets.

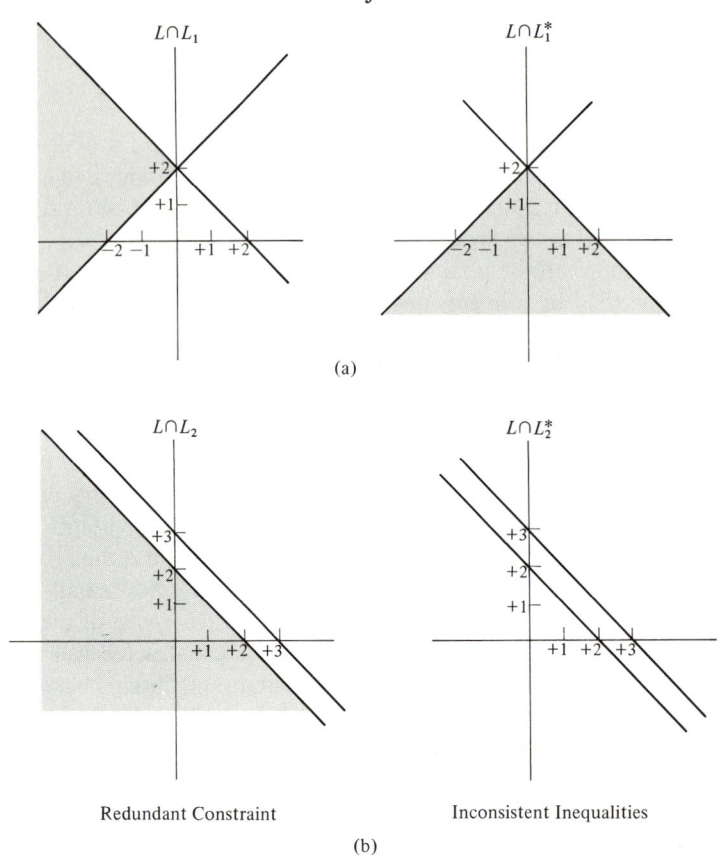

Redundant Constraint   Inconsistent Inequalities

(b)

neither open nor closed since it may include some but not all of its boundary points. An example is the set $N = \{(x, y): x, y \in R, x \geq 0 \text{ and } y > 0\}$.

Since half spaces are sets, new sets can be obtained from the intersection of two or more half spaces. Such sets are of particular interest since they specify the points that are members of both sets simultaneously, and therefore satisfy the inequality constraints that define both sets. Figure 1.9(a) illustrates $L \cap L_1$ and $L \cap L_1^*$ while Figure 1.9(b) illustrates $L \cap L_2$ and $L \cap L_2^*$ where:

$$L = \{(x, y): x, y \in R \text{ and } y \leq 2 - x\}$$
$$L_1 = \{(x, y): x, y \in R \text{ and } y \geq 2 + x\}$$
$$L_1^* = \{(x, y): x, y \in R \text{ and } y \leq 2 + x\}$$
$$L_2 = \{(x, y): x, y \in R \text{ and } y \leq 3 - x\}$$
$$L_2^* = \{(x, y): x, y \in R \text{ and } y \geq 3 - x\}$$

In $L \cap L_2$ shown in Figure 1.9(b), the set of points that satisfies the restriction given for set L is the same as the set which belongs to $L \cap L_2$; that is, the intersection of L and L_2 is just L. In this case, the inequality restriction given by L_2 is redundant. Furthermore, for $L \cap L_2^*$ also illustrated in Figure 1.9(b), there is no

Sec. 1.6]   POINT SETS   21

Figure 1.10

A Strictly Bounded Set.

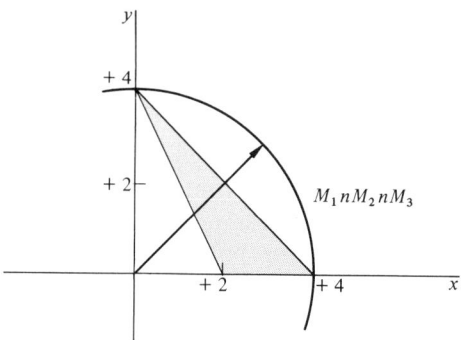

set of points that simultaneously satisfies the restrictions given in L and L_2^*; that is, the intersection of L and L_2^* is the empty set. When this is the case, the inequalities are said to be inconsistent. Note that if the weak inequality ($\leq$) in the rule for L were reversed, then the intersection of L and L_2 would consist of the infinite strip between (and including) the two lines. Finally, if the rule for the set L_1 had involved only a strict inequality ($>$), the set $L \cap L_1$ would not have included the set of points given by $\{(x, y): x \leq 0, y \leq 2 \text{ and } y = 2 + x\}$; the line segment in Figure 1.9(a) beginning at (0, 2) and extending down to the left would be excluded.

In each of the nonempty sets described so far, the set appears to extend without bound in at least one direction. When this is true, the set is a strictly *unbounded* set. More precisely, a set is strictly *bounded* if it lies within (is a subset of) a circle whose center is at the origin and which has a finite radius.[8] Figure 1.10 illustrates a strictly bounded set that can be inscribed in a circle with a radius of 4 units (or, for that matter, any radius ≥ 4) centered at the origin. This set is described by the intersection of the following three sets:

$$M_1 = \{(x, y): x, y \in R \text{ and } y \leq 4 - x\}$$
$$M_2 = \{(x, y): x, y \in R \text{ and } y \geq 4 - 2x\}$$
$$M_3 = \{(x, y): x, y \in R \text{ and } y \geq 0\}$$

The set illustrated in Figure 1.10 is both closed and strictly bounded. Can you distinguish between strictly bounded and closed when these terms are applied to sets?

Sets also may be bounded on some side. For example, the set given by $L \cap L_1^*$ in Figure 1.9(a) is bounded from above (by 2, since all points in the set $L \cap L_1^*$ have y coordinates of 2 or less). The value 4 for y constitutes a least upper bound (l.u.b.) to the set $M_1 \cap M_2 \cap M_3$ in R. The largest of lower boundary points is known as the greatest lower bound (g.l.b.).

The notion of boundedness also applies to line segments. Suppose that the letter x is used to stand for any point on a line segment and that a and b are real

---

[8] More generally for sets in n-dimensional space an n-dimensional sphere is substituted for the circle.

numbers, $a \leq b$, then the following definitions are given (note the use of the semicolon in this notation):

1. Closed interval $[a; b] = \{x: a \leq x \leq b\}$
2. Open interval $(a; b) = \{x: a < x < b\}$
3. Half-closed interval $[a; b) = \{x: a \leq x < b\}$
   or $(a; b] = \{x: a < x \leq b\}$

Observe that a closed interval is a closed set since it contains its boundary points. An open interval consists only of interior points and is thus an open set. The half-closed interval is neither closed nor open since it contains some but not all (one but not both) of its boundary points. Intervals are just particular sets of real numbers.

Convex Sets

Intuitively, a convex set is a set with no holes or indentations in it. If p_1 and p_2 constitute any two points in a convex set, then any point on the line segment connecting these two points is in the convex set. Figure 1.11(a) illustrates four convex sets while Figure 1.11(b) illustrates four sets that are not convex. In each of the latter four cases, two points have been located in the set for which the connecting straight line does not lie entirely within the set.

Figure 1.11

Convex and Nonconvex Sets.

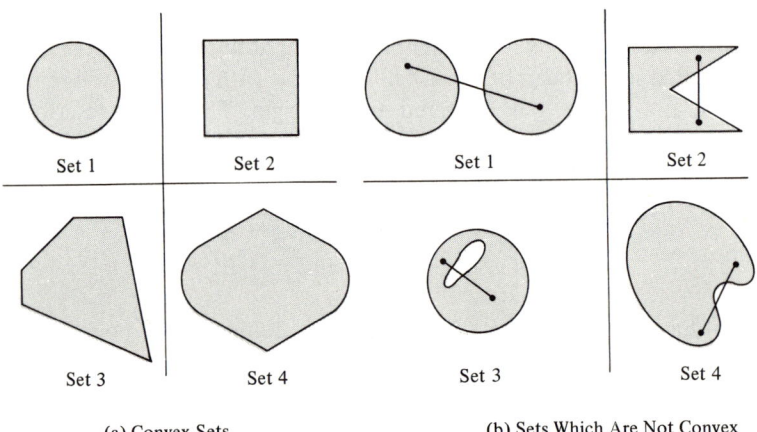

Definition—Convex Set: *A set X is convex if, for any two points p_1, p_2 in the set, the line segment joining these points is also in the set.*[9]

An examination of the boundary points of the convex sets in Figure 1.11 indicates that there are two types of boundary points. Some boundary points lie

---

[9] By convention, a set consisting of a single point is convex.

Sec. 1.7] *SUMMARY* 23

"between" other boundary points while others do not. That is, some boundary points lie on line segments constructed between two other boundary points while others do not. Boundary points that do not lie on any line segment between two other points in a convex set are called *extreme* points. Examples include the circumference of a circle and the vertices of a square or triangle. Finally, it is noted that a convex set need not be closed or bounded. The open half space K in Figure 1.8 is an example.

1.7 Summary

This chapter began with the concept of a set: a well-defined collection of distinct elements. "Well-defined" means that it is known whether or not a particular element belongs to a set or not. The adjective "distinct" means that no two elements in a set are identical. Two sets are said to be equal if they have the same elements.

Often the order in which the elements of a set are selected is important to a problem. When the order is both known and important, the set is written as an n-tuple. An ordered pair is the special case where n is equal to two.

Certain special sets are of particular importance in set theory. Among these are the null, or empty, set; the set of all elements relevant to a particular problem, the universe set; the Cartesian product set of two sets, the set of all ordered pairs of the two sets; and subsets. A subset of a given set is a set whose elements, if any, are elements of the given set. Any collection of subsets of a given set whose intersection is empty and whose union again yields the given set forms what is called a partition of the given set. A partition, then, is a set. In addition, the set of all subsets of a given set is called the power set of the given set.

Several operations may be defined on a group of sets. The most important are set union, the set of elements included in one or more of the sets; and set intersection, the set of elements common to all sets in the group.

Similar to the set operations just discussed, is the idea of matching elements of different sets. Where this matching is such that each element in one set is matched to a single element in a second set and vise versa, a one-to-one correspondence is established between the two sets, and the sets are said to be equivalent.

The most important use in this chapter of the ideas developed so far is in the definition of a function. A function is a set of ordered pairs in which each and every element in one set (called the domain set) is matched by some rule to a single element in another set (called the range set). The range set need not be exhausted by the matching or mapping process. A function is a subset of a Cartesian product set. The rule for a function is commonly given by an equation. This equation may provide all the additional information needed to determine the set of ordered pairs, the function, if the domain and range sets are understood to be the set of real numbers. However, the reader must be careful not to confuse the pairing rule with the function: they are not the same.

A point set is an ordered n-tuple of points. A point set specified by a linear inequality is a half space. Half spaces are called open half spaces if they do not include their boundary points and closed half spaces if they do (the boundary points are a subset of the point set). The closed half space has two types of points, boundary points and interior (nonboundary points); an open half space is made up of only interior points. The sets made up of the intersections of two or more half spaces satisfy more than one inequality. A set may be strictly bounded; that is, all of its points may fall within (be a subset of) a circle of finite radius centered at the origin. A convex set is a set with no holes or indentations; a straight line connecting any two points in a convex set lies entirely within the set. Boundary points of a convex set that do not lie on a line segment connecting other boundary points are called the extreme points of the set.

To quote Professor Spivey at this point:[10]

> The reader may now have the impression that we have proliferated definitions merely for the sport of it, getting further and further away from possible applications in the process, but it turns out that these concepts are of fundamental importance to much of both pure and applied mathematics.

For example, the set-theoretic approach is a very useful and intuitive path to understanding both calculus and linear algebra. Furthermore, the ideas underlying many new and important techniques including mathematical programming and critical path analysis are perhaps most easily grasped if the basic ideas of set theory are used in their development. In short, needs for these new definitions and concepts will develop at various and occasionally unexpected times.

Key Concepts

Concept	Reference Section
Set	1.1
Element	1.1
Ordered Set or n-tuple	1.1
Set Equality	1.1
The Empty or Null Set	1.2
The Universe or Population Set	1.2
Subsets and Proper Subsets	1.2
Partitions	1.2
Mutually Exclusive or Disjoint	1.2
The Power Set	1.2
Set Union	1.3
Set Intersection	1.3
Set Difference	1.3
Set Complement	1.3
One-to-One Correspondence	1.4
Equivalence	1.4
Pair	1.4
Cartesian Product Set	1.4
Relation	1.4

[10] A. W. Spivey, *Linear Programming and the Theory of the Firm*, New York: Macmillan, 1960, p. 23.

Key Concepts (cont.)

Concept	Reference Section
Function	1.5
Domain	1.5
Range	1.5
Mapping	1.5
Set Functions	1.5
Point Sets	1.6
Half Spaces	1.6
Boundary Points	1.6
Interior Points	1.6
Convex Sets	1.6
Extreme Points	1.6

Appendix 1A: The x-Element Technique

In this chapter Venn diagrams are used to illustrate various useful relationships. Another means of verifying expressions given in set form is the x-element technique. The procedure works as follows for the *equality relation*. Suppose that the set equality $A = B$ is under consideration. The relation is analyzed to see if an arbitrary element, x, which is included in the set A on the left-hand side of the relation is included in set B on the right-hand side of the relation. Then the same procedure is repeated for an arbitrary element on the right-hand side. It is necessary to examine the right-hand side, because although all elements on the left are included on the right, some element on the right may not be included on the left. Let x be this arbitrary element. Thus, x is a variable over the set in the sense that it can stand for any member of the set. The technique is illustrated for the associative law, $A \cup (B \cup C) = (A \cup B) \cup C$.

1. Let $x \in (A \cup (B \cup C))$; then either $x \in A$ or $x \in (B \cup C)$, by the definition of set union.
2. If $x \in A$, then $x \in (A \cup B)$ by the definition of set union. This implies that $x \in ((A \cup B) \cup C)$.
3. If, on the contrary, $x \notin A$ but $x \in (B \cup C)$, then either $x \in B$ or $x \in C$, by the definition of set union.
4. If $x \in B$, then $x \in (A \cup B)$ which implies that $x \in ((A \cup B) \cup C)$.
5. If $x \notin B$, but $x \in C$, then $x \in$ (any set) $\cup C$ by the definition of set union. In particular, $x \in ((A \cup B) \cup C)$, since $(A \cup B)$ is just some set.

Therefore if x is an element on the left-hand side of this relation, it is an element on the right-hand side. It is now necessary to show that if x is an element included on the right-hand side of the relation, then it is included on the left. This is left to the reader as an exercise.

This appendix concludes with an example in which all elements included on the left-hand side are included on the right-hand side, but the reverse is *not* true. Let A be the set of positive integers less than 10, B be the set of positive integers less than 8, and C be the set of positive integers less than 12. The set A is *not* equal to the set $B \cup C$ although every element in A is an element in $B \cup C$. This is because there are elements in $B \cup C$, 11 and 10, which are not in A. The set A is a subset of $B \cup C$ however. (Can you show this using the x-element technique?) Any of the set relationships discussed in Chapter 1 may be established using this technique.

Appendix 1B: Number Sets

The sets of real numbers, R, and ordered positive integers, I^+, are used extensively in this book. Figure 1B.1 interrelates various sets including the set R and the ordered set I^+.

The ordered set I^+ of positive integers or natural numbers is best thought of as the set of common counting numbers, $I^+ = (1, 2, \ldots, n, \ldots)$. There is no end to the listing.

Figure 1B.1

Number Sets.

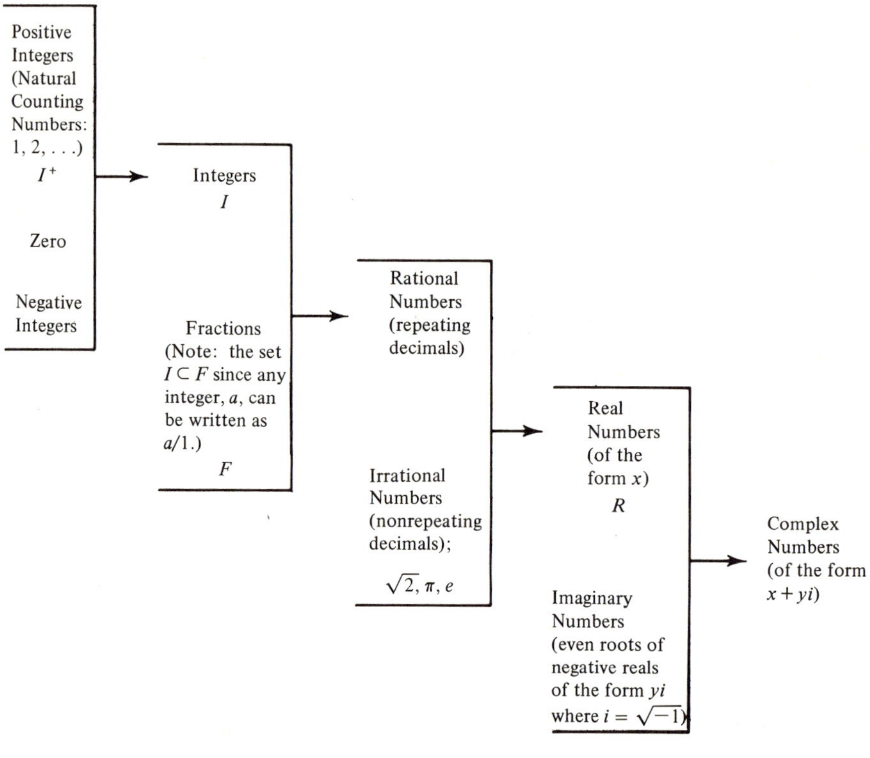

```
Positive
Integers
(Natural
Counting
Numbers:
1, 2, . . .)
    I⁺         →    Integers
                        I
    Zero
                                        Rational
   Negative                             Numbers
   Integers         Fractions          (repeating
                   (Note: the set      decimals)
                   I ⊂ F since any
                   integer, a, can                      Real
                   be written as                        Numbers
                   a/1.)                                (of the
                        F                               form x)
                                        Irrational          R
                                        Numbers                         Complex
                                        (nonrepeating                   Numbers
                                        decimals);                      (of the form
                                        √2, π, e                        x + yi)
                                                        Imaginary
                                                        Numbers
                                                        (even roots of
                                                        negative reals
                                                        of the form yi
                                                        where i = √(−1))
```

Appendix 1C: The Distinction Between a Function and an Equation

Let X be the set of real numbers, and let Y also be the set of real numbers. Let y be an element of the set Y and x be an element of the set X. Then the rule $y = 5x$ maps each x in X to a single value y in Y. This rule can be used to describe a function, call it f, in the product set X cross Y. The ordered pairs of this function f are too numerous to be listed, but the function f can be described in full by

$$f = \{(x, y): y = 5x, x \in X = R, y \in Y = R\}. \tag{1C.1}$$

If any of the information inside the braces were excluded, the description of the function would be incomplete. It may seem harmless, in this case, to exclude the symbolic sentence $y \in Y = R$ from the description. This equation sentence says: "y is an element of the

APPENDIX 1C: DISTINCTION BETWEEN A FUNCTION AND AN EQUATION

set Y which is equal to the set of real numbers." If this part of the description of f is excluded, equation (1C.1) becomes

$$f = \{(x, y): y = 5x, x \in X = R\}. \tag{1C.2}$$

Suppose that this description of the function f is used, is f an onto function? If the range set is R, the answer is yes. But suppose that the range is the set $C = \{z + bi:$ where $z \in R, b \in R; i = \sqrt{-1}\}$. The set C is called the set of complex numbers. (See Appendix 1B where a complex number is shown as the sum of a real plus an imaginary number.) If the range set were C, that is, if it is intended that $y \in C$, then f would be an into function only. The function f maps the elements of X *onto* the subset of C for which $b = 0$. This book uses only the real numbers; complex numbers will not be considered. Therefore, the second and more abbreviated description of f will not be confusing.

Suppose now that an even more abbreviated description of f is considered:

$$f = \{(x, y): y = 5x\}. \tag{1C.3}$$

There is now an even greater danger of confusion. Consider the following set, which is a function:

$$g = \{(1, 5), (2, 10)\}. \tag{1C.4}$$

Is g the same function as f? The answer is no. The elements $(3, 15)$, and $(.5, 2.5)$ are elements of f as defined by equation (1C.1), but they are not elements of g, so the functions are distinct. The distinctness is due to the fact that the domain of f as defined earlier is the set of real numbers, while the domain g is the set $\{1, 2\}$. However, the distinctness of f from g is not obvious given only the abbreviated description that $f = \{(x, y): y = 5x\}$. Nevertheless, abbreviations are useful if they do not lead to confusion. When it is understood clearly that the domain of x is the set of real numbers, the above abbreviated description for the function f should not be confusing.

The student may ask: If the description of f has been abbreviated to this extent, why not carry the process one step further? What information is conveyed by the symbol (x, y) in the description of f? If the symbol (x, y), is eliminated, the following description of the function f remains:

$$f = \{y = 5x\}. \tag{1C.5}$$

This is not acceptable. The equation $y = 5x$ is an equation and not a set while f is being used to stand for a function, a set of ordered pairs. The following sentence however would be acceptable: "The function whose rule is $y = 5x$." This sentence describes the function, but now the name of the function f is eliminated. An alternative form is $y = f(x) = 5x$. This form tells us that there is a function, called f, whose domain and range are understood to be the set of real numbers, and whose rule is $y = 5x$.

Even if the domain set is understood to be the set of real numbers, however, there is still one possible avenue for confusion that could result from specifying a function in abbreviated form; that is, by giving only its rule, $y = 5x$, or only its rule and name, $y = f(x) = 5x$. For example, consider the following full description of a function h:

$$h = \{(y, x): y = 5x, x \in X = R, y \in Y = R\}.$$

Note that this function has the same rule as f; that is, $y = 5x$. But h and f are not identical. The ordered pair $(1, 5)$ is an element of f, but it is not an element of h; the pair (x, y) from f is replaced with (y, x) in h. Similarly, $(1, 0.2)$ is an element of h but not an element of f. The distinction between h and f may be clearer if it is noted that the ordered pairs of f can all be written in the form $(x, 5x)$ while the ordered pairs of h can all be written in the form $(5x, x)$. To avoid this possibility of confusion the symbol x is usually used for the elements in the domain set of the function, whereas the symbol y is commonly used to indicate the element in the range set. Thus, when the rule $y = 5x$ is given to specify a function, the function f is intended. Usually, if reference to the function h is desired, its rule would be described by $y = x/5$, and the ordered pairs would be written (x, y).

In this chapter, sets have been denoted by capital letters unless they were functions. In later chapters, it is desirable to specify functions using small or capital letters. Thus the form $y = f(x)$ is used where f stands for the function or $y = F(x)$ where F stands for the function.

Problems

1. Under what conditions would two ordered sets be equal?
2. Is the following true?
$$\{2, 4\} = \{\{2\}, \{4\}\}.$$
3. Suppose that a market researcher is testing a new product. For each potential customer interviewed, he determines:
    (a) Whether the potential customer prefers the new product to the leading competitive product.
    (b) The income of the potential customer.

    Suppose that the set of all possible results is partitioned into the sets {prefer new, income high}, {prefer new, income low}, {prefer other, income high}, and {prefer other, income low}. Can you think of any uses that this partition might have in helping the company to decide where and how to advertise and market its product?
4. Suppose $S^{**} \subseteq S$ and $S \subseteq S^{**}$ where S and S^{**} are *any* sets, what do you conclude?
5. Can you show that $\emptyset$ is unique?
6. If $S = \{x, y, z\}$; $K = \{\{x, y\}, \{x, z\}, \{x, y, z\}\}$, indicate which of the following hold (all parts of each statement must be true before it can be said that the statement holds) and indicate why any statement is false.

    (a) $x \in S$; $x \in K$ \qquad (e) $\{x, y, z\} = \{y, x, z\}$
    (b) $S \in K$; $\{x\} \in S$ \qquad (f) $S = K$
    (c) $\{x, y\} \subseteq K$; $\{x, y\} \subseteq S$ \qquad (g) K is a partition of S
    (d) $\{\{x, y\}\} \subseteq K$ \qquad (h) $K \subseteq S$

7. (a) If $S^* \subseteq S \subseteq L$, is $S^* \subseteq L$? (That is, is the subset relation transitive?)
    (b) Is $S \subseteq S^*$? (That is, is the subset relation reflexive?)
8. Consider the set $A = \{1, 2, 3\}$. Write out the set 2^A.
9. Contrast the sets that make up the elements of the power set for S with the sets that are the elements of the set of all ordered sets that can be constructed from S. (Try the set $S = \{1, 2\}$.)
10. Show a set B as a proper subset of a set A using the Venn diagram technique.
11. Indicate, using a Venn diagram, the union and intersection of three overlapping sets.
12. Given the statements: (a) $A \cup \emptyset = A$; (b) $A \cap \emptyset = \emptyset$. Are they true or not? Why? (These statements are known as identity laws.)
13.[11] Illustrate, using the Venn diagram technique, that
    (a) $A \cap (B \cap C) = (A \cap B) \cap C$ \qquad (An associative law)
    (b) $A \cap (B \cup C) = (A \cap B) \cup (A \cap C)$ \qquad (A distributive law)
14. Illustrate $M \cap K = \emptyset$ using a Venn diagram.
15. (a) Give an example in which $M \cup K = \emptyset$ and $M \cap K = \emptyset$.
    (b) What is implied if only $M \cup K = \emptyset$ is true?
    (c) What is implied if $M \cap K = \emptyset$ only is true?

---

[11] Problems 13 and 14 are somewhat difficult, but serve as a good means for the reader to test his understanding of set relations.

PROBLEMS

16. Is it true that $A \cap A' = \emptyset$ and $(A')' = A$? Why or why not?
17. Illustrate $A - B'$ where A is the universe set, using the Venn diagram technique.
18. Illustrate (a) $(A \cup B)' = A' \cap B'$, and (b) $(A \cap B)' = A' \cup B'$. These two relationships are known as de Morgan's laws.
19. Illustrate $M - K = \emptyset$ and $M - K = M$. Can $M - K = K$? Why or why not?
20. If $M - K = K - M$, what can you conclude?
21. (a) Show $B = (A \cap B) \cup (B - A)$.
    (b) Write B in terms of $A \cap B$ and a complement of $(A \cap B)$.
    (c) Write B in terms of B' only.
    (d) Write B using $(A \cap B)$ and $(A' \cap B)$ only.
22. Define the set $A \triangle B$, the symmetric difference between sets A and B, to be equal to $(A \cap B') \cup (A' \cap B)$. Illustrate this set with a Venn diagram where $A \cap B \neq \emptyset$. Is $A \triangle (B \cap C) = (A \triangle B) \cap (A \triangle C)$ true?
23. If for some collection of sets E_i, $i = 1, \ldots, n$, it is known that $E_i \subseteq A$; $E_i \cap E_j = \emptyset$ for $j = 1, \ldots, n$ and $j \neq i$; and $E_1 \cup E_2 \cup \cdots \cup E_n = A$, what conclusion(s) can be drawn?
24. Verify the argument below by writing the hypotheses and conclusions in set terms and showing that hypotheses 1 and 2 imply the conclusions. Also show this using Venn diagrams.

    Hypotheses: 1. All products of the X Company have annual sales of at least 100,000 dollars.
    2. No product with sales of 100,000 dollars or more is unsuccessful.

    Conclusion: No product of the X Company is unsuccessful.

    (If one wished to argue with this line of reasoning, where could he begin?)
25. The report of the inspector on an assembly line showed the following for 100 units:

Item	Defect	Number of pieces
1	strength (defect A)	50
2	flexibility (defect B)	20
3	radius (defect C)	30
4	A and B	15
5	A and C	8
6	B and C	12
7	A, B, and C	5

    The report was returned. Why? What is your estimate of the number of pieces with both radius and strength defects, but with no flexibility defect (a) given that item 7 is correct? (b) given that items 1, 2, 3, 4, 5, and 6 are correct and that the number of units with defect B only is 1?
26. A factory committee consists of three management members and two union members. For a proposal to be passed on from the committee to the stockholders for consideration, a simple majority of the members must vote for the proposal. Each member has one vote.

    (a) How would you describe in set terms a group that could veto any proposal brought before it? How many such groups are there in the present situation? List them.
    (b) If, in addition to the original problem statement, one member has the power to veto any decision by his no vote, list or define in set terms the groups that could veto a proposal. How many groups would there be?

27. In a survey of consumer buying intentions conducted on 1000 families at the beginning of year 19—, 200 of the families interviewed indicated that they would purchase a new kitchen range during the coming calendar year. The other 800 families said they did not intend to purchase a new range. At the end of the year a resurvey of the same 1000 families showed that of those who indicated that they would purchase a new range, 180 did so while 200 of those who had no initial purchase intentions did so.
   (a) What proportion of families acted in accordance with their expressed intentions at the beginning of the year?
   (b) Do the data suggest that if a family intends to purchase a stove it usually does so? Why?
   (c) Do the data suggest that the intentions of people in a group to purchase are a good means of estimating the total purchase that will be made by that group?

28. An advertising agency is provided with the following information concerning the audience for each of three magazines.

Magazine	Audience (thousands)
M_1	1000
M_2	800
M_3	500
M_1 and M_2	160
M_1 and M_3	100
M_2 and M_3	80
M_1, M_2, and M_3	16

How many different individuals make up the total audience of the three magazines together?

29. What can be said about the number of elements in equivalent sets?

30. Are equivalent sets necessarily equal? Why?

31. Is there a natural pairing between equal sets that makes them equivalent?

32. Is an ordered pair an ordered set? Why or why not?

33. Write an ordered pair using braces { } only.

34. Use the notion of an ordered pair to define a one-to-one correspondence.

35. Let $S = \{2, 4, 6, 8\}$. The power set 2^S is a set of subsets. The Cartesian product $S^2 = S \times S$ is a set of ordered pairs.
   (a) How many elements are there in *each member* of $S \times S$?
   (b) List all the subsets of S (elements of 2^S) that contain two elements. Are there ordered pairs in $S \times S$ for each of these subsets? Are there any elements in S^2 for which there is no 2-element subset in 2^S?

36. Is the set R^4 a Cartesian product set? Why or why not?

37. Is the function of S into M given at the beginning of Section 1.5 the only function of S into M possible? Why? If not, give two others.

38. Is a function a subset of the product set, if both are properly defined?

39. Is a function a relation? Is the converse true?

40. Is a function a one-to-one correspondence? Is the converse true?

41. Does Figure 1.6 illustrate a mapping from B into A? Why?

42. Does Figure 1.6 illustrate a function from A onto B? Why?

43. Suppose that a line is added from $3 \in A$ to $20 \in B$ in Figure 1.6. Are any functions defined?

PROBLEMS   31

44. If the present line from $3 \in A$ to $15 \in B$ in Figure 1.6 were taken out and redrawn from $3 \in A$ to $10 \in B$, would any functions exist? Why or why not?

45. Is an onto function an into function? Why or why not?

46. Extend the notion of a function to three sets, C, D, and H, by giving an example.

47. From the product set $S \times S$ illustrated in Figure 1.5, select a subset that exhausts the domain (that is, for each element s of S, there is an ordered pair that contains s as its first element), and which is a relation, but not a function.

48. From the elements of the relation selected in answering problem 47, select a subset so that the subset is a function. Is your function onto or into?

49. Can $S \times S$ be treated as the universe set for the function defined by your answer to problem 48? If not, specify an appropriate universe set. Find the complement of the function. Is the complement of the function itself a function?

50. For the following question, consider the product set $C \times D$ where $C = \{1, 2, 3, 4\}$ and $D = \{a, b, c, d, e\}$. Listed below are some descriptions of subsets. For each description, either find a subset that meets the description, or state why it is impossible to do so.

(a) This subset of $C \times D$ is an *onto* function from C *onto* D.
(b) This subset of $C \times D$ is a function C into D and also a function from C *onto* a *subset* of D.
(c) This subset of $C \times D$ is an *onto* function, and it also is a *one-to-one* function, from C *onto* a *subset* of D.
(d) This subset of $C \times D$ is an *onto* function from C *onto* a *subset* of D, but it is not a *one-to-one* function.
(e) This subset of $C \times D$ is an *into* function from C *into* D, and it is also a *one-to-one* function.

51. Let $X = R$ and $Y = R$ where X is the domain set and Y is the range set. For each of the following rules specify whether the rule defines a function or a relation from X into Y. If a function is *not* defined from X into Y, specify the maximum domain possible in the set of real numbers, R, and the subset of the range used in the mapping for which a function is defined. Also give the subset of the range used in the mapping for all functions of X into Y defined by the rule as given. It is useful to graph the rule in most cases.

Rule	Function or Relation	Maximum Real Domain	Subset of the Range Used in the Mapping
a. $y = 6x$	Function	R	R
b. $y = 1/x$			
c. $y = x^3$			
d. $y = x^2$			
e. $y = x^4 - x^2$			
f. $y = (x-1)/(x+1)$			
g. $y^2 - x = 0$	Relation	$\{x: x \in R, x \geq 0\}$	$\{y: y \in R, y \geq 0\}$ or $\{y: y \in R, y \leq 0\}$ or both
h. $y^2 - x^2 = 6$			

52. Suppose that $y = 2x$ where $x \in A$, $y \in B$, and $A = B = I^+$. Is there a function defined from A into B? From B into A? Why?

53. Specify a simple function by using a rule in graphic, verbal, and tabular form, respectively.

54. A few special functions are given below; give an example for each one:
    (a) Set function: The domain is a collection of sets.
    (b) Real-valued function: The range is a subset of the real numbers.
    (c) Constant function or constant: All elements in the domain are mapped into a single element in the range; that is, only one element in the range set is used in the mapping.
    (d) Absolute-value function: The domain is the set of real numbers and the equation is $f(x) = |x|$ where $|x| = x$ if $x \geq 0$ and $= -x$ if $x < 0$, that is, if $x = -3$, $f(-3) = |-3| = -(-3) = 3$.
    (e) Identity function: A set is mapped onto itself.

55. Given: $Q = \{(x, y): x = 2y, x \in A, y \in B\}$, $A = B = \{1, 2, 3, 4\}$. Is the set Q a function? If not, why not? If so, give the set of ordered pairs and specify the range and domain of the function.

56. Let the set $J = \{j_1, j_2, \ldots, j_i, \ldots, j_n\}$ be a set of n jobs, and let $T = \{t_1, t_2, \ldots, t_l, \ldots, t_n\}$ be a set of times.
    (a) Define and write the function from J into T which assigns to each job the time needed to perform that job only. Call this set $M = \{m_i\}$.

    Now consider a new set $K = \{k_1, k_2, \ldots, k_n\}$ where k_i is the set of jobs that must be completed before job j_i can be begun.

    (b) Define a function P, for project, from K into M. Write P in set terms. (This may be easier to do with specific values given for the various terms. If you have trouble, try using the specific values given in part (c).)
    (c) Suppose that $n = 4$. The times to complete jobs 1 through 4 are 1, 2, 3, and 4, respectively; $k_1 = \emptyset$; $k_2 = k_3 = \{j_1\}$; and $k_4 = \{j_1, j_2, j_3\}$. Write P and determine the shortest time needed to complete the project.
    (d) Suppose that the time for job 2 is 1 instead of 2, then what is the shortest time needed to complete the project?

57. An oil wildcatter must decide whether to drill for oil at a specific location or sell his drilling rights. He may sell immediately or, alternatively, he may take seismographic recordings that yield information about the subsurface structure on the location and make his decision after examining the results. The information obtained is not perfect, however. Sometimes oil is found although no favorable subsurface structure was indicated and vice versa. Suppose that the following facts are known:
    (a) Sale price of drilling location (regardless of seismograph recordings) before a well is drilled: $100,000.
    (b) Cost of seismograph recordings: $10,000.
    (c) Cost of drilling: $40,000.
    (d) Value of a producing well: $500,000.
    (e) Value of a dry well: $0.

    Set up a function from the set of possible outcomes of the wildcatter's decisions into the set of real numbers such that each point is mapped to its money value. For example, if

Symbol	Meaning	Symbol	Meaning
s	sell	fs	favorable structure
tr	take recordings	nf	structure not favorable
nr	take no recordings	o	oil
d	drill	no	no oil

then two results are:
    (a) (s, $100,000) sell immediately, and
    (b) ((tr, fs, d, no), $-$ $50,000) in which recordings are taken, favorable structure is indicated, a well is drilled, but no oil is located.

Figure 1.12

Decision Tree: Wildcatter Problem

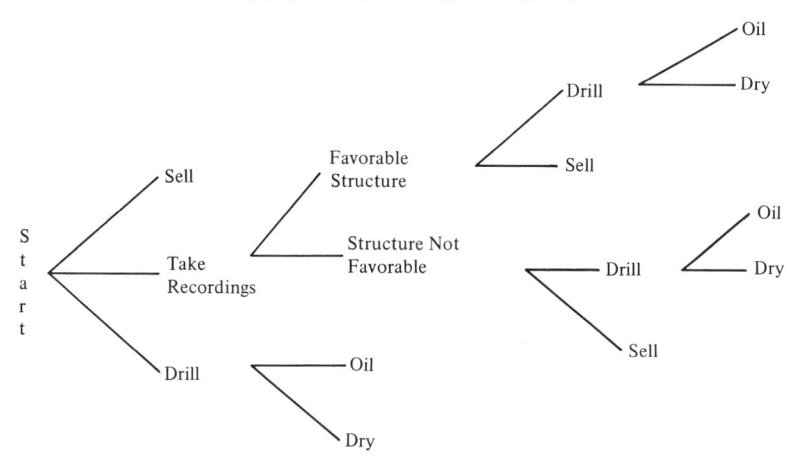

```
                                                    ,Oil
                                          ,Drill <
                                          /         ` Dry
                   Favorable             /
           ,Sell   Structure     /      ----- Sell
          /       /             /
         /       /                              ,Oil
  Start <- Take            Structure Not    ,Drill <
         \  Recordings      Favorable      /      ` Dry
          \       \                        /
           \       \                       \
            `Drill  \----- Oil              ` Sell
                     \
                      ` Dry
```

The above possibility list (decision tree diagram) may be helpful.

58. The following sets, which describe intervals of real numbers, are in common usage and the following simplified notation is common.

A closed set: $[a; b] = \{x: x \in R; a \leq x \leq b; a < b \in R\}$
An open set: $(a; b) = \{x: x \in R; a < x < b; a < b \in R\}$
Half-open sets: $[a; b) = \{x: x \in R; a \leq x < b; a < b \in R\}$
$(a; b] = \{x: x \in R; a < x \leq b; a < b \in R\}$

Write the following intervals using this simplified notation.

(a) $(a; b) \cup \{a\} \cup \{b\}$
(b) $[a; b) \cup [b; c]$
(c) $(a; b] \cap [b; c)$

59. (a) Is a convex set necessarily strictly bounded?
(b) Is a convex set necessarily closed?

60. (a) Is a straight line a convex set?
(b) Is $y = x^2$ a convex set?

61. Is a closed half space a convex set?

62. See if you can form two convex sets whose intersection is neither the null set nor convex, and whose union is not convex.

63. Is the set of real numbers a strictly bounded set? Is it convex?

64. A firm makes two products, X and Y, from two raw materials, M_1 and M_2. The firm has available 14 units of M_1 and 12 units of M_2. It also has available 16 hours of labor. If for each unit of X it takes 2 units of M_1, 1 unit of M_2, and 1 unit of time while for each unit of Y it takes 1 unit of M_1, 1 unit of M_2 and 2 units of time, indicate graphically the possible amounts of X and Y that can be produced and where the optimum profit point may be.

65. Given sets:

$A = \{1, 2, 3, 4\}$, $B = \{\text{aardvarck, heffalump}\}$,
$C = \{a, b, \{c\}\}$, $D = \{1, 4, 9, 16\}$, and $E = \{c\}$.

(a) Indicate whether the following statements are true or false:
  (i) $E \subset C$
  (ii) $\{1, 2\} \subset A$
  (iii) $\{\{a, b\}, \{\{c\}\}\}$ is a partition of C
(b) Indicate which of the following sets are legitimate functions and change those that are not so that they are functions.
  (i) $f = \{(x, y): y = x^2, x \in A, y \in D\}$ where A is the domain set and D is the range set.
  (ii) $f = \{(\text{aardvarck}, \{c\}), (\text{heffalump}, \{c\})\}$, where B is the domain set and C is the range set.
  (iii) $f = \{(c, 1), (c, 2)\}$, where E is the domain set and A is the range set.
(c) Form the following sets:
  (i) The power set of B
  (ii) The Cartesian product set $B \times E$
  (iii) A one to one function using A and D
  (iv) $A \cap D$

CHAPTER 2

Limits and Continuity

This chapter deals with a concept fundamental to the development of both differential and integral calculus, the concept of a limit. This concept is basic to understanding the meaning of the slope of a curve, the measurement of the area under a curve, the value of irrational numbers, and many other important relationships useful in the development of mathematical models for important situations and the solution of managerial problems.

The limit of a special function, the sequence, is discussed in Section 2.2 and a procedure is presented for determining whether the function has a limit. The concept and the procedure are then extended to more general functions and a number of illustrations are discussed. The limit concept is then used in the development and definition of the continuity of a function.

2.1 Limits

It is relatively easy to find the limits of simple functions, and because more complicated functions can often be divided into groups of simple functions, the limits for these too can be found by applying the basic rules for taking limits to each member of the groups. Also, it is generally possible to identify readily those cases in which a limit does not exist; therefore, in practice, the evaluation of limits usually does not present a serious problem for the analyst.

Assume, for example, that the regional sales manager of a firm has reached the following judgment about assigning salesmen to a given territory: Each salesman assigned can exploit one half of the *untapped* sales potential of the territory. This judgment is based on the manager's knowledge of the potential customers, their locations throughout the territory, and the attractiveness of customers in the territory to his competitors. The expected contribution of each additional

Table 2.1

Proportion of Sales Potential Contributed by Each Additional Salesman and Cumulative Total (Per Cent)

Salesman Number	Proportion of Total Potential Added by the Salesman	Cumulative Proportion of Potential Sales[a]
1	.5000000	.5000000
2	.2500000	.7500000
3	.1250000	.8750000
4	.0625000	.9375000
5	.0312500	.9687500
6	.0156250	.9843750
7	.0078125	.9921875
⋮		
n	$(0.5)^n$	$\sum_{i=1}^{n} (0.5)^n$
⋮		

[a] The Greek letter capital sigma $\sum$ is the symbol used to indicate summation. Thus $\sum_{i=1}^{n} (0.5)^n$ means $(0.5)^1 + (0.5)^2 + \cdots + (0.5)^n$. The letter i is known as the variable of summation.

salesman is presented in Table 2.1. With the assignment of *enough* salesmen, the regional manager can be assured of realizing *essentially all* the sales potential of the territory. But there *is* a limit to the sales he can expect; he can expect no more than 100 per cent of the potential. Furthermore he can never (quite) attain that level, since each new salesman adds only one half the remaining potential. Thus, no matter how many salesmen are added, some small percentage of potential cannot be obtained; 100% of potential cannot be reached.

The last paragraph contains the essence of the concept of a limit: first, that the limit, in this case 100 per cent of the potential, never needs to be reached for it to be a limit; second, that it is possible to approach the limit as closely as desired, in this case merely by assigning "enough" salesmen to the territory. For example, if the regional manager wanted at least 95 per cent of the potential sales of the territory how many salesmen would he need to assign? From Table 2.1 it can be seen that it is not until the fifth salesman is assigned that the total sales of the company will reach 95 per cent of potential sales, and then they should exceed the goal and reach almost 97 per cent. By making the number of salesmen large enough, it is possible to approach the limit as closely as desired.

Table 2.1 also contains a second example of a limit. The proportion of the total potential added by an additional salesman becomes smaller and smaller and eventually is almost zero as the number of salesmen becomes large. Thus, although each saleman always makes a positive contribution to sales in the territory, the contribution approaches zero as a limit.

Other examples of limits are the maximum speed for an airplane designed with aluminum skin (beyond a particular speed atmospheric friction would melt the skin), the minimum coefficient of friction of a ball bearing (it can approach but not reach zero), and the minimum level of unemployment attainable in the economy (some persons will always be in the process of changing jobs and some others of seeking work upon entrance to the work force). Intuitively, a limit can be thought of as a particular value that a phenomenon or process approaches and could potentially, although not necessarily, reach.

2.2 Sequences

The relationship between the number of salesmen assigned to the territory and the proportion of the sales potential achieved is a special type of function known as a *sequence*.

> ***Definition—Sequence.*** *A function f with domain variable n over the ordered set (or a subset) of positive integers I^+ is known as a sequence. For each $n \in I^+$ the sequence has a particular value y_n specified by the rule of the sequence.*

Actually, two different sequences have been defined in the illustration given in Section 2.1. One has as its domain the ordered positive integers $I^+ = (1, 2, 3, \ldots, n, \ldots,)$ and as domain variable n, the number of the individual salesman assigned.

Its rule is: the contribution $f(n)$ to total sales by *that* individual salesman is found from the equation,
$$f(n) = (\tfrac{1}{2})^n = (0.5)^n.$$
For example, the contribution of the second salesman ($n = 2$) is $(0.5)^2 = 0.25$. In set terms, this is the sequence f_1:
$$f_1 = \{(n, y_n): n \in (I^+), y_n = (\tfrac{1}{2})^n\}.$$
The second sequence necessarily has the same domain; the ordered positive integers $I^+ = (1, 2, 3, \ldots, n, \ldots,)$ but the domain variable n now signifies the *total* number of salesmen assigned to the territory. Its rule is: the cumulative contribution of all the salesmen through the nth assigned to the territory is found from the equation,
$$f(n) = \sum_{i=1}^{n} (\tfrac{1}{2})^i = \sum_{i=1}^{n} (0.5)^i.$$
That is, the cumulative contribution of the first and second salesmen is equal to the sum of the contributions of each; that is $(\tfrac{1}{2})^1 + (\tfrac{1}{2})^2 = (0.50) + (0.25) = 0.75$ (for $n = 2$). In set terms this sequence f_2 is written,
$$f_2 = \left\{(n, y_n): n \in (I^+), y_n = \sum_{i=1}^{n} (\tfrac{1}{2})^i\right\}$$

The limit of f_1 is the particular value L_1 approached in the range as the domain variable n becomes very large and approaches infinity. Although ∞ itself is not a number and therefore n can never equal ∞, the notation used to suggest limit values or values to be tested as limits is generally written in the form
$$\lim_{n \to \infty} y_n = L_1. \tag{2.1}$$
If y_n can be made as close as desired to L_1 by choosing a large enough value N of n, then L_1 is the limit of f_1.

In Section 2.1 the limit of the sequence f_1 is stated to be zero; after the number of salesmen in the territory has become large, the next or nth salesman assigned makes a contribution to total sales, $f_1(n) = (\tfrac{1}{2})^n$, that is very small and the contribution of each additional salesman becomes successively smaller. Similarly, the limit of the sequence f_2 is 1; the *sum* of the (successively smaller) contributions of individual salesmen becomes increasingly close to 100% of potential sales as n grows larger. These implications are now developed in a more rigorous fashion.

A particular sequence $f(n)$ typically takes on a different value for each n, but in many cases successive values of the sequence, y_n, tend to cluster around some fixed value L. If there is such a number L, and if y_n is close to the number L for all values of n greater than a particular number N, then L is the limit of $f(n)$. For y_n to be "close" in the sense desired here, the inequality
$$|y_n - L| < \epsilon \tag{2.2a}$$
must hold for a small positive number ϵ and for all $n > N$.

In the case of the sequence f_1, the inequality $|y_n - L_1| = |y_n - 0| = (y_n - 0) < \epsilon$ must hold. Because for f_1, y_n is always greater than zero, the absolute value sign

Sec. 2.2]   SEQUENCES   39

has been dropped and the requirement for a limit becomes, for any $\epsilon > 0$,

$$(y_n - 0) < \epsilon \qquad (2.2b)$$

which is equivalent to $y_n < \epsilon$. For *any* desired $\epsilon > 0$, it must be possible to select N such that inequality (2.2b) is satisfied.

This suggested procedure is now tested for the numerical values in the salesman example. The value of L_1, which is tested as the limit for f_1 as $n \to \infty$, is the value approached by y_n as $n \to \infty$; that is, $(\frac{1}{2})^n = 0$. For an arbitrarily selected value ϵ_1, of .05 it can be seen from Table 2.1 that $n \geq 5$ gives a value y_n such that the inequality holds:

$$y_5 = .03125 < .05.$$

For an ϵ_2 of .01 a domain value $n \geq 7$ makes inequality (2.2b) hold:

$$y_7 = .0078125 < .01.$$

The definition of a *limit of a sequence* is now formally stated.

>   **Definition—The Limit of a Sequence:**   *The sequence f with domain variable $n \in (I^+)$ has a limit L (to the value of its range variable y_n) if for any $\epsilon > 0$ there exists a value N of the domain variable n such that*
>
> $$|L - y_n| < \epsilon \qquad (2.2a)$$
>
> *for all $n > N$. The limit L is generally written* $\lim_{n \to \infty} y_n = L$.

This test can be applied also in the case of f_2. The test value for the limit as $n \to \infty$ is the range value as n becomes very large. For f_2

$$\lim_{n \to \infty} y_n = \lim_{n \to \infty} \sum_{i=1}^{n} (\tfrac{1}{2})^i = 1.00.$$

The variable n can never reach ∞, but this is immaterial. A limit value is a value *approached*, but *not* necessarily reached. The sum above can never reach one, nor can $\lim_{n \to \infty} (\tfrac{1}{2})^n$ ever reach zero although it can approach zero as closely as one might wish. The test for the limit L_2 shows that $f_2(n)$ can approach 1.00 as closely as one might wish.

Using the same arbitrarily small ϵ_1 and ϵ_2 as in the earlier illustrations—but this time using the test inequality in the form (2.2a)—for $\epsilon_1 = .05$, $N = 5$ works:

$$.03125 = |1.00 - .96875| = \left| L_2 - \sum_{i=1}^{5} (.5)^i \right| < .05;$$

for $\epsilon_2 = .01$, $N = 7$ works:

$$.0078125 = |1.00 - .9921875| = \left| L_2 - \sum_{i=1}^{7} (.5)^i \right| < .01.$$

---
[1] Here the Greek letter ϵ is used to mean "a small finite distance," and is different from the symbol $\in$ used to mean "an element of a set" in Chapter 1 as well as in this chapter.

The $\lim\limits_{n\to\infty} \sum\limits_{i=1}^{n} (\frac{1}{2})^i = L_2$, as this procedure illustrates for any ϵ.

The procedure is now illustrated with other examples.

Example 2.1

Given the sequence f_3 with rule $y = f(n) = (n + 2)/n$, select a number to be tested as the limit L_3 for the sequence and test it to see if L_3 is the limit of the sequence as $n \to \infty$. The values of the sequence are:

$$y_1 = \frac{3}{1}, \quad y_2 = \frac{4}{2}, \quad y_3 = \frac{5}{3}, \ldots, \quad y_n = \frac{n+2}{n}, \ldots$$

The number to be tested as the limit should be the value y_n approaches for n very large. As $n \to \infty$, y_n approaches the value 1 (since 10,002/10,000 is close to 1 and 100,002/100,000 is even closer). To test 1 as the limit for any small value ϵ, a domain value N must be found such that

$$|1 - y_n| < \epsilon \quad \text{for } n > N.$$

Because the value y_n is always greater than 1, $|1 - y_n|$ equals

$$(y_n - 1).$$

Then, by the rule for y_n,

$$y_n - 1 = \left(\frac{n+2}{n} - 1\right)$$

and since $1 = n/n$, this can be stated in terms of a common denominator,

$$y_n - 1 = \left(\frac{n+2}{n} - \frac{n}{n}\right) = \frac{n+2-n}{n} = \frac{2}{n}.$$

In order for the value $L_3 = 1$ to be the limit of this sequence, it must be possible to choose a value N of n such that the (modified) test value $2/n$ is smaller than any designated ϵ.

Assume that ϵ is specified as $1/100$. Can $2/n$ be made smaller than $1/100$ by choosing N large enough; that is, can a value N be chosen such that $2/n < (1/100)$ for all $n > N$? Simplifying by cross multiplication gives

$$200 < n.$$

Therefore, by choosing $N = 200$ this inequality will hold for $n > N$. What if ϵ were specified to be $1/1,000,000$? Again by choosing $N = 2,000,000$, the inequality will hold. That is, can a value N of n be chosen such that $2/n < (1/1,000,000)$ for all $n > N$? Again cross multiplying, $2,000,000 < n$. For $N = 2,000,000$, the difference between L_3 and $y_{2,000,001+}$ is less than $1/1,000,000$. The general rule[2] then is choose N such that $2/n < \epsilon$ or $2/\epsilon < n$ for any $n > N$.

---

[2] From this inequality, the value of N for f_3 can be calculated for any ϵ desired. The calculation to determine the N for an ϵ of $1/100$ is as follows: $n > 2/\epsilon$; $n > 2/(1/100)$; $n > 200$. Since N takes on only integer values, N must be 200 and the inequality holds for $n > N = 200$, or $n = 201$ or more. In the second test for $\epsilon = 1/1,000,000$, $n > 2/\epsilon$; $n > 2/(1/1,000,000)$; $n > 2,000,000$; therefore N must be 2,000,000 and the inequality holds for $n > N = 2,000,000$, or for $n = 2,000,001$ or more.

Sec. 2.2] SEQUENCES

The sequence f_3 is a discrete function with values only at particular points, as shown in Figure 2.1. Successive values of y_n approach closer and closer to 1 from above. It is possible to approach 1 as closely as desired by choosing n large enough, but y_n will always be greater than 1; the limit $L_3 = 1$ can only be approached, not reached.

Figure 2.1

The Sequence f_3 for Selected Values, n.

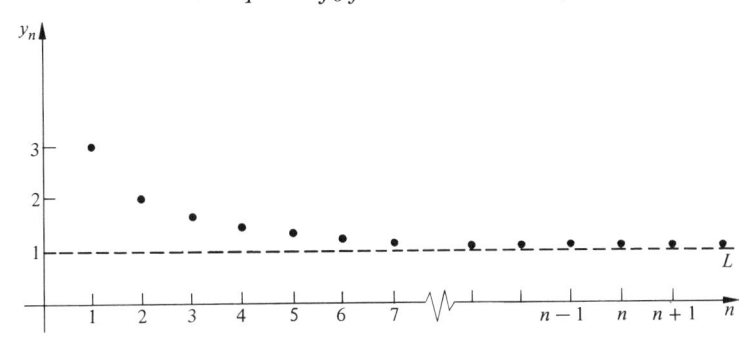

Example 2.2

Given the sequence f_4, with rule $y_n = n/(n + 2)$, select a number to be tested as the limit L_4 for the sequence, and test it to see if L_4 is the limit of f_4 as $n \to \infty$. The values of f_4 are

$$y_1 = \frac{1}{3}, \quad y_2 = \frac{2}{4}, \quad y_3 = \frac{3}{5}, \ldots, \quad y_n = \frac{n}{n+2}, \ldots$$

As $n \to \infty$, it appears that $y_n \to 1$. This value will be used for the test-limit L_4. For any designated ϵ, a domain value must be found such that $|1 - y_n| < \epsilon$. Since y_n is always less than 1, the absolute value sign can be dropped, giving $1 - y_n < \epsilon$. Substituting for y_n and simplifying as before gives

$$1 - \left(\frac{n}{n+2}\right) < \epsilon$$

or since $1 = (n + 2)/(n + 2)$,

$$\left(\frac{n+2}{n+2} - \frac{n}{n+2}\right) < \epsilon$$

or

$$\frac{n+2-n}{n+2} < \epsilon$$

or

$$\frac{2}{n+2} < \epsilon.$$

By cross multiplying

$$\frac{2}{\epsilon} < (n + 2).$$

Thus for $\epsilon = 1/100$,

$$\frac{2}{\epsilon} < (n + 2)$$

gives

$$\frac{2}{1/100} < (n + 2)$$
$$200 < n + 2$$
$$198 = N < n.$$

Thus $N = 198$ and for $n = 199$ or greater, the difference between L_4 and y_{199+} is less than $1/100$. For $\epsilon = 1/1,000,000$

$$\frac{2}{1/1,000,000} < (n + 2)$$
$$2,000,000 < n + 2$$
$$1,999,998 = N < n.$$

Thus, for $n = 1,999,999$ or greater, the difference between L_4 and $y_{1,999,999+}$ is less than $1/1,000,000$.

Because the function f_4 is also a sequence, its values are discrete. It approaches the limit, $L_4 = 1$, only from below, since its value is always less than 1. The function f_4 is shown graphically in Figure 2.2.

A sequence may approach a limit by alternating values on either side of a limit and approaching that limit as well as only from above or only from below as shown in Examples 2.1 and 2.2. (Can you find the rule for such a function?)

The discussion thus far should not be interpreted to suggest that every sequence has a limit, because there are many sequences that do not.[3] Some increase without limit as in

$$f_5 = \{(n, y_n): n \in (I^+), y_n = 2n\}.$$

Others decrease without limit, as in

$$f_6 = \{(n, y_n): n \in (I^+), y_n = (-3n + 3)\}$$

and some merely alternate in value, as in

$$f_7 = \{(n, y_n): n \in (I^+), y_n = 2 + (-1)^n\}.$$

---

[3] Sequences that do have a limit are known as *convergent* sequences, and they converge to the limiting value L. A sequence is convergent if, for large enough n, *all* the numbers in the sequence are close to some fixed number (the limit). The ϵ test of closeness is used in a formal definition of a convergent sequence. The value L for a sequence that converges from below is known as the least upper bound (l.u.b.) of the sequence, and for a sequence that converges from values above, the greatest lower bound (g.l.b.). If it is difficult to identify the limit to a sequence; Cauchy's convergence criterion can be used as a test of convergence; if the numbers of a sequence get close (enough) to *each other*, the sequence is convergent. Again the formal statement of the criterion is stated in terms of an ϵ test.

Figure 2.2

The Sequence f_4 with Rule $y = n/(n+2)$.

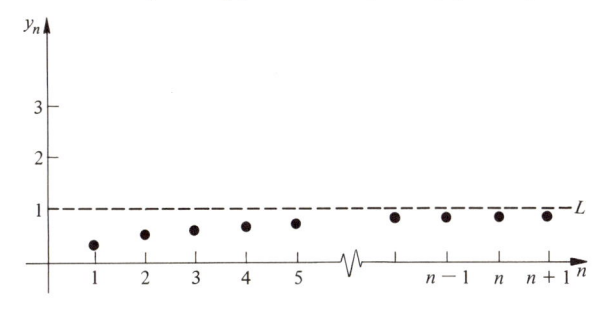

2.3 The Limit of a Function (General)

The discussion in Sections 2.1 and 2.2 dealt with sequences, functions having as domain the ordered set of positive integers. In this section the discussion is expanded to cover functions having as their domain the set of all real numbers, or a subset thereof. The main difference between the limit of a sequence and the limit of a function is that it is frequently necessary to discover whether a limit to the function exists for a given value in the domain, not just for values of $n > N$ in the domain, as in the case of the sequence. Also, in the case of a sequence, n, the domain variable, varies over the *ordered* set of positive integers; therefore $n \to \infty$ implies n increasing only.

For the case of the general function $x \to a$ implies that x approaches the value a in the domain from *either* direction, left to right or right to left. For a value L to be a limit to the function f as $x \to a$ (that is, as x approaches the value a in the domain from *either* direction) it must be possible for y to be brought as close to L as desired by choosing x close enough to a in either direction, but not equal to a. In more precise symbols, if the difference between L and $f(x)$ can be made less than an arbitrarily small number $\epsilon > 0$ by choosing x close to a on either side of a; that is, in the interval $(a - \Delta x_1) < x < (a + \Delta x_2)$, where Δx_1 and Δx_2 are small distances on the left and right of the point a, then L is the limit of f; that is, if

$$|L - f(x)| < \epsilon \qquad (2.2c)$$

for x such that $(a - \Delta x) < x < (a + \Delta x)$, then L is the limit of f as $x \to a$.

The symbols $|L - f(x)| < \epsilon$ may be translated into a more useful form by algebraic manipulation. (The reason for the change will become clear from the discussion of Figures 2.3–2.5.) If $L - f(x)$ is positive in inequality (2.2c), then $|L - f(x)| = L - f(x)$, and the absolute value sign can be dropped to give $L - f(x) < \epsilon$. Adding $f(x)$ to both sides of this inequality gives $L < \epsilon + f(x)$. Subtracting ϵ from both sides gives $L - \epsilon < f(x)$. If $L - f(x)$ is negative in inequality (2.2c), then $f(x) - L$ is positive and $|L - F(x)| < \epsilon$ can be written as $f(x) - L < \epsilon$. Adding L to both sides gives $f(x) < L + \epsilon$. Thus, $|L - f(x)| < \epsilon$ can be replaced by two inequalities, one for each possible algebraic value of

$|L - f(x)|$. The two inequalities are combined to obtain

$$(L - \epsilon) < f(x) < (L + \epsilon)^4$$

which is the form used in the definition of a *limit of a function*.

Definition—Limit of a Function: *Given the function*

$$f = \{(x, y), x \in R, y \in R, y = f(x)\},$$

then the value of the function approaches a limit L as $x \to a$, a value in the domain, if, for any arbitrarily small positive number ϵ, the inequality

$$(L - \epsilon) < f(x) < (L + \epsilon) \qquad (2.3)$$

can be made to hold by choosing any value of x in an interval $(a - \Delta x_1) < x < (a + \Delta x_2)$, except perhaps $x = a$.

If inequality (2.3) holds for the values of x in the specified interval, then $f(x)$ has the limit L as $x \to a$.

The requirements stated in the definition are illustrated in Figure 2.3 for the function f with rule $y = f(x) = 2x$. Assume that the question has been asked, "Does f with rule $y = 2x$ have a limit as $x \to 3$?" First, one must determine an appropriate value to test as the limit. The best starting place is to try $f(a)$, that is, $f(3)$. For this function $y = f(3) = 2(3) = 6$. The value 6 can be used as the test limit value of f as $x \to 3$.

Given the requirement that $f(x)$ be within ϵ distance of L, it is possible by choosing x in the interval $(a - \Delta x) < x < (a + \Delta x)$ for $f(x)$ to fall within the interval

$$(L - \epsilon) < f(x) < (L + \epsilon).$$

The arbitrary value of ϵ illustrated in Figure 2.3 is 0.5. By choosing a value of $x > 3$ but close to 3, $f(x)$ is made to fall within the interval between 6.0 and $6 + 0.5 = 6.5$. Any value of x such that $3.00 < x < 3.25$ will suffice. For example, for $x = 3.24$, $f(x) = f(3.24) = 6.48 < 6.50$. Thus $\Delta x_2 = 0.25$ identifies the upper end of the appropriate interval.[5] Testing for a limit from above is sometimes called "testing for a right-hand limit."

---

[4] An illustration may help the reader see that these expressions are equivalent. The two values $|6 - 4|$ and $|4 - 6|$ are equal—both are 2 in absolute value. Thus $|6 - 4| < 3$ is a correct statement. By dropping the absolute value sign this may be rewritten

$$-3 < (6 - 4) < +3,$$
$$-3 < 2 < +3,$$

or redoing this for $|4 - 6| < 3$ and dropping the absolute value sign:

$$-3 < (4 - 6) < +3$$

or

$$-3 < -2 < +3.$$

[5] For a value of $a = 3$ in the domain in Figure 2.3 and an ϵ of 0.5 it is shown that $2.75 < 3 < 3.25$ is the range of appropriate domain values. These are symbolically $a - \Delta x_1 < a < a + \Delta x_2$. The value of Δx that satisfies this relationship is 0.25 on both sides of a, so the appropriate interval is $a \pm \Delta x = 3 \pm 0.25$.

Figure 2.3

Function f with Rule $y = f(x) = 2x$ with Limit L as $x \to 3$.

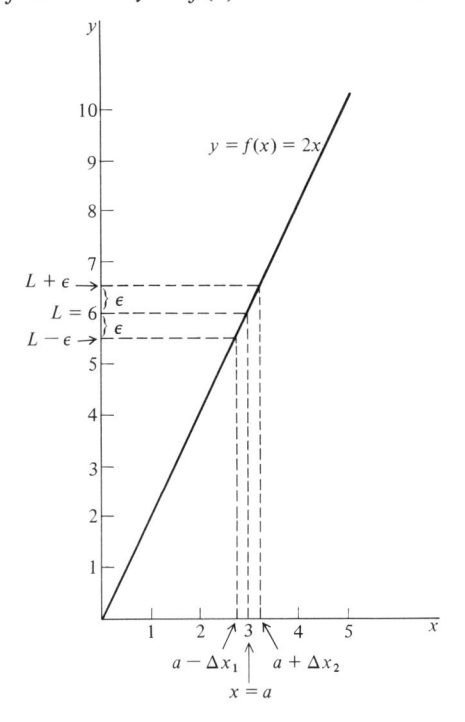

Similarly for the ϵ of 0.5 and values of $x < 3$, $(6 - 0.5) = 5.5$ is the lower limit of the interval within which $f(x)$ may fall: Any value of x such that $2.75 < x < 3.00$ will satisfy the requirements for the existence of the limit for $\epsilon = 0.5$ on the lower (left-hand) side (the left-hand limit), that is, $\Delta x_1 = 0.25$ also. For the value of $x = 2.76$, $y = f(2.76) = 2(2.76) = 5.52$; therefore, $(L - \epsilon) = (6 - 0.5) = 5.50 < 5.52 = f(a - \Delta x_1)$. For any $\epsilon > 0$ it will always be possible to find an interval for values of x such that $(L - \epsilon) < f(x) < (L + \epsilon)$ for any x (except perhaps $x = a$) in the interval $(a - \Delta x_1) < x < (a + \Delta x_2)$. Thus the value $L = 6$ *is* the limit of f as $x \to 3$:

$$\lim_{x \to 3} 2x = 6.$$

In this example, $\Delta x_1 = \Delta x_2$; however, this is not necessarily the case, as illustrated in Figure 2.4. Here $\Delta x_1 > \Delta x_2$, and as before, the function g (whose rule has not been given) does have a limit as $x \to (a = 3.3)$,

$$\lim_{x \to 3.3} g(x) = 6.$$

It is very important to note that nothing in this chapter has been said about the value of the function at $x = a$, except that this is a useful value to use as a test value for the limit. There is no requirement that $f(x)$ be defined at the point $x = a$ for a limit to exist as $x \to a$.

Figure 2.4

Function g with Rule y = g(x).

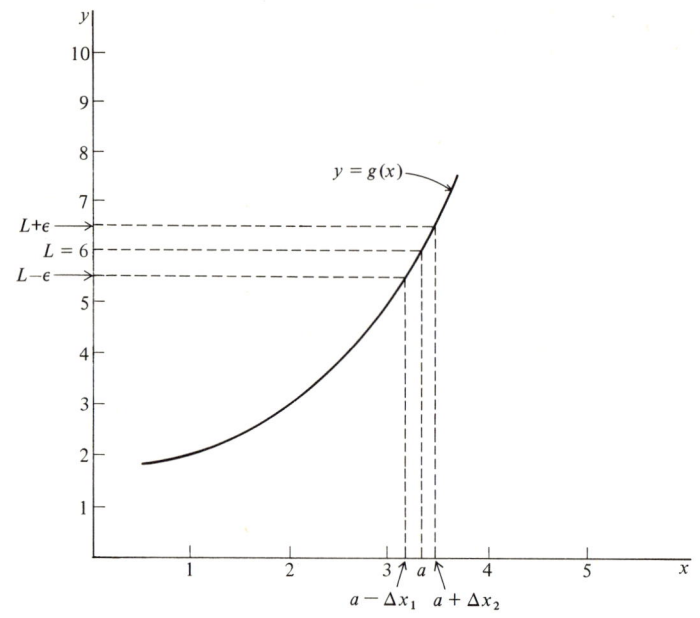

The requirements of the definition of a limit for a function at a point can be fulfilled by the following procedure:

1. Test the function for a left-hand limit L at the point a (a fulfillment of the ϵ test when moving from left to right in the domain).
2. Test the function for a right-hand limit L' at a (a fulfillment of the ϵ test when moving from right to left).
3. If both L and L' exist *and* if $L = L'$, then the function has a limit at the point a.

This procedure is that applied in carrying out the test for Figures 2.3 and 2.4 with the value $f(a)$ as the test value for the limit. The procedure for determining the limit of a function at a point is now illustrated further in a series of examples.

Example 2.3

Given the function f with rule $v = f(z) = z^2$, does the function approach a limit as $z \to a$, where a is a value in the domain?

The value $f(a)$ can be tested as the candidate limit; that is, $v_0 = f(a) = a^2$ can be tested as the limit L. This test is left as an exercise to the student, but it is valid that the left-hand limit (LHL) is

$$\text{LHL} = L = a^2,$$

Sec. 2.3] THE LIMIT OF A FUNCTION (GENERAL)

that the right-hand limit (RHL) is

$$RHL = L' = a^2,$$

and that

$$L = L' = a^2.$$

Therefore a^2 is the limit; $\lim_{z \to a} z^2 = a^2$. The same function is now tested for a limit elsewhere.

Example 2.4

Given the function f with rule $v = f(z) = z^2$, does the function approach a limit as $z \to \infty$ (z goes to infinity)?

No test limit can be suggested. There is no limit to f as $z \to \infty$; z increases without limit and $v \to \infty$. Thus the same function may have a limit at a given point in its domain, but, as the domain variable approaches infinity, the function may increase (or decrease) without limit. The value of the function may be unbounded.

Example 2.5

Given the function f with rule $v = (z^2 - 4)/(z - 2)$, does f have a limit as $z \to 2$? First the value $f(a) = f(2)$ can be used as a test limit at $z = 2$:

$$v = \frac{4 - 4}{2 - 2} = \frac{0}{0}.$$

Thus f is indeterminate at $z = 2$. The test limit $f(a)$ appears not to hold; however, it is possible to factor the right-hand side of the equation for the rule for f as follows:

$$v = \frac{(z + 2)(z - 2)}{(z - 2)}.$$

Thus, except at $z = 2$, at which the denominator becomes zero,

$$v = (z + 2).$$

The value $f(a) = f(2)$ in this simplified form of the rule for f may be tried as the candidate value of the limit; $f(2)$ is now

$$v_0 = (2 + 2) = 4.$$

The function f is shown graphically in Figure 2.5. There are values of f at every point *except* at $f(a)$. For any ϵ, say .001, the candidate limit $z = 4$ is a left-hand limit L, as shown in step 1 below.

Step 1. By selecting a value of $z > (a - \Delta z_1) = (2 - .0005) = 1.9995$, then $f(a - \Delta z_1) > 3.999 = (4.000 - \epsilon)$. Thus $L = 4$.

Step 2. By selecting a value of $z < (a + \Delta z_2) = (2 + .0005) = 2.0005$, then $f(a + \Delta z_2) < 4.001 = (4.000 + \epsilon)$. The same would hold for any other ϵ. The number 4 is also a right-hand limit L'.

Figure 2.5

Function f with Rule $v = (z^2 - 4)/(z - 2)$ (*that is undefined at* $z = 2$).

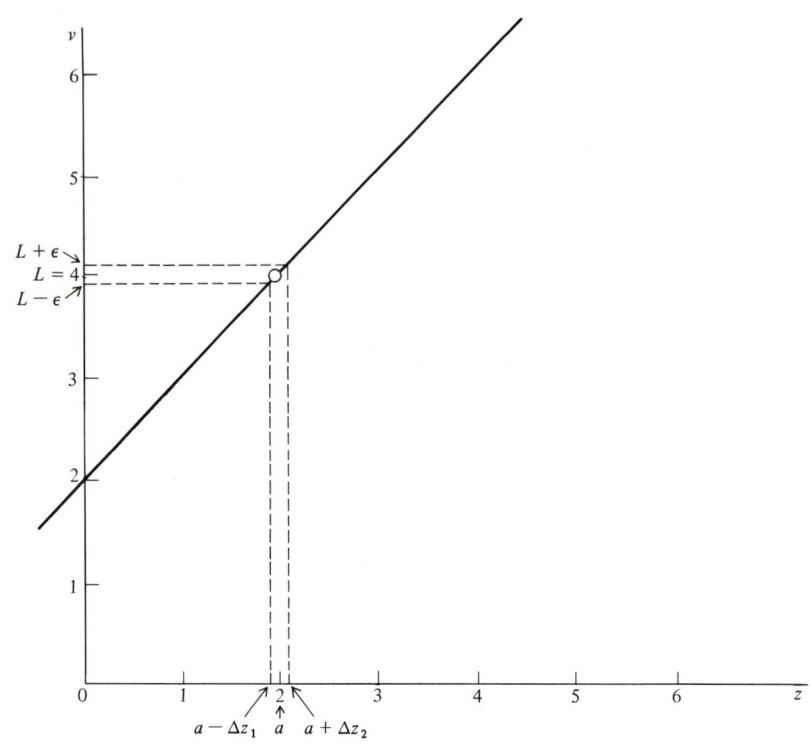

Step 3. The two limits L and L' are equal, therefore,

$$\lim_{z \to 2} \frac{z^2 - 4}{z - 2} = 4$$

despite the fact that, at the point $z = 2$,

$$\frac{z^2 - 4}{z - 2} = \frac{0}{0}.$$

Example 2.6

In this example a very interesting function is presented. Assume that the capacity output of a given machine is 1000 units per month, and that each machine costs $500 to purchase. The function that describes the machine-costs for different output levels each month is the function f with the following rule:

$$y = f(x) = \begin{cases} \$\ 500 \text{ for } 0 < x \le 1000 \\ \$1000 \text{ for } 1000 < x \le 2000 \\ \$1500 \text{ for } 2000 < x \le 3000 \\ \text{and so on.} \end{cases}$$

Figure 2.6
Cost of Machines for Different Output Levels.

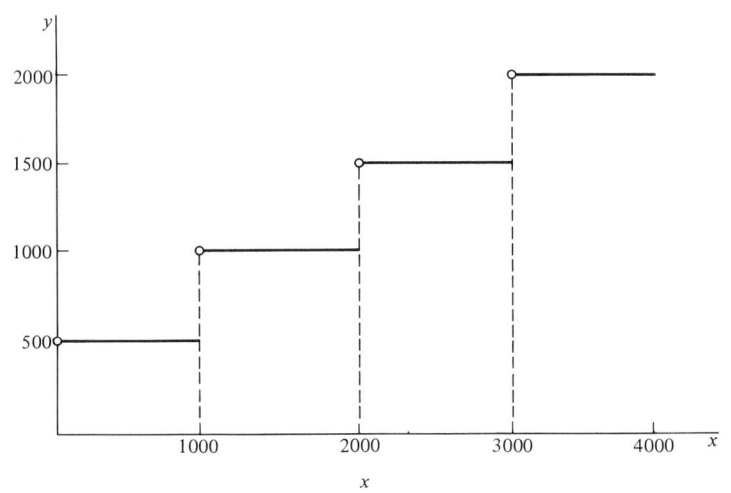

Does the function have a limit as $x \to 2000$? The function is shown graphically in Figure 2.6. As $x \to 2000$ from the left, the function f has a value of $1000 for all values up to and including 2000 units; thus the candidate limit of $1000 is established. It meets the requirements for a left-hand limit (it is equal to $1000 everywhere in the interval $1000 < x \leq 2000$); however, there is *no* value of x to the right of $x = 2000$ for which the value $f(x) = \$1000$. In fact, for every value $2000 < x \leq 3000$, the value of f equals $1500, so that f has a right-hand limit L' equal to $1500. The results of the limit test would show the following.

1. f has a left-hand limit $L = \$1000$.
2. f has a right-hand limit $L' = \$1500$.
3. However, $L \neq L'$.

The function f does *not* have a limit as $x \to 2000$.

2.4 Theorems on Limits

Several important theorems on limits are presented here without proof. The first theorem deals with the limit of a constant. Briefly stated, it says that the limit of a constant is that constant. Formally it is stated as follows:

Theorem 2.1—The Limit of a Constant: *Given a function f with rule $y = f(x) = k$, where k is a constant, then the limit of the function at any point in the domain is equal to that constant, k. That is,*

$$\lim_{x \to a} f(x) = k. \tag{2.4}$$

Thus, given a function with rule $y = f(x) = 3$, the limit of the function at any point in the domain is 3.

The second theorem deals with the limit of the algebraic sum of two (or more) functions. Briefly stated it says: The limit of the algebraic sum of two (or more) functions equals the algebraic sum of their respective limits.

Theorem 2.2—The Limit of the Algebraic Sum of Two (or More) Functions: *Given the functions f and g with rules $y = f(x)$ and $u = g(x)$ defined over a common domain, then the limit of the sum $f(x) \pm g(x)$ as the domain variable x approaches a value a is equal to the algebraic sum of their respective limits as $x \to a$. That is,*

$$\lim_{x \to a} (f(x) \pm g(x)) = \lim_{x \to a} f(x) \pm \lim_{x \to a} g(x). \tag{2.5}$$

Thus given the functions f, g, and h with common domain $x \in R$ and rules

$$y = f(x), \quad u = g(x), \quad w = h(x)$$

and limits as $x \to a$ as follows:

$$\lim_{x \to a} f(x) = A$$
$$\lim_{x \to a} g(x) = B$$
$$\lim_{x \to a} h(x) = C$$

then

$$\lim_{x \to a} (f(x) + g(x) - h(x)) = \lim_{x \to a} f(x) + \lim_{x \to a} g(x) - \lim_{x \to a} h(x)$$

or

$$= (A + B - C).$$

The next theorem deals with the limit of a product of two (or more) functions. Its brief statement is: The limit of a product of two or more functions is equal to the product of their respective limits.

Theorem 2.3—The Limit of the Product of Two (or More) Functions: *Given the functions f and g with rules $y = f(x)$ and $u = g(x)$ defined over a common domain, then the limit of the product of the functions as $x \to a$ in the domain is equal to the product of their respective limits as $x \to a$*

$$\lim_{x \to a} f(x) \cdot g(x) = \left(\lim_{x \to a} f(x)\right)\left(\lim_{x \to a} g(x)\right). \tag{2.6}$$

Given the same illustrative functions f, g, and h, with limits A, B, and C, respectively, then

$$\lim_{x \to a} (f(x) \cdot g(x) \cdot h(x)) = \left(\lim_{x \to a} f(x)\right)\left(\lim_{x \to a} g(x)\right)\left(\lim_{x \to a} h(x)\right) = (A)(B)(C).$$

Sec. 2.4] THEOREMS ON LIMITS

An immediate corollary of this theorem is that the limit of a product of two functions, the first with a constant value k, is equal to the product of their limits; that is, it is equal to k times the limit of the second function. This is formally stated in Corollary 1 to Theorem 2.3.

Corollary 1 to Theorem 2.3—The Limit of a Constant Times the Value of a Function: Given the functions f and g with rules $y = f(x) = k$ and $u = g(x)$, the limit of $f(x) \cdot g(x)$ as $x \to a$ in the domain is

$$\lim_{x \to a} (f(x) \cdot g(x)) = k \cdot \lim_{x \to a} g(x) = kL, \tag{2.7}$$

where L is $\lim_{x \to a} g(x)$.

A second corollary is that the limit of a product of a function with itself one or more (n) times is equal to the product of its limit n times. That is,

Corollary 2 to Theorem 2.3: The limit of the nth product of a function is equal to the nth product of the limit to the function; that is,

$$\left[\lim_{x \to a} f(x) \right]^n = L^n. \tag{2.8}$$

The next theorem is for the quotient of two functions, which briefly stated says that the limit of a quotient is equal to the quotient of the limits. Formally stated, it is as follows:

Theorem 2.4—The Limit of a Quotient of Two Functions: Given the functions f and g with rules $y = f(x)$, $u = g(x)$ defined over a common domain, then the limit of the quotient $f(x)/g(x)$ as $x \to a$ in the domain, is equal to the quotients of the limits, provided the $\lim_{x \to a} g(x)$ in the denominator is not zero.

$$\lim_{x \to a} \left(\frac{f(x)}{g(x)} \right) = \frac{\lim_{x \to a} f(x)}{\lim_{x \to a} g(x)}, \quad \lim_{x \to a} g(x) \neq 0. \tag{2.9}$$

For the illustrative functions f and g, with limits as $x \to a$ of A and B, respectively, this means that

$$\lim_{x \to a} \frac{f(x)}{g(x)} = \frac{\lim_{x \to a} f(x)}{\lim_{x \to a} g(x)} = \frac{A}{B}.$$

A function must have a limit at a given point in order for the function to be continuous at that point. This necessary but insufficient condition for continuity is discussed in Section 2.5. The test for the existence of the limit has its greatest usefulness in the test for continuity, since applications to many management problems require that functions be continuous.

2.5 Continuity

The next problem to be dealt with is that of the continuity of a function. When is a function continuous in an interval $[a; b]$? An intuitive answer might be: "When it is possible to draw a graph for the function over that interval without lifting the pencil from the paper." The functions in Figures 2.3 and 2.4 satisfy this criterion over their full domains; those in Figures 2.5 and 2.6 do not. In Figure 2.6, for example, it is necessary to lift the pencil from the paper at each point that is a multiple of 1000; the function does not have a limit at any of the points at which x is equal to a multiple of 1000. It may be inferred that, when a function does not have a limit at a point, it is not continuous at that point. Thus the existence of a limit at a point is a necessary, but not sufficient, condition for continuity at that point.

In Figure 2.5 for the function with rule

$$v = \frac{z^2 - 4}{z - 2}$$

another problem is indicated. This function does have a limit, even at the point $z = 2$. However, since the function is not defined at $z = 2$, a pencil would have to be lifted from the paper in drawing the curve for that point. The function must also be defined at a point in order to be continuous at the point. This is another necessary but not sufficient condition for continuity. The necessary and sufficient conditions are stated in the following definition:

> *Definition—Continuity at a Point:* *A function f with rule $y = f(x)$ is said to be continuous at a point $x = a$ in the domain if and only if the function has a limit L as $x \to a$, the function exists (is defined) at a, and the value of the function $y = f(x)$ is equal to L at a. This may be expressed in symbols as*
>
> $f(x)$ *is continuous at* $x = a$ *if and only if* $\lim_{x \to a} f(x) = L$,
> $f(a)$ *exists (is defined), and*
> $f(a) = L.$

In other words, once it has been determined that a function has a limit L as $x \to a$, then two additional conditions are necessary for the function to be continuous at $x = a$:

 1. $f(a)$ must exist, and
 2. the value $f(a)$ must be equal to the limit L.

It is now possible to define the continuity of a function in an interval.

> *Definition—Continuity of a Function:* *A function f with rule $y = f(x)$ is said to be continuous in an interval $a \leq x \leq b$, if it is continuous at every point in the interval.*

Sec. 2.5] CONTINUITY 53

The graphs of Figure 2.7 illustrate various ways in which functions can fail

Figure 2.7

Noncontinuous Functions.

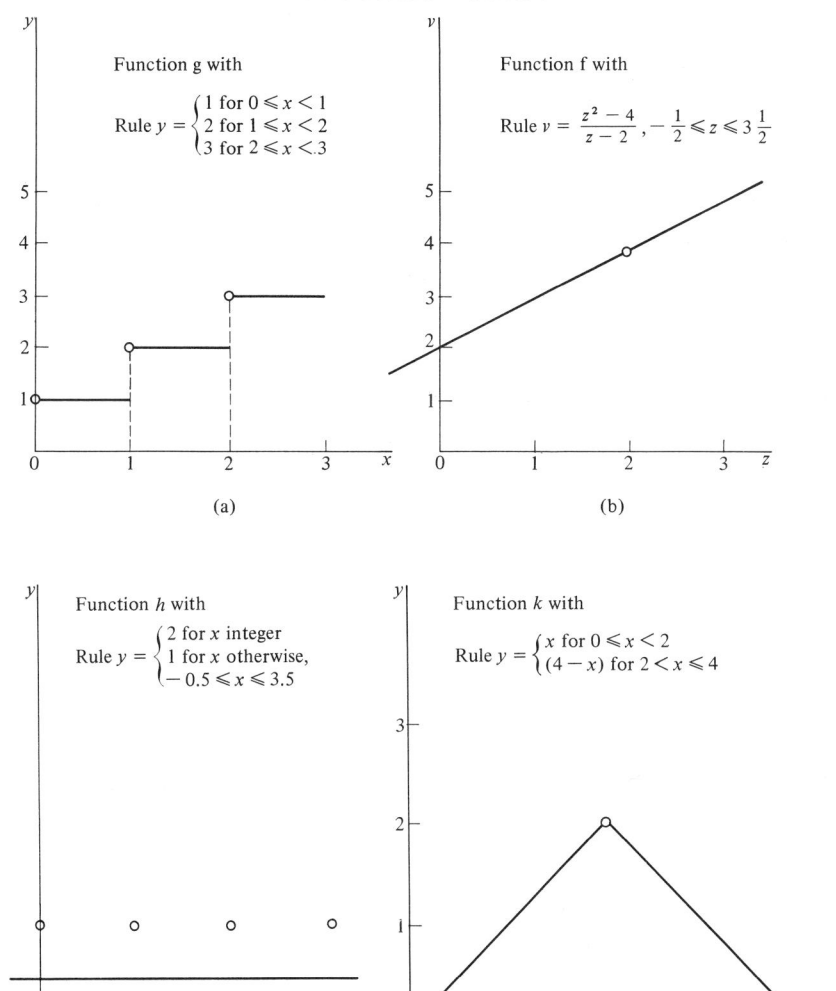

Function g with

Rule $y = \begin{cases} 1 \text{ for } 0 \leqslant x < 1 \\ 2 \text{ for } 1 \leqslant x < 2 \\ 3 \text{ for } 2 \leqslant x < 3 \end{cases}$

Function f with

Rule $v = \dfrac{z^2 - 4}{z - 2}, -\dfrac{1}{2} \leqslant z \leqslant 3\dfrac{1}{2}$

(a) (b)

Function h with

Rule $y = \begin{cases} 2 \text{ for } x \text{ integer} \\ 1 \text{ for } x \text{ otherwise,} \\ -0.5 \leqslant x \leqslant 3.5 \end{cases}$

Function k with

Rule $y = \begin{cases} x \text{ for } 0 \leqslant x < 2 \\ (4 - x) \text{ for } 2 < x \leqslant 4 \end{cases}$

(c) (d)

the tests for continuity in an interval. The function g in Figure 2.7(a) is not continuous in the interval $0 \leq x \leq 3$ since at $x = 0, 1, 2, 3$ it does not have a limit (following the reasoning used in the discussion of Figure 2.5). The function f (Figure 2.7(b)) has a limit at every point in the interval, but it is not continuous since at the point $z = 2$, f is not defined because it reduces to 0/0. The function h (Figure 2.7(c)) is not continuous, although it can be shown to have a limit at every point in the domain and is defined at every point. Nevertheless, it is not continuous, because at each integer value in the domain, the value is not the value of the respective limits at those points. Figure 2.7(d) shows another function which is continuous at every point with the exception of the point $x = 2$, again because the point $x = 2$ has been left out of the definition of the function. Had it been included, the function v would have been continuous.

2.6 Summary

A sequence is a function whose domain is an ordered subset of the positive integers. The sequence approaches a limit L if for any arbitrarily specified small positive value ϵ there exists a domain integer N such that the value of the sequence $f(n)$ can be within ϵ of L for $n > N$; i.e., $|L - f(n)| < \epsilon$ for all $n > N$. The definition of a limit for a general function is quite similar. A function f has a limit L at a point a in the domain if the value of the function can be made as close as desired to L by choosing values of the domain variable x close to a. Stated more precisely, the value L is the limit value of the function f at a point a if for any positive ϵ, no matter how small, the difference $|L - f(x_0)|$ can be made less than ϵ by choosing x_0 close to a.

For functions with a single variable over a domain of the real numbers (or a subset thereof), the procedure for determining if a limit exists at a point a in the domain can be reduced to three steps: determining (1) if a left-hand limit exists as $x \to a$ from the left, (2) if a right-hand limit exists as $x \to a$ from the right, and (3) if so, if the two limits are equal. If the test fails at any of the three steps, the function does not have a limit as $x \to a$. It is noted that no test for a limit makes any requirement of $f(a)$ itself. However, the most convenient value to test as the value L is the value $f(a)$; in many cases in which f has a limit, $f(a)$ *is* that limit. Several theorems for the algebraic manipulation of limits are presented and illustrated, and several specific examples of determining limit values are presented.

The three-step procedure for verifying that a test value is a limit is useful also in determining if a function is continuous at a point in the domain, because the limit L must exist at a point a in order for the function to be continuous at that point. The two conditions required for continuity in addition to the three that must be fulfilled for the function to have a limit as $x \to a$ are: (4) $f(a)$ must exist and be finite, and (5) $f(a)$ must be equal to L, the limit value of f. If any one of the five conditions is not met, the function is not continuous at a. In order for a function to be continuous in an interval it must be continuous at every point in the interval. Several examples of continuous and noncontinuous functions are presented in the chapter.

Key Concept

Concept	*Reference Section*
Sequence	2.2
Limit of a Sequence	2.2
Convergent Sequence	2.2

Limit of a Function   2.3
Test Limit Value   2.3
Right-Hand Limit   2.3
Left-Hand Limit   2.3
Limit of a Sum   2.4
Limit of a Product   2.4
Limit of a Quotient   2.4
Continuity of a Function at a Point   2.5
Continuity of a Function   2.5

Problems

1. Let $\{x_n\}$ assume the following values:

$$x_1 = \tfrac{3}{1}, \quad x_2 = 0, \quad x_3 = \tfrac{5}{3}, \quad x_4 = \tfrac{1}{2}, \quad x_5 = \tfrac{7}{5}, \quad x_6 = \tfrac{2}{3},$$

$$x_n = \frac{n+2}{n}, \quad (n, \text{ odd}), \quad x_n = \frac{n-2}{n}, \quad (n, \text{ even}).$$

(a) What value of n will make $|L - x_n|$ less than $\epsilon = 1/1000$?
(b) Graph the behavior of $\{x_n\}$ in this case.

Find the following limits:

2. $\lim\limits_{x \to \infty} \dfrac{24}{7 + x^3}$

3. $\lim\limits_{x \to \infty} \dfrac{3 + x^5}{13 + 3x^5}$

4. $\lim\limits_{x \to \infty} \dfrac{x^4}{x^3 + 23x^2 + 10x}$

5. $\lim\limits_{x \to \infty} \dfrac{1}{x^2}$

6. (a) $\lim\limits_{x \to 0} f_1(x) = \dfrac{1}{x^2}, \quad x > 0$

   (b) $\lim\limits_{x \to 0} f_2(x) = \dfrac{1}{x}, \quad x > 0$

   (c) $\lim\limits_{x \to 0} f_3(x) = \dfrac{1}{x^{(1/2)}}, \quad x > 0$

   Which function increases most rapidly?

7. $\lim\limits_{x \to -3} f(x) = \dfrac{x^2 + 7x + 12}{x + 3}$

   (*Hint:* Factor the numerator.)

8. $\lim\limits_{x \to \infty} f(x) = 1 + (-1)^x$

9. $\lim\limits_{x \to \infty} f(x) = 1 + \dfrac{(-1)^x}{x}$

Find the limit and graph the function for the functions with the following rules.

10. $\lim\limits_{x \to 2} f(x) = \dfrac{1}{x - 2}$

11. $\lim\limits_{x \to 3} f(x) = \begin{cases} 0 \text{ if } x \text{ is an odd integer,} \\ 1 \text{ otherwise.} \end{cases}$

12. $\lim\limits_{x \to 3} f(x) = \dfrac{x^2 - x - 6}{x - 3}$

13. $\lim\limits_{x \to +1} f(x) = x(x - 1)$

Given the rules for functions, evaluate the limits at the points indicated.

14. $\lim\limits_{x \to 0} \dfrac{x^3 + 2x}{1 - x^2}$

15. $\lim\limits_{x \to \infty} \dfrac{2x^2 - 1}{x^2 + 1}$

16. $\lim\limits_{x \to 0} \dfrac{x^2 + 2x}{x - 2x^2}$

17. $\lim\limits_{x \to \infty} \dfrac{2x^3 + 3x^2 + 4}{3x^3 - 4x^2 + 30}$

18. $\lim\limits_{x \to 0} \dfrac{(1 + x)^2 - 1}{x}$

19. $\lim\limits_{x \to 2} \dfrac{x^2 - 4x + 4}{x - 2}$

20. $\lim\limits_{x \to -2} \dfrac{x^2 - 2x - 8}{x^2 - x - 6}$

21. $\lim\limits_{x \to \infty} \left(x + \dfrac{1}{x} \right)$

22. $\lim\limits_{x \to 1} \dfrac{x^2 - 1}{x - 1}$

23. $\lim\limits_{x \to 0} \dfrac{1 - (1 + x)^3}{x}$

Continuity.

24. Discuss the continuity of Problems 10 to 13.

25. Show that $y = \dfrac{x - 3}{x^2 - 16}$ is continuous at $x = 3$ and at $x = 5$. Find any point of discontinuity and sketch the graph.

PROBLEMS

26. Show that $y = x^2 - 4x + 5$ is continuous at points where $x = -1, 0, 3, 6$. Are there any points of discontinuity?

Determine which of the following functions (problems 27 to 29) is continuous and/or discontinuous and where.

27. $f(x) = 4x^3 - 5x^2 - 2x + 1$

28. $f(x) = 4 - |x|$

29. $f(x) = \begin{cases} 2 & \text{for } 1 < x < 3 \\ 3x + 1 & \text{elsewhere.} \end{cases}$

30. Find y such that $f(x)$ is continuous at $x = 2$ when

$$f(x) = \begin{cases} \dfrac{x^2 + 2x - 8}{x - 2} & \text{for } x \neq 2 \\ y & \text{for } x = 2. \end{cases}$$

31. Is it possible for a function to have a limit at a point if the function is discontinuous at that point? Can the function be discontinuous if the limit exists at that point?

In the following problems (nos. 32 to 37), find the points of discontinuity of each of the functions whose rules are given. Make a sketch of the functions in the neighborhood of the discontinuities.

32. $y = \dfrac{2}{x + 1}$

33. $y = \dfrac{x^2 - 2x + 1}{x - 1}$

34. $y = \dfrac{x + 4}{x^2 - 16}$

35. $y = \dfrac{1}{\sqrt{x}}$ (for the positive root only)

36. $y = \dfrac{x + 3}{x^2 + 7x + 12}$

37. $y = \begin{cases} |x|, & |x| < 1 \\ x^2 - 1, & |x| > 1 \end{cases}$

38. If $\lim\limits_{u \to a} g(u) = g(a)$, and $g(a)$ exists and is finite, is the function $g(u)$ continuous?

39. A bright young management technologist was attempting to approximate a demand curve with a particular function. The instructions for use of the computer program specified that any function so used had to be continuous over the full range for which it was used. Are there any points in the interval $0 \leq x \leq 10$ at which the following functions cannot be used? If so, why? Do the functions have a limit at any points identified?

(a) $y = \dfrac{(x - 2)^3}{x^2 - 4x + 4}$

(b) $y = \dfrac{x^2 - 25}{x + 5}$

40.

Fill in the proper spaces below for functions with rules as indicated.

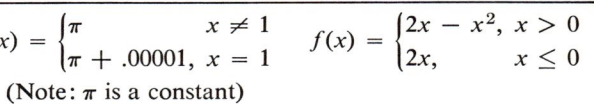

$$f(x) = \begin{cases} x^2 - 2x, & x \neq 1 \\ -2, & x = 1 \end{cases} \qquad f(x) = \begin{cases} 2 - |x|, & -1 \leq x < 3 \\ |x|, & -2 < x < -1 \end{cases} \qquad f(x) = \begin{cases} \pi, & x \neq 1 \\ \pi + .00001, & x = 1 \end{cases} \qquad f(x) = \begin{cases} 2x - x^2, & x > 0 \\ 2x, & x \leq 0 \end{cases}$$

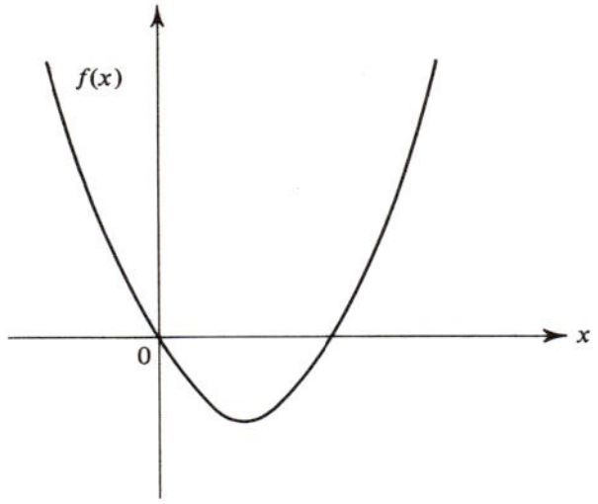

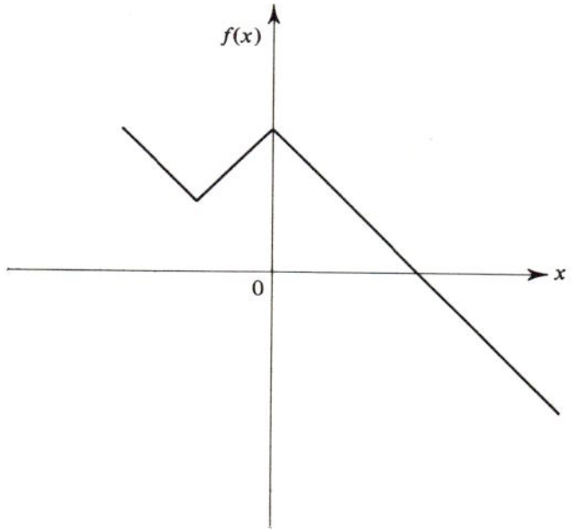

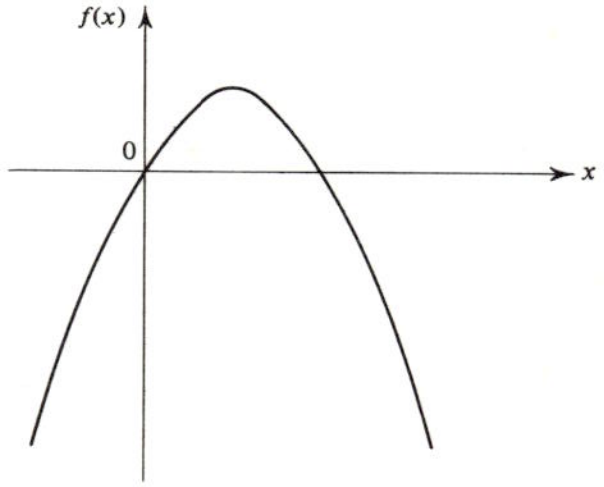

(Note: π is a constant)

Existence of limits	Equals	Does not exist	Equals	Does not exist	Equals	Does not exist	Equals	Does not exist
$\lim_{x \to 0} f(x) =$								
$\lim_{x \to 1} f(x) =$								
Continuity	Cont.	Not cont.	Cont.	Not cont.	Cont.	Not cont.	Cont.	Not cont.
At $x = 0$, $f(x)$ is								
At $x = 1$, $f(x)$ is								

CHAPTER

3

The Derivative, Maxima and Minima

This chapter deals with a new function f', called a derivative function (or, in general usage, just "the derivative"). This function allows the manager to determine how rapidly the range value y of a given function f changes for a very small change in the domain variable x at particular points in the domain. For example, given that a manager has developed a function from which he can predict the level of profits for any given level of sales, he can use the derivative to help him predict the level of sales at which his profits reach their maximum. He can also answer such questions as "Are profits increasing as rapidly as sales this month?"

Given a function f with rule $y = f(x)$, a new function f', the derivative function, can be defined over the same domain, with rule $y' = f'(x)$. This function is interpreted intuitively in Section 3.1 and then more formally in later sections of the chapter. The general (definitional) form for the rule of the derivative function at a point x_0 is

$$y' = \lim_{\Delta x \to 0} \frac{f(x_0 + \Delta x) - f(x_0)}{\Delta x}.$$

It requires the use of the limit process for a quotient involving the rule for the original function f. Stated as a set, the function f is

$$f = \{(x, y); y = f(x)\}$$

and its derivative function f' is

$$f' = \left\{(x, y'); y' = \lim_{\Delta x \to 0} \frac{f(x + \Delta x) - f(x)}{\Delta x}\right\}.$$

The derivative function f' is a function of x defined for all x for which the limit $\lim_{\Delta x \to 0} \frac{f(x + \Delta x) - f(x)}{\Delta x}$ exists. The use of x rather than x_0 indicates that the rule holds for all x in the domain for which $\lim_{\Delta x \to 0} \frac{f(x + \Delta x) - f(x)}{\Delta x}$ is defined. If f' exists, its value at a particular point x_0 is the value of y' the derivative function at the point x_0. The value y' of the derivative function at a given point is also known as the instantaneous rate of change of f at that point.

The process of deriving rules for the derivative function for given functions is called the process of *differentiation*. Several examples of differentiation are given in Sections 3.2 and 3.3, for functions with specific rules, such as $y = f(x) = x^n$, specific rules for f' are derived in Section 3.4. Sections 3.6 and 3.7 indicate how important characteristics of the derivative are used to help to identify the points in an interval at which the function reaches its maximum and its minimum values.

3.1 General Discussion

The meaning of the derivative may be clarified by a familiar analogy. Assume that a manager on his way to a board of director's meeting glances down to his speedometer to see how fast he is going at that particular instant. This information is useful if the road on which he is traveling has a posted speed limit and especially if he is late for his meeting, since it may help to answer such questions as "Will I be able to pass the car ahead without exceeding the speed limit?" or "What average speed must I make in order to arrive at the meeting on time?"

The speedometer provides an indication of the *instantaneous rate of change* of the position of the manager's car relative to the road. This is important information in relation to the maximum permissible rate. From two other devices—the odometer, which measures the number of miles covered over the road, and his

Sec. 3.2] THE DERIVATIVE 61

watch, which tells him how long he has been enroute—the manager can determine his average rate of change, the number of miles he would cover in an hour's travel. It will be shown that the instantaneous rate of change is the limit of the average rate of change as the time interval over which the average is calculated becomes very small (approaches zero).

It may be useful to examine other characteristics of the instantaneous rate of change. Consider three situations in which a vehicle is moving. In situation (a), the vehicle is moving at a constant 30 miles per hour (mph) for an hour; in situation (b), it is accelerating from 20 to 40 mph at a constant rate of acceleration over an hour; in situation (c), it is accelerating from 0 to 60 mph at a constant rate over an hour. In each case the vehicle covers 30 miles in that time, thus its average rate of change is 30 mph. In case (a), illustrated in Figure 3.1(a), the car is going

Figure 3.1

Three Functions Representing Different Instantaneous Rates of Change.

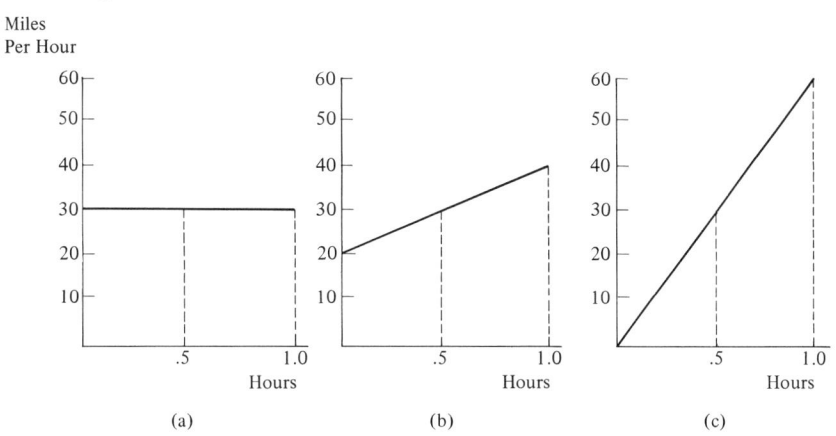

30 miles per hour at each instant. The average rate of change and the instantaneous rate of change over the road are the same at each instant of time. However, in cases (b) and (c), only at *one* point in time—and it is important to recognize that the domain variable is stated in hours—is the *instantaneous* rate of change equal to 30 miles per hour—at the one-half-hour point. This is shown in Figure 3.2 by the fact that the tangents to the curves in that figure all have the same slope at the half-hour point in the domain. In Section 3.2, the tangent to a curve at a point is defined, and the slope of the tangent at the point is shown to be the instantaneous rate of change of the function at that point; the instantaneous rate of change is then defined to be the derivative of the function at the point.

3.2 The Derivative

Consider situation (c) graphed in Figure 3.3 showing the total distance covered by an automobile starting from rest and reaching 60 mph at the end of an hour by

Figure 3.2

Curves Showing the Total Distance Covered in Each of the Three Situations Above, plus Tangents to the Curves at the Half-Hour Point.

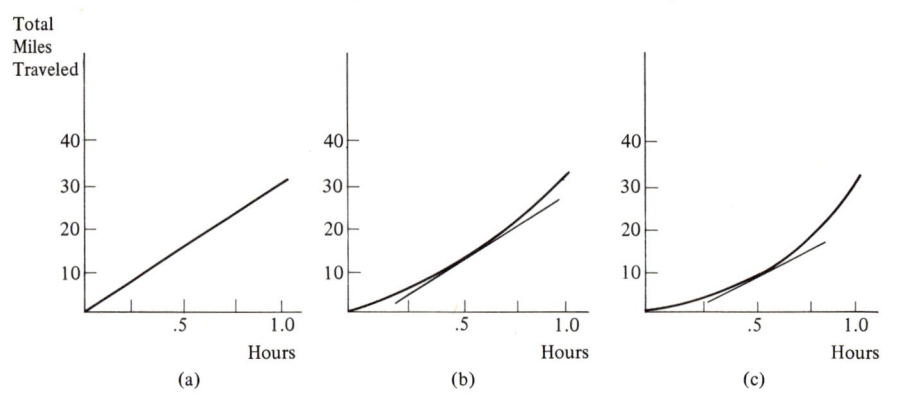

(a) (b) (c)

accelerating at a constant rate. Over the hour this auto achieves an average rate of change, an average speed, of 30 mph and thus covers a total of 30 miles during the hour's travel. Figure 3.3 is a blown-up version of Figure 3.2(c) with lines 1 and 2 plus identifying labeling added to the diagram.

For the discussion of this figure it may be helpful to refresh a few concepts. Recall from plane geometry that a secant is a line connecting two points on a continuous curve and not intersecting the curve between these points. Lines 1 and 2, and in a sense line 3 also, in Figure 3.3 are secants. (Lines 2 and 3 have been extended beyond the curve to aid in visualizing the discussion.)

Starting from point A in Figure 3.3, secant 1 has been drawn connecting point A to point B_1. Then the second secant has been drawn from point A to B_2, (and extended beyond B_2 also) a point closer to A than is point B_1. It can be seen in Figure 3.3 that the point B_i moves closer and closer to the point A, and the secant drawn from A to B_i approaches line 3 as a limit. The line 3 is defined to be the *tangent* to the curve at the point A.

> **Definition—Tangent to a Curve:** *Given the graph of a function f whose rule is $y = f(x)$ and the secant connecting a point A to another point B on the graph of f, the limit position of the secant as $B \to A$ on the curve is known as the* tangent *to the curve at the point A.*

The point A corresponds to the point $x_0 = \frac{1}{2}$ hour in the domain of the function f in Figure 3.3. The slope of secant 1 in Figure 3.3 is the average rate of change of f from the point A to the point B_1, measured by:

$$\frac{\Delta y_1}{\Delta x_1} = \frac{f(x_0 + \Delta x_1) - f(x_0)}{(x_0 + \Delta x_1) - (x_0)}.$$

The numerical value of the average rate of change is

$$\frac{\Delta y_1}{\Delta x_1} = \frac{30 - 7.5}{1 - 0.5} = \frac{22.5}{0.5} = 45 \text{ mph,}$$

Sec. 3.2]   THE DERIVATIVE   63

Figure 3.3

Average Rates of Change and the Instantaneous Rate of Change at Point A.

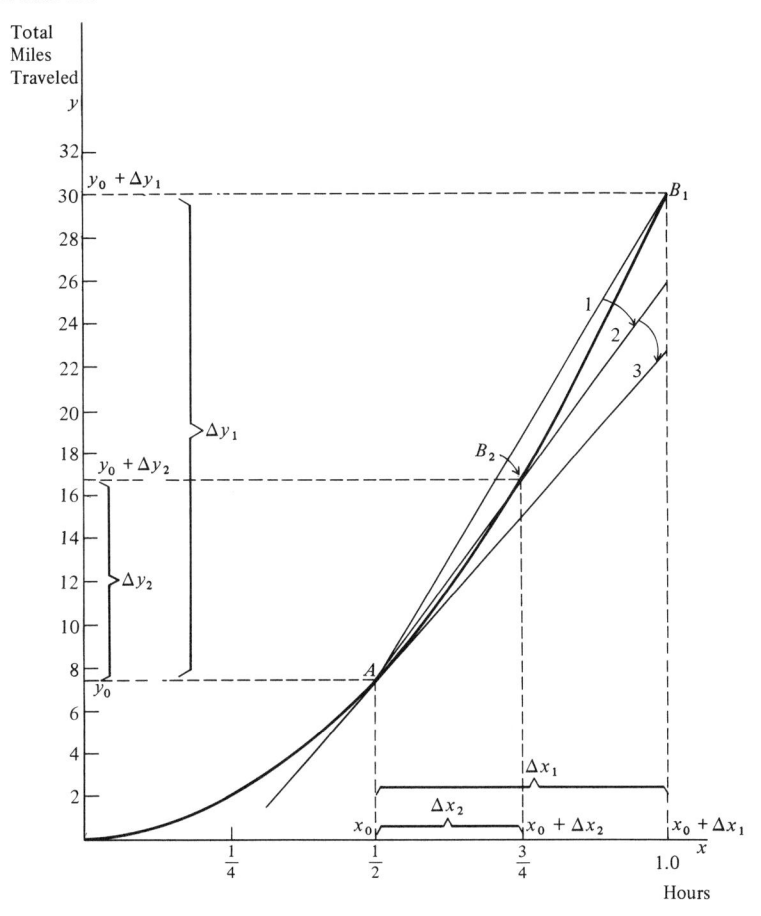

as read from Figure 3.3. The slope of secant 2 is the average rate of change of f from the point A to the point B_2, measured by

$$\frac{\Delta y_2}{\Delta x_2} = \frac{f(x_0 + \Delta x_2) - f(x_0)}{(x_0 + \Delta x_2) - (x_0)}.$$

Its numerical value is

$$\frac{\Delta y_2}{\Delta x_2} = \frac{16.875 - 7.500}{0.75 - 0.50} = \frac{9.375}{.25} = 37.5 \text{ mph}.$$

In each case the numerical value of the average rate of change is equal to the *slope* of the secant and is calculated from the coordinate values at two points on each secant (as read from Figure 3.3). As $B_i \to A$, both Δy and Δx approach 0; however, the limit of the average rate of change remains finite. It is possible to verify that this is true from the fact that secant 3 does have a finite slope even though Δy and Δx approach zero. The limit of the average rate of change to the curve $f(x)$ as Δx approaches zero is the instantaneous rate of change of f at the point x_0:

Definition—Instantaneous Rate of Change: *Given the function f with rule $y = f(x)$ and its average rate of change*

$$\frac{\Delta y}{\Delta x} = \frac{f(x_0 + \Delta x) - f(x_0)}{\Delta x},$$

over some interval

$$x_0 \leq x \leq (x_0 + \Delta x)$$

then the limit of the average rate of change of f as $\Delta x \to 0$,

$$\lim_{\Delta x \to 0} \frac{\Delta y}{\Delta x} = \lim_{\Delta x \to 0} \frac{f(x_0 + \Delta x) - f(x_0)}{\Delta x}, \qquad (3.1)$$

is the instantaneous rate of change of f at x_0.

The instantaneous rate of change of f at the point A is equal to the slope of the tangent to f at the point A. This slope can be read from the graph in Figure 3.3 to be

$$\frac{22.5 - 7.5}{0.5} = 30 \text{ mph.}$$

(This is calculated between points A and C on that figure, a rise of 15 over a run of 0.5.) Thus the instantaneous rate of change is 30 mph at the point A, the point corresponding to the half-hour point in the domain of f. The slope of the tangent to the curve at A is equal to the instantaneous rate of change of f at A. This limit is defined to be the derivative of y with respect to x at $x = x_0$.

Definition—The Derivative: *Given a function f with rule $y = f(x)$, then the function f' whose rule*

$$y' = \lim_{\Delta x \to 0} \frac{f(x + \Delta x) - f(x)}{\Delta x} \qquad (3.1)$$

is derived from the rule of f, is called the derivative function *of f, and is defined for those values of x at which the above limit exists.*

The derivative function f'—or for convenience, the derivative—may not exist (may not have a value) at every point in the domain of f. The limit defined by (3.1) must exist, in order for f' to exist at a point. It is important to recognize that the rule for the *derivative function f'* is defined as a limit—the limit stated in equation (3.1)—and that this limit is quite different from the limit to the original function f at the same domain point as illustrated numerically on page 65.

Generally, applications of the derivative involve functions that do have a derivative at every point in the relevant portion of their domain. Such functions are said to be *differentiable* over that portion of their domain.

Sec. 3.2] THE DERIVATIVE

Definition—Differentiable Function: *A function f is said to be differentiable in an interval if, at every point in that interval, the limit given by equation (3.1) exists.*

Another way of expressing differentiability is to say that a function is differentiable in an interval if its derivative exists at every point in that interval. A function that is differentiable in an interval meets the definition of a continuous function given in Chapter 2. (Can you see why?) But a continuous function may not be everywhere differentiable. This is discussed further in the next paragraph following the new definition for a tangent.

It is now possible to give a more precise definition of the tangent to a curve at a point.

Definition—The Tangent to a Curve at a Point: *Given a function f with rule $y = f(x)$, then the line passing through a point P on the graph of f corresponding to a point x_0 in the domain of f, and having slope equal to the value of the derivative function f' at x_0 is the* tangent *to f at the point P.*

It is important to note that two *different* limits are being discussed. The first limit is $\lim_{x \to x_0} f(x)$ at x_0, and the second limit (3.1), $\lim_{\Delta x \to 0} \frac{f(x + \Delta x) - f(x)}{\Delta x}$ at x_0. This latter limit may be referred to as the "derivative limit." In Figure 3.3 the $\lim_{x \to 0.5} f(x) = 7.5$, while the derivative limit, $\lim_{\Delta x \to 0} \frac{f(x + \Delta x) - f(x)}{\Delta x} = 30$ at $x = 0.5$. If f, the function itself, does not have a limit at a point x_0, it cannot be continuous at x_0; if it is not continuous at x_0, it cannot have a derivative. There are also continuous functions for which the derivative function does not exist. As shown in Figure 3.4, the derivative function f' does not exist at points at which the tangent to f is vertical, such as x_1, or at points such as x_0 at which the left-hand derivative limit (the limit (3.1) moving toward x_0 from left to right) is equal to a (the slope of the tangent to the right of the point x_0) and the right-hand derivative limit are not equal. The right-hand derivative limit (again limit (3.1)) is equal to b, but $b \neq a$. (This is similar to the case shown in Figure 2.7(d).) Other cases similar to those in Chapter 2, in which the limit for a function did not exist at a point, are points at which the derivative limit, equation (3.1), would not exist. A function f must be continuous in order for it to be differentiable, but continuity of f does not insure that f', the derivative limit, will exist at a given point.

There are many symbols for the value of the derivative f' of the function f. Some of these are:

$$\frac{dy}{dx}$$
$$\frac{d}{dx}(y)$$
$$y'$$
$$f'(x).$$

Figure 3.4

Functions Not Differentiable at x_0 and x_1, Respectively.

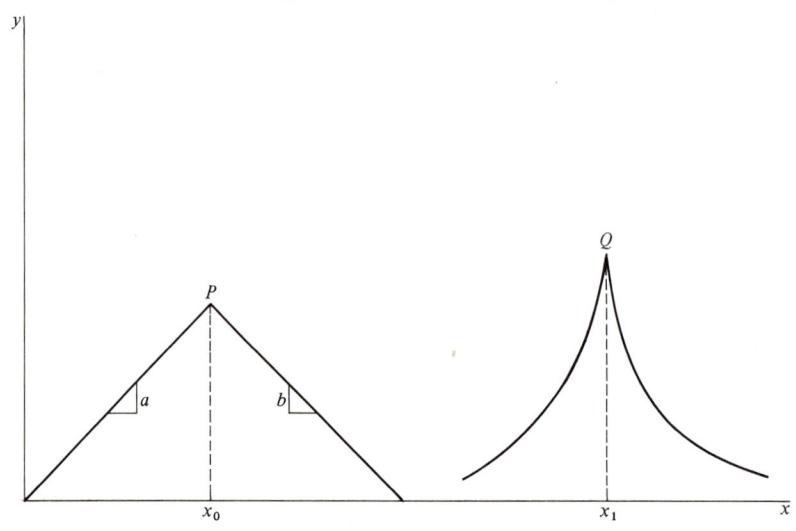

These symbols are in common use in the literature of mathematics and are used interchangeably in the pages to follow.

3.3 Differentiation

Calculating the value of the derivative of a function at a given point is called *differentiation*. The process of differentiation using the formal definition can be divided into four steps.

1. Given $y = f(x)$, the rule for f, calculate the value y_0 at $x = x_0$. Then assume Δx, an increment to x, and calculate $y_0 + \Delta y$, the new value of $f(x + \Delta x)$.
2. Determine Δy by subtracting $y_0 = f(x_0)$ from $y_0 + \Delta y = f(x_0 + \Delta x)$.
3. Divide Δy by Δx to get $\Delta y / \Delta x$.
4. Find the limit of this fraction as $\Delta x \to 0$.

This process is illustrated with another example, this time in algebraic form only.

Example 3.1

The manager studying the growth of demand for a new electronic part concludes that its growth is exponential. He hopes to produce enough of the parts to meet expected demand, and has developed a procedure for doing so which depends on predictions of the rate at which sales are changing at key times. The two times that are of most interest to him are the second and third years. The rule for the function f that he feels is appropriate is $y = x^2$. He needs to find the rule for f' and evaluate it when $x = 2$ and $x = 3$.

Solution:

Substituting $x + \Delta x$ for x gives $y + \Delta y$. Then

1. $y + \Delta y = (x + \Delta x)^2$.

Subtracting the value of y from $y + \Delta y$ gives

2. $y + \Delta y - y = \Delta y = (x^2 + 2x\,\Delta x + \Delta x^2) - x^2$.

Dividing both sides of the equation by Δx gives

3. $\dfrac{\Delta y}{\Delta x} = \dfrac{2x\,\Delta x + \Delta x^2}{\Delta x} = 2x + \Delta x$.

Taking the limit of both sides as $\Delta x \to 0$ gives

4. $\lim\limits_{\Delta x \to 0} \dfrac{\Delta y}{\Delta x} = \lim\limits_{\Delta x \to 0} (2x + \Delta x) = 2x$

since the second term on the right goes to zero in the limit. The value of f' then is $2x$ and

$$\frac{dy}{dx} = 2(2) = 4 \quad \text{at } x = 2$$

and

$$\frac{dy}{dx} = 2(3) = 6 \quad \text{at } x = 3.$$

The manager can now go on to make his production decision based on the rate of change of sales at the key times.

It is apparent that this process is an extremely clumsy one. Section 3.4 shows that the rule for the derivative function can be derived, once and for all, for large classes of functions. Once the rules have been derived, the values of the derivative can be determined at any point merely by substituting the values of the domain variable into the rule for the derivative. A few rules, derived for important types of functions, make it possible to find the derivatives for large classes of functions. Before dealing with these, one more topic is introduced, the differential of a function.

Throughout the remainder of this chapter and frequently in later chapters, the concept of a differential is important, for example, in the discussion and proofs in Section 3.4, especially that for the "chain rule." It may be helpful therefore to clarify the difference between an increment to y, that is Δy, and the differential of y; that is dy. The differential is also useful in its own right, since under proper conditions the differential dy is a close approximation to the change in the value of a function Δy, and may be used in estimating Δy as illustrated in some of the problems at the end of the chapter.

Reversing the logic of the procedure used to develop the derivative function f', it is possible to make use of the derivative in approximations for the function f. The concept of the *differential* is formally defined as follows:

Definition—The Differential dy: *Given the function f differentiable in an interval, the differential dy of the function is defined by the formula*

$$dy = f'(x)\,dx = \frac{dy}{dx}\,dx \tag{3.2}$$

where $dx \equiv \Delta x$, a small increment in x, and where dx is known as the differential of x.

The significance of the differential as compared with an increment Δy in y can be seen from Figure 3.5. The increment Δy is the change in y corresponding to a

Figure 3.5

Function f with Rule $y = f(x)$ Showing Increments Δx in x and Δy in y as Well as the Differentials, dx and dy.

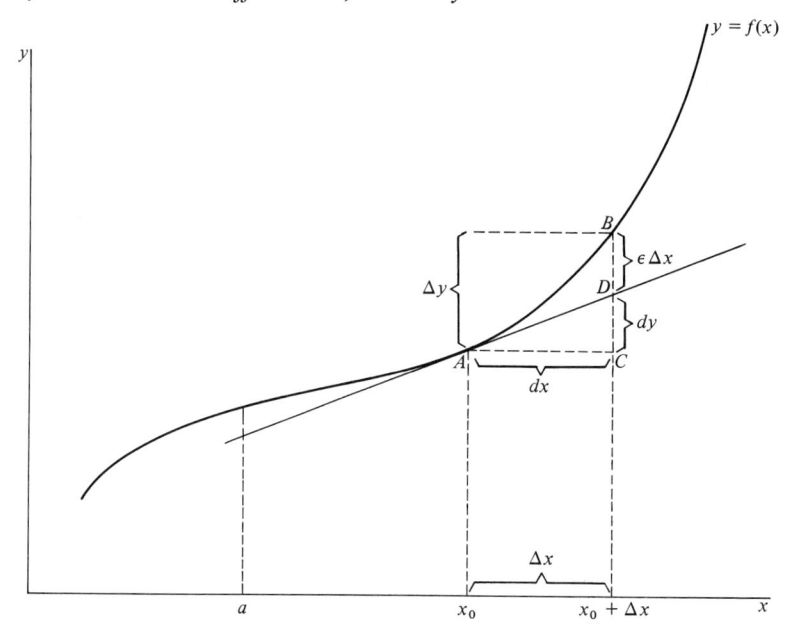

change in x of Δx. The differential dy is the change in y that is obtained by assuming that y changes at a constant rate equal to the value of the derivative of y at the point x_0. Thus, if Δx is small, dy could be used to approximate Δy in the neighborhood of x_0. By the definition of a derivative as a limit, the following equation can be stated:

$$\lim_{\Delta x \to 0} \frac{\Delta y}{\Delta x} = f'(x) + \epsilon$$

where $\lim_{\Delta x \to 0} \epsilon = 0$. In the limit, the change in y divided by the change in x is the derivative of y with respect to x. For finite increments in x, the quotient $\Delta y/\Delta x = CB/AC$ in Figure 3.5 differs from $dy/dx = CD/AC$ by a small error

Sec. 3.3] DIFFERENTIATION

term $\epsilon\,(\Delta x)$. Given

$$\frac{\Delta y}{\Delta x} = f'(x) + \epsilon$$

cross multiplying gives

$$\Delta y = f'(x)\,\Delta x + \epsilon\,\Delta x$$

where $\epsilon\,\Delta x = BD$ in Figure 3.5. But because $\Delta x \equiv dx$, the equation for Δy can be rewritten as

$$\Delta y = f'(x)\,dx + \epsilon\,\Delta x$$

or

$$\Delta y = dy + \epsilon\,\Delta x.$$

This is shown geometrically in Figure 3.5. In an interval near x_0, dy may be used as a close approximation to Δy; in that region $\epsilon\,\Delta x$ is quite small.

The derivative dy/dx, although not a quotient, frequently behaves as if it were a quotient, as here; that is, given

$$\frac{dy}{dx} = \frac{dy}{dx},$$

cross multiplying by dx would give

$$dy = \frac{dy}{dx} \cdot dx$$

which is by definition the differential of f. A similar result applies in the case of the chain rule, Rule 3.6, which is presented later in the chapter. Intuitively, it may be helpful to think of dy/dx as a quotient, although rigorously this is not the case; it is the limiting value of a quotient.

Example 3.2 illustrates the use of the differential to approximate the function f with rule $y = f(x) = 4x^2$ in the neighborhood of the point $x = 1$.

Example 3.2

Find Δy and dy for $x = 1$ of the function f with rule $y = f(x) = 4x^2$, and show the error $\epsilon\,\Delta x = \Delta y - dy$ as a function of dx. What is $\epsilon\,\Delta x$ for $dx = .2$ and $.1$?

Solution:

$\dfrac{dy}{dx} = 8x$ (since as shown in Example 3.1, $\dfrac{d}{dx}(x^2) = 2x$ and $4(2x) = 8x$)
$dy = 8x\,dx$
$\Delta y = f(x + dx) - f(x) = 4(x + dx)^2 - 4x^2$
$\quad = 4\bigl(x^2 + 2x\,dx + (dx)^2\bigr) - 4x^2 = 8x\,dx + 4(dx)^2$
so $\epsilon\,\Delta x = \Delta y - dy = 8x\,dx + 4(dx)^2 - 8x\,dx = 4(dx)^2$.
If $dx = .2$ then $\epsilon\,\Delta x = 4(.04) = .16$
If $dx = .1$ then $\epsilon\,\Delta x = 4(.01) = .04$
Thus the approximation of y is closer for dx small. (Note that in this case $\epsilon\,\Delta x$ is independent of x.)

3.4 General Differentiation Rules

General rules for the derivative functions of some basic functions are now developed and discussed. Once such rules have been established, differentiation by the four-step process just illustrated can be superseded by the more efficient technique of applying the rules. A few proofs have been included in this section, so that the reader can observe the procedure for developing the rules for the derivative functions. These proofs may also help provide the reader with an intuitive understanding of the meaning of the derivative. In each case the rule for a derivative is stated as an average rate of change for the class of function in question, the resulting expression simplified, and its limit taken to arrive at the expression of the instantaneous rate of change. Several other rules are stated without proof in the body of the chapter, and some of the proofs are included in Appendix 3A. The same procedure is used in these proofs, but the simplification steps are somewhat more complicated.

Three rules for derivative functions that are similar to the rules for limits are readily stated and proved. Briefly stated, the first of these rules is: A function f with constant value in an interval does not change, that is, *the value of the derivative of a constant is zero*. The graph of such a function in an interval is shown in Figure 3.6.

Figure 3.6

Function f, Which Is Constant in an Interval.

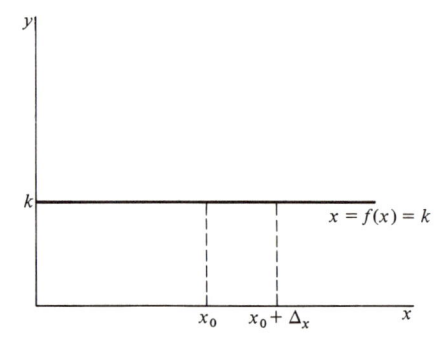

Rule 3.1: Given the function f with rule $y = f(x) = k$, which is differentiable in an interval, then the rule for the derivative function f' is

$$y' = 0. \tag{3.3}$$

Proof:

Given $y = f(x) = k$, then at $x = x_0$

$$y' = \lim_{\Delta x \to 0} \frac{f(x_0 + \Delta x) - f(x_0)}{\Delta x}$$

Sec. 3.4] GENERAL DIFFERENTIATION RULES

by definition, but $f(x_0) = k$, and also $f(x_0 + \Delta x) = k$, so

$$y' = \lim_{\Delta x \to 0} \frac{k - k}{\Delta x} = \lim_{\Delta x \to 0} \frac{0}{\Delta x}$$

$$y' = 0 \qquad \text{Q.E.D.}[1]$$

Stated briefly, the second rule is: The rule for the derivative of a function g that is equal to a constant k times the value of another function f is that constant times the value of the derivative of f; that is, *the derivative of a constant times a function is the constant times the derivative of that function*. Figure 3.7 illustrates a specific case of such a pair of functions f and g.

Figure 3.7

Functions f and g for the Case in Which
$y = f(x) = x$ *and* $u = k[f(x)] = kx = x/2$.

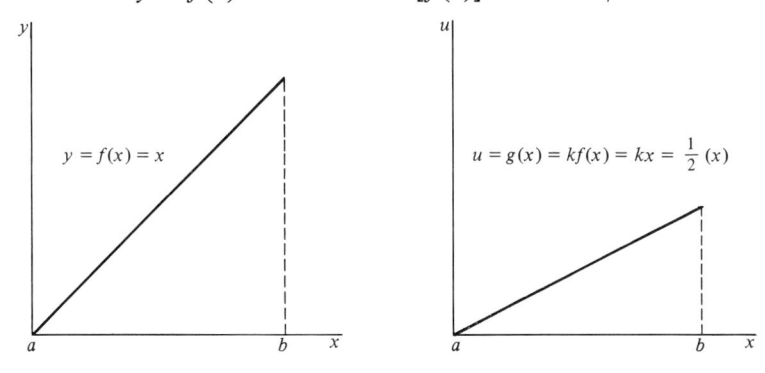

Rule 3.2: Given the function f with rule $y = f(x)$ and the function g with rule $u = g(x) = k[f(x)]$, both differentiable in an interval, then the rule for the derivative function g' of g is

$$u' = k[f'(x)]. \tag{3.4}$$

Proof:

Given $u = g(x) = k[f(x)]$, then at $x = x_0$

$$u' = \lim_{\Delta x \to 0} \frac{g(x + \Delta x) - g(x)}{\Delta x}$$

by definition, but $g(x) = k[f(x)]$ and $g(x + \Delta x) = k[f(x + \Delta x)]$, so by substitution

$$u' = \lim_{\Delta x \to 0} \frac{k[f(x + \Delta x)] - k[f(x)]}{\Delta x} = \lim_{\Delta x \to 0} k \left[\frac{f(x + \Delta x) - f(x)}{\Delta x} \right].$$

---

[1] Q.E.D. is an abbreviation of a Latin term *quod erat demonstrandum*, which freely translated means "that which was sought to be proved has been proved."

By the rules for limits, the limit of a constant times the value of a function is equal to the constant times the limit of that function (Corollary 1 to Theorem 2.3); thus,

$$u' = k \left[\lim_{\Delta x \to 0} \frac{f(x + \Delta x) - f(x)}{\Delta x} \right].$$

But the factor inside the brackets is the rule for the derivative of f, thus

$$u' = k[f'(x)] = k(y'). \qquad \text{Q.E.D.}$$

Again the proof follows from the definition, algebraic manipulation, and simplification.

The rule for the derivative function h' of the algebraic sum of two functions f and g with rules $y = f(x)$ and $u = g(x)$ differentiable in the interval $a \leq x \leq b$, is the sum $f'(x) \pm g'(x)$; that is *the value of the derivative of a sum of two (or more) functions is equal to the sum of the values of their respective derivatives.*

Rule 3.3: *Given two (or more) functions f and g, with rules $y = f(x)$ and $u = g(x)$ differentiable in a common interval, and the function h with rule $v = f(x) \pm g(x)$, then the rule for the derivative function h' is given by*

$$v' = \frac{d}{dx}[f(x) \pm g(x)] = f'(x) \pm g'(x). \tag{3.5}$$

Proof:

Given $v = f(x) \pm g(x)$, then

$$v' = \frac{d}{dx}[f(x) \pm g(x)] = \lim_{\Delta x \to 0} \frac{[f(x + \Delta x) \pm g(x + \Delta x)] - [f(x) \pm g(x)]}{\Delta x}.$$

Collecting terms gives:

$$v' = \frac{d}{dx}[f(x) \pm g(x)] = \lim_{\Delta x \to 0} \frac{[f(x + \Delta x) - f(x)] \pm [g(x + \Delta x) - g(x)]}{\Delta x}.$$

Since the limit of a sum is equal to the sum of the limits, this can be written as

$$v' = \frac{d}{dx}[f(x) \pm g(x)] = \lim_{\Delta x \to 0} \frac{f(x + \Delta x) - f(x)}{\Delta x} \pm \lim_{\Delta x \to 0} \frac{g(x + \Delta x) - g(x)}{\Delta x}.$$

By definition, these terms are:

$$f'(x) \pm g'(x),$$

thus

$$v' = f'(x) \pm g'(x). \qquad \text{Q.E.D.}$$

Rule 3.4 is the rule for the derivative function of a polynomial function. Stated in brief terms, the value of the derivative of the function f with rule $y = f(x) = x^n$ is $y' = nx^{(n-1)}$; that is, *the derivative of x^n is $nx^{(n-1)}$*.

The derivation of this rule for the derivative requires only the standard steps used in each case thus far. In this instance, however, the simplification steps are somewhat complex, as shown in Appendix 3A.

Sec. 3.4] GENERAL DIFFERENTIATION RULES

Rule 3.4: *Given the function f with rule $y = f(x) = x^n$, differentiable in an interval, then the rule for the derivative function f' is*

$$y' = nx^{n-1} \qquad (3.6)$$

for rational n.[2]

It is now possible to compare how much easier differentiation is by using Rule 3.4 than the four-step procedure for differentiation used for the function with rule $y = x^2$ in Example 3.1.

Example 3.3

Given the function f with rule $y = x^2$, find the rule for f' by using Rule 3.4

$$y' = \frac{d}{dx}(x^2) = 2x^{(2-1)} = 2x.$$

The value of the derivative function can be calculated at $x = 2$ and $x = 3$, as was done in Example 3.1. Again the result is $2(2) = 4$ and $2(3) = 6$, respectively.

Two more examples are presented to illustrate some of the rules discussed thus far.

Example 3.4

Given the function f with rule $y = f(x) = x^3 - 2x + 4$, find the rule for f' by use of Rules 3.1 to 3.4;

$$y = x^3 - 2x + 4$$
$$y' = 3x^{(3-1)} - 2x^{(1-1)} + 0 = 3x^2 - 2.$$

Example 3.5

Given the function f with rule $y = x^5 + 17x^3 - x + 4$, find the rule for the derivative f' by use of Rules 3.1 to 3.4;

$$y = x^5 + 17x^3 - x + 7$$
$$y' = 5(x)^{(5-1)} + 17(3)x^{(3-1)} - (1)(x)^{1-1} + 0$$
$$y' = 5x^4 + 51x^2 - 1.$$

Rule 3.5 is the rule for the derivative function h' of a function h that is equal to the product of two other functions f and g,

$$h'(x) = F(x) \cdot g'(x) + g(x) \cdot f'(x);$$

that is *the derivative of a product of two functions is equal to the first times the derivative of the second, plus the second times the derivative of the first.* Its proof (Appendix 3A) is also somewhat complex at the simplification step.

---

[2] The rule $y' = nx^{n-1}$ also applies for irrational exponents, although the proof in Appendix 3A has not been extended to show this.

Rule 3.5: *Given the function f with rule $y = f(x)$ and g with rule $v = g(x)$, differentiable over an interval, and the function h with rule $u = [f(x) \cdot g(x)]$, the rule for the derivative function h' is given by*

$$u' = \frac{d}{dx}[f(x) \cdot g(x)] = f(x) \cdot g'(x) + g(x) \cdot f'(x). \tag{3.7}$$

For the rule to be valid the definition of a derivative function requires that

$$h' = f(x) \cdot \lim_{\Delta x \to 0} \frac{g(x + \Delta x) - g(x)}{\Delta x} + g(x) \lim_{\Delta x \to 0} \frac{f(x + \Delta x) - f(x)}{\Delta x}.$$

The proof in Appendix 3A begins with the definition of h' and ends in the form just stated. Example 3.6 illustrates how this rule is used.

Example 3.6

Given the function h with rule $v = 4x^2(7x + 4)$, find v', the rule for the derivative function. Because the rule for h involves the product of terms from two other functions, $f(x) = 4x^2$ and $g(x) = 7x + 4$, its derivative is equal to "the first times the derivative of the second plus the second times the derivative of the first":

$$\begin{aligned}
v' = h'(x) &= 4x^2 \frac{d}{dx}(7x + 4) + (7x + 4)\frac{d}{dx}(4x^2) \\
&= 4x^2(7 + 0) + (7x + 4)4(2x) \\
&= 4x^2(7) + (7x + 4)8x \\
&= 28x^2 + 56x^2 + 32x \\
&= 84x^2 + 32x \\
&= 4x(21x + 8).
\end{aligned}$$

The same result could have been achieved by differentiating h after its rule had been multiplied through to give $v = 4x^2(7x + 4) = 28x^3 + 16x^2$. (Try it.)

Rule 3.6, the *chain rule*, deals with the rule for a derivative function of "a function of a function." That is, the rule gives the value of the instantaneous rate of change of f for changes in the domain variable of g where f is a function with rule $y = f(u)$, whose domain variable u itself is the *range* variable for the other function, g, whose rule is $u = g(x)$. The set statements of the two functions are:

$$\begin{aligned} f &= \{(u, y), \ y = f(u)\} \\ g &= \{(x, u), \ u = g(x)\}. \end{aligned}$$

Illustrative functions are shown graphically in Figure 3.8 for particular rules for f and g to demonstrate the meaning of a composite function. A change in the domain variable x of g in that figure brings a change in u which, in turn, leads to a change in y. Given the rules illustrated in Figure 3.8, a change in x from 2 to 4 leads to a change in u from 1 to 2; in turn, this brings about a change in y from 1 to 4:

$$y = f(u) = f(g(x)) = \left(\frac{x}{2}\right)^2 = \frac{x^2}{4}$$

as can be seen in Figure 3.8.

Sec. 3.4] GENERAL DIFFERENTIATION RULES 75

Figure 3.8

The Functions f with Rule $y = f(u)$, where u is the Value of the Function g, and g with Rule $u = g(x)$.

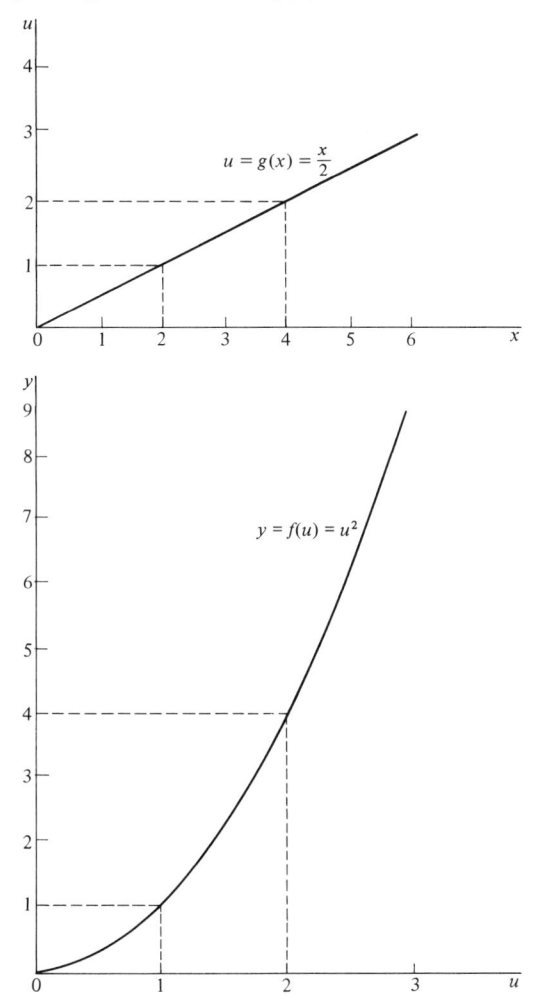

The following relationships are also valid:

$$\Delta y = f(2) - f(1) = 4 - 1 = 3$$
$$\Delta u = g(4) - g(2) = 2 - 1 = 1$$
$$\Delta x = 4 - 2 = 2$$

also

$$\frac{\Delta y}{\Delta x} = \frac{\Delta y}{\Delta u} \cdot \frac{\Delta u}{\Delta x} = \frac{3}{1} \cdot \frac{1}{2} = \frac{3}{2}.$$

The relationships of the increments in y, u, and x are used in the proof of the chain rule given in Appendix 3A. The chain rule, one of the most useful rules of differentiation, is best remembered in terms of the equation form of the statement of the rule.

> **Rule 3.6—The Chain Rule:** *Given the function f with rule $y = f(u)$, whose domain variable is the range variable of another function g with rule $u = g(x)$ differentiable over intervals $a \leq x \leq b$ and $c \leq u \leq d$, respectively, then the rule for the derivative function of f in the domain of g is given by*
>
> $$\frac{dy}{dx} = \frac{dy}{du} \cdot \frac{du}{dx}. \tag{3.8}$$

Given the illustrative functions f and g with rules

$$y = f(u) = u^2, \quad u = g(x) = \frac{x}{2},$$

the chain rule, equation (3.8),

$$\frac{dy}{dx} = \frac{dy}{du} \cdot \frac{du}{dx}$$

becomes

$$\frac{dy}{dx} = \frac{d}{du}(u^2) \cdot \frac{d}{dx}\left(\frac{x}{2}\right) = (2u)\left(\frac{1}{2}\right) = u.$$

But $u = x/2$ and thus

$$\frac{dy}{dx} = \frac{x}{2}.$$

In this particular illustration the same result is obtained by substituting for u in $f(u)$ and differentiating directly by Rule 3.4. (Try it.) But this is not always possible.

Example 3.7

Given the function f with rule $y = (2x^2 + 4)^{(1/3)}$, is it possible to find dy/dx? The rule for the derivative function cannot be determined directly. There is no rule for doing so. But, if a substitution is made such that $u = (2x^2 + 4)$, then the rule for f is changed to the form $y = u^{(1/3)}$, and Rule 3.4 for dy/du applies. However, dy/dx is wanted not dy/du. This can be found from Rule 3.6, because $y = f(u) = u^{(1/3)}$, where $u = g(x) = (2x^2 + 4)$; therefore, by Rule 3.6,

$$\frac{dy}{dx} = \frac{dy}{du} \cdot \frac{du}{dx},$$

$$\frac{dy}{du} = \frac{1}{3} u^{(-2/3)},$$

$$\frac{du}{dx} = 4x;$$

Sec. 3.4] GENERAL DIFFERENTIATION RULES

thus,
$$\frac{dy}{dx} = \frac{1}{3} u^{(-2/3)} \cdot (4x).$$

Substituting for u in the first term gives
$$\frac{dy}{dx} = \frac{1}{3} (2x^2 + 4)^{(-2/3)}(4x).$$

The chain rule can be used to extend Rule 3.4 generally. This is shown in Example 3.8. The result of this example does not need to be stated formally as an additional rule; the chain rule is always available for use in extensions of the rules formally developed.

Example 3.8

Given the function f with rule $y = f(u) = u^n$ and the function g with rule $u = g(x)$ differentiable over corresponding intervals, then

$$\frac{d}{dx}(y) = \frac{d}{du} u^n \frac{du}{dx} = nu^{n-1} g'(x). \tag{3.9}$$

By the chain rule, Rule 3.6,
$$\frac{dy}{dx} = \frac{dy}{du} \cdot \frac{du}{dx}.$$

By Rule 3.1,
$$\frac{dy}{du} = nu^{n-1}.$$

By definition
$$\frac{du}{dx} = g'(x).$$

Substituting these values gives:
$$\frac{dy}{dx} = nu^{n-1} g'(x).$$

Rule 3.7 is derived through use of Rules 3.5 and 3.6. It provides an approach to differentiating a function whose rule involves the quotient of rules for other functions.

Rule 3.7: *Given the functions f and g with rules $y = f(x)$ and $u = g(x)$, differentiable in an interval and the function h with rule $v = f(x)/g(x)$, also differentiable in the interval, the rule for the derivative function h' is given by*

$$v' = \frac{d}{dx}\left(\frac{f(x)}{g(x)}\right) = \frac{g(x) \cdot f'(x) - f(x) \cdot g'(x)}{[g(x)]^2}, \quad g(x) \neq 0. \tag{3.10}$$

Proof:[3,4]

Rewrite $h(x)$ from the quotient form,

$$v = h(x) = \frac{f(x)}{g(x)}$$

to the product form,

$$h(x) = f(x) \cdot [g(x)]^{-1} = [g(x)]^{-1} \cdot f(x).$$

By Rules 3.5 and 3.3, the rule for the derivative function is given by

$$h'(x) = v' = ([g(x)]^{-1} \cdot f'(x)) + (f(x) \cdot (-1)[g(x)]^{-2} \cdot g'(x)).$$

Rewriting this in terms of a quotient gives

$$h'(x) = v' = \frac{f'(x)}{g(x)} - \frac{f(x) \cdot g'(x)}{[g(x)]^2}.$$

Rewriting these over a common denominator gives

$$h'(x) = v' = \frac{g(x) \cdot f'(x) - f(x) \cdot g'(x)}{[g(x)]^2}. \qquad \text{Q.E.D.}$$

Example 3.9

Given the function f with rule $y = (x - 2)/x^2$, find the rule for the derivative function f'. The rule is

$$y' = \frac{x^2(1) - (x - 2)(2x)}{x^4} = \frac{x^2 - 2x^2 + 4x}{x^4} = \frac{4x - x^2}{x^4} = \frac{4 - x}{x^3}.$$

Rule 3.5 could be used if the problem is rewritten as a product. Rewriting gives

$$y = \frac{(x - 2)}{x^2} \quad \text{as} \quad y = (x - 2)x^{-2}.$$

The rule y' is

$$y' = (x - 2)(-2x^{-3}) + x^{-2}(1) = -2x^{-2} + 4x^{-3} + x^{-2} = 4x^{-3} - x^{-2}$$
$$= \frac{4}{x^3} - \frac{1}{x^2} = \frac{4 - x}{x^3}.$$

3.5 Higher Order Derivatives

So far the differentiation process has been applied only once for a given function; that is, the derivative function f' has been determined from the original

---

[3] A mnemonic device to remember this formula is as follows:

$$\frac{d}{dx}\left(\frac{hi}{ho}\right) = \frac{ho[d(hi)] - hi[d(ho)]}{ho \cdot ho}$$

where "*hi*" is the numerator, "*ho*" is the denominator, "*d(hi)*" and "*d(ho)*" are their respective derivatives and *ho[d(hi)]* is read "*ho de hi.*"

[4] This rule can be proved directly from the definition as are Rules 3.5 and 3.6 in Appendix 3A. A proof relying on the multiplicative rule and the chain rule has been used here.

Sec. 3.6] CRITICAL POINTS

function f. The result of this process is called the *first derivative* function f'. It is also possible, however, to repeat the process by deriving the instantaneous rate of change of f'. The new derivative function f'' is called the second derivative function of f. The process can be repeated in turn to derive third, and successive, derivatives f''', f'''', and so on, which are called higher order derivatives. In set terms, the second derivative function can be written

$$f'' = \left\{(x, y''); y'' = \lim_{\Delta x \to 0} \frac{f'(x + \Delta x) - f'(x)}{\Delta x}\right\}.$$

This is an application of the definition of a derivative to the first derivative function f'. The same qualifications for the existence of the limit apply for higher order derivatives as applied to the first derivative. The symbols often used for higher order derivatives are f'' for the function, and y'', $(d^2y)/(dx^2)$, and $y'' = f''(x)$ for its rule.

Given a function that expresses the total miles traveled as a function of the time enroute, it has been shown that the first derivative function shows the speed of travel at any given time, that is the instantaneous rate of change in miles covered. The second derivative is the rate of acceleration at any instant; it tells whether speed is increasing or decreasing at any moment of time; it is the rate of change of the rate of change. If the speed is constant, the second derivative is zero; there is no change in the rate of change (the speed) of travel.[5]

The procedure for finding the rule for the second derivative is now illustrated for a simple case.

Example 3.10

Given the function f with rule $y = (2x^2 + 4)^5$, find rules for the first and second derivative functions of f. These rules are:

$$y' = \frac{d}{dx}(2x^2 + 4)^5 \qquad y'' = \frac{d^2y}{dx^2} = \frac{d}{dx}\left(\frac{dy}{dx}\right) = \frac{d}{dx}[20x(2x^2 + 4)^4]$$

$$y' = 5(2x^2 + 4)^4(4x) \qquad y'' = 20x \cdot 4(2x^2 + 4)^3 4x + (2x^2 + 4)^4(20)$$

$$y' = 20x(2x^2 + 4)^4; \qquad y'' = 320x^2(2x^2 + 4)^3 + 20(2x^2 + 4)^4.$$

The first and second derivatives can now be used in the identification of extreme points of a function in an interval. The discussion centers on continuous functions, and a number of complications are avoided.

3.6 Critical Points

It is now possible to use the derivative function in finding the maximum and minimum values of a function—the maximum profit that can be achieved by a company, for example, or minimum costs. More than one point may meet the criteria for a maximum or minimum in a given interval, but only one can be *the*

---

[5] This brief discussion may be useful in understanding the presentation given in the introduction to Chapter 7.

maximum and one *the* minimum. This value of the function is called the *global* maximum (or minimum). The other points are known as relative maxima and minima. This section shows that one set of maxima and minima correspond to those points at which the value of the derivative function equals zero. But this is the case for both maxima and minima; therefore, it is necessary to look at the second derivative of the function in order to distinguish between the two by observing the "rate of change of the rate of change." Another set of maxima and minima for the function may occur at the end points of the interval at which the first derivative may not be zero. These must also be investigated if all the maxima or minima for the function are being sought. The global maximum (minimum) can be identified only by comparing *all* the relative maxima (minima) and selecting the largest (smallest). Finally, there is a set of points for which the value of the first derivative equals zero, but the value of the function is neither a maximum nor a minimum, and another set at which the first derivative does not exist. The value of the second derivative generally permits the first set of such cases to be distinguished. The formal definition of relative maxima (minima) is as follows:

Definition—Relative Maximum (Minimum): *Given the function f with rule $y = f(x)$ differentiable in an interval $a \leq x \leq b$, f has a relative maximum (minimum) at x_0 if there exists a neighborhood containing x_0 such that $f(x_0) \geq f(x)$, $(\leq f(x))$ for every x in the neighborhood.*[6]

Points in the domain for which the function reaches relative maxima and minima are defined as extreme points, a subset of the set of critical points.[7] The classification *critical points* includes, in addition to extreme points, those points at which the derivative function is not defined plus another class of points known as inflection points (discussed in Section 3.8).

Definition—Critical Point: *Given f a function whose rule is $y = f(x)$, defined at every point in the interval $a \leq x \leq h$, the critical points of the function are those values of x for which one or more of the following applies:*

$y' = 0$, *the value of the first derivative function is zero,*
$x = a$ *or* $x = h$,
y' *is not defined.*

In set terms, if the interval is $X = \{x: a \leq x \leq h\}$, then the set of critical points, C, is a subset of X; specifically,

---

[6] For the endpoints to be maxima and minima this definition must be modified to apply in half neighborhoods of x_0. For the beginning point in the interval the half neighborhood is the right-hand neighborhood of x_0; for the ending point of the interval, the left-hand neighborhood.

[7] The mathematics of extreme and critical points becomes very involved. A full discussion is beyond the scope of this book. The discussion here is limited specifically to functions with one domain variable and to those aspects that are most useful to the manager. The discussion is extended to functions with more than one domain variable in Chapter 5.

Sec. 3.6] CRITICAL POINTS 81

$C = \{x: x \in X \text{ and } f'(x) = 0, \text{ or } x = a, \text{ or } x = h, \text{ or } f'(x) \text{ is not defined}\}$.

Some of the possible types of critical points are illustrated in Figure 3.9.

Definition—Extreme Points: *Given the critical points of a function f as just defined, the extreme points of f are the set of all relative maxima and minima.*

The extreme points are a subset of the critical points including all relative maxima or minima.

Figure 3.9

The Function f with Rule $y = f(x)$.

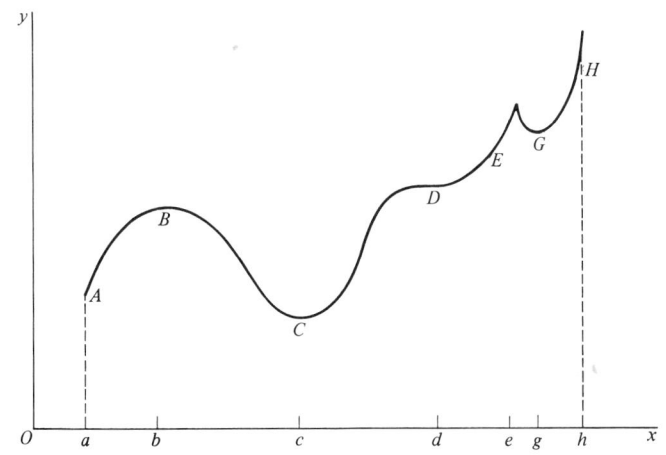

The relevant interval in the domain is $a \leq x \leq h$. For x equal to b, c, and g, the function f reaches relative maxima and minima (the value of f' is equal to zero). The points at which x equals a and h are endpoints of the interval that can be seen to be a relative minimum and maximum, respectively. At $x = e$, the function f' is not defined,[8] a case similar to those in Figure 3.4. At $x = d$, the value of the first derivative is zero, but the function has neither a relative maximum nor a relative minimum (it is an inflection point). Thus the set of critical points is

$$C = \{a, b, c, d, e, g, h\}.$$

The extreme points $T \subset C$ are

$$T = \{a, b, c, e, g, h\}, \text{ but not } d.$$

---

[8] This type of critical point has been identified to make the reader aware of it, since in a practical problem it could prove troublesome; however, it will not be discussed further; it seldom arises.

3.7 Maxima and Minima

Here the assertion that there is an important relationship between the slope of a function at a point in a given subinterval and whether there is a relative maximum or minimum of the function at that point is investigated in detail.

Figure 3.10

The Tangent to the Function f at $x = x_0$.

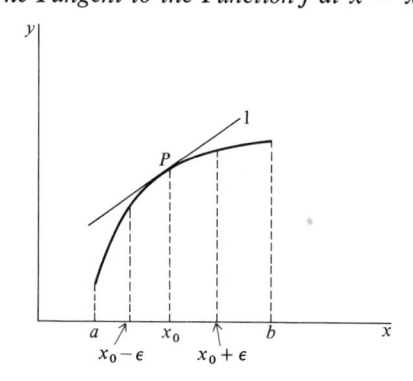

Given the function f in Figure 3.10, if one tangent lies above the function at $x = x_0$, then more than one tangent must also lie above f to the left of x_0 and to the right of x_0. It is possible for an infinite number of points on either side of x_0 (the neighborhood of x_0) to lie below tangent 1. A tangent at any of these points, for example, $x_0 - \epsilon$ or $x_0 + \epsilon$, where ϵ is a very small change from x_0, must also lie above f. (These tangents are not drawn in Figure 3.10.) In order for the function f to lie below the tangent 1 at $x = x_0$ in some neighborhood of x_0, (the absolute value of) the slope of the tangent at $x_0 - \epsilon$ must be greater than that at x_0, which in turn must be greater than that at $x_0 + \epsilon$. In other words, the *rate of change of the slope* of the curve must be negative.

A function whose tangent lies everywhere above the curve in an interval is said to be concave downward in that interval. This theorem is stated without proof

> **Theorem 3.1:** *Given a function f with rule $y = f(x)$, if f'' exists and is negative (positive) in an interval $a \leq x \leq b$, then the graph of f is concave downward (upward) at each point in the interval. The tangent of f lies everywhere above (below) the graph of f in the interval.*

The significance of this theorem for maximum and minimum values of a function is discussed in the next paragraphs. The graph of the function that is concave upward or downward forms the boundary of a convex point-set; a line connecting two points within the concave set lies everywhere within the set. The tangent is the limit of the line connecting a point on the boundary with another point on the boundary as the second point approaches the first.

Sec. 3.7]   MAXIMA AND MINIMA   83

Figure 3.11

Tangents to the Graph of the Function f.

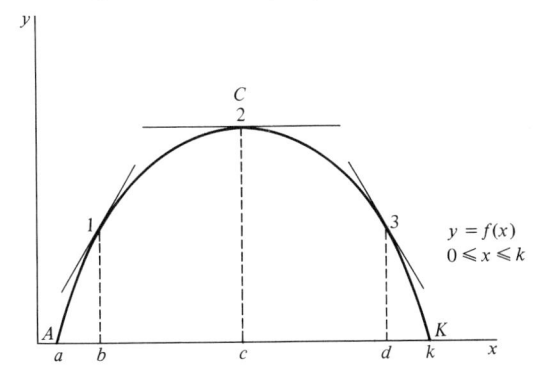

Figure 3.11 shows the graph of a function that is concave downward. It reaches a maximum value at point C. The slope of tangent 1 at $x = b$ is positive; the rate of change for the function is positive. At point $x = d$, the slope of tangent 3 to the curve at D is negative; the rate of change of the function is now negative. At the point $x = c$, the tangent 2 is horizontal, its slope is 0. At this point the rate of change of the function is zero and the function can be seen to reach a relative maximum. (It increased steadily to c, flattened out and began to decrease.) The relative maximum is also the global (only) maximum in the interval in this case.[9]

The sufficient condition for a function to be concave downward in an interval is that the value of f'' is negative. (This condition is not a necessary condition since it is possible for the value of f'' to be zero and further tests to confirm that it is concave downward.) For the value of the function that is concave downward to be a maximum, the first derivative must be zero. This is formally stated in Theorem 3.2 which is presented without proof.

Theorem 3.2: *Given the function f' with rule $y = f(x)$ differentiable in the interval $a \leq x \leq b$, the value of the function is a (relative) maximum at a point $x = x_0$ if*

1. *the value of f' is zero and*
2. *the value f'' is negative*

at $x = x_0$. The value of f is a (relative) minimum at $x = x_0$ if

1. *the value of f' is zero and*
2. *the value f'' is positive.*

Thus a necessary condition for a maximum or minimum value for f *within* an interval is that where $f'(x)$ exists, $f'(x_0) = 0$.

---
[9] This is an intuitive statement of the intermediate value theorem which is not presented formally.

The statement of the theorem is in reverse order from the discussion in the preceding paragraph; the critical point is identified through the first derivative test (first-order condition) and the distinction between maxima and minima (and later inflection points) made through the second-order conditions. Note that it is possible to have a relative maximum or minimum without the above being satisfied: at a or h the endpoints of the interval in Figure 3.9, or at e, a point at which the first derivative is not defined.

Figure 3.12 shows that at a minimum point, such as that at h, the slope of the tangent at H, is also zero, but the tangents to the curve in some subinterval around h all lie below the curve. In the neighborhood of the minimum point, the slope of the tangents have gone from negative to zero to positive. The rate of change in the slope of the tangent is positive.

Figure 3.12

The Function f with Several Extreme Points.

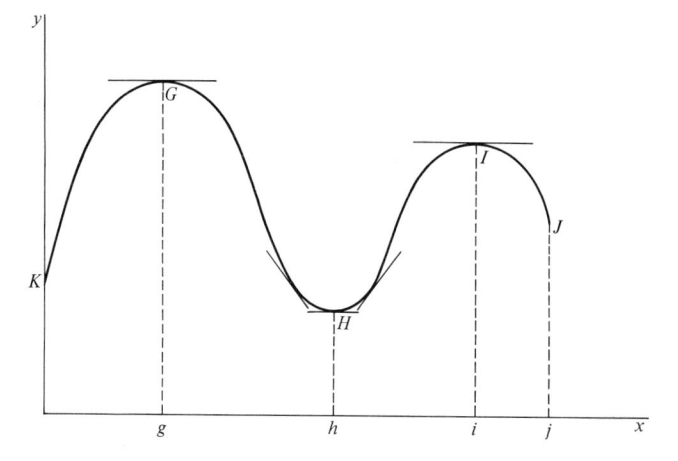

For a maximum value of a function at $x = x_0$, the tangents of the curve all lie *above* the curve in a subinterval around x_0. For a minimum value, the tangents in the subinterval all lie below. Analytically the maxima can be distinguished from the minima by the value of the second derivative of f at $x = x_0$. When the value $y'' = f''(x_0)$ is negative, it can be seen from the discussion of the characteristics of a tangent that the function reached a maximum; when $y'' = f''(x_0)$ is positive, the function reached a minimum at x_0.

A procedure for finding the extreme points for a function f with rule $y = f(x)$ follows. First, the necessary (although not sufficient) first-order (first derivative) conditions for maxima and minima within an interval must be investigated as follows:

1. Find the rule for the first derivative function.
2. State the rule as an equation and set it equal to zero.
3. Solve the equation to find the values x_i of the domain variable for which the value of the first derivative function is zero.

Sec. 3.7]   MAXIMA AND MINIMA

4. Substitute these domain values into the rule for f to find the values of f at each point.

Then further steps that in conjunction with the first order conditions constitute sufficient conditions for maxima and minima, the second-order conditions are:

5. Find the rule for the second derivative function.
6. Solve it for $y'' = f''(x)$ at each respective value of x_i found in Step 3.
7. If the value of y'' is negative, the extreme point is a maximum; if it is positive, the extreme point is a minimum.[10]
8. Solve for the value y at $x = a$ and $x = b$, the endpoints of the interval.
9. Compare all maxima to find the one with the largest value of y. This is the global maximum for the interval. Compare the minima to determine the global minimum.

These steps are now illustrated with an example.

Example 3.11

Given the function f with rule

$$y = \frac{x^3}{3} - \frac{5}{2}x^2 + 4x + 2, \quad (0 \leq x \leq 10)$$

find the extreme points in this interval.

The rule for the first derivative function f' is found and set equal to zero,

$$f'(x) = y' = \frac{dy}{dx} = x^2 - 5x + 4 = 0.$$

The equation is solved for x:

$$(x - 1)(x - 4) = 0$$
$$x = 1, 4.$$

The apparent extreme points occur at $x = 1$ and $x = 4$. The value of the function f at $x = 1$ and $x = 4$ is found by substituting these values of x into the rule for f

at $x = 1$
$$y = \frac{x^3}{3} - \frac{5}{2}x^2 + 4x + 2$$
$$y = \tfrac{1}{3} - \tfrac{5}{2} + 4 + 2$$
$$y = \tfrac{2}{6} - \tfrac{15}{6} + \tfrac{24}{6} + \tfrac{12}{6} = \tfrac{23}{6} = 3\tfrac{5}{6}$$

at $x = 4$
$$y = \frac{x^3}{3} - \frac{5}{2}x^2 + 4x + 2$$
$$y = \tfrac{64}{3} - \tfrac{80}{2} + 16 + 2$$
$$y = \tfrac{128}{6} - \tfrac{240}{6} + \tfrac{108}{6} = -\tfrac{4}{6} = -\tfrac{2}{3}.$$

[10] The value y'' may also be zero. This case is discussed in Section 3.8.

The rule for the second derivative function of f is

$$y'' = \frac{d^2y}{dx^2} = 2x - 5.$$

Its value at the points $x = 1$ and $x = 4$ is:

$$2x - 5 = 2 - 5 = -3 \quad \text{at } x = 1,$$
$$2x - 5 = 8 - 5 = 3 \quad \text{at } x = 4.$$

The rate of change of the slope is negative at $x = 1$, the value of the function at the extreme point is a maximum; at $x = 4$ it is positive, and the value is a minimum.

The values of f at the endpoints must also be determined to see if they are relative maxima or minima:

at $x = 0$,
$$y = 2$$
at $x = 10$,
$$y = \tfrac{1000}{3} - \tfrac{500}{2} + 40 + 2 = 333\tfrac{1}{3} - 250 + 42 = 125\tfrac{1}{3}.$$

The graph of the function is presented in Figure 3.13. Given the extreme points of the function, the curve is relatively simple to complete. In the interval $0 \leq x \leq 10$,

Figure 3.13

Sketch of the Function f from Example 3.9.

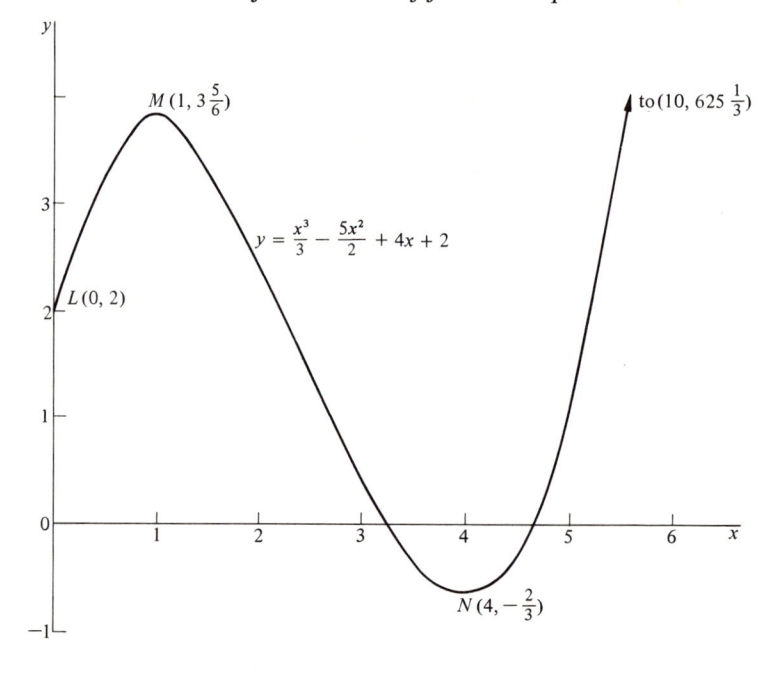

the function is increasing (although at a decreasing rate) everywhere to the left of $x = 1$, reaches a maximum at $x = 1$ (the point M) and then decreases between $x = 1$ and $x = 4$. At $x = 4$ (the point N) the function reaches a minimum and then increases everywhere to the right of $x = 4$. The endpoint (0, 2) (point L) is a relative minimum. The point $(10, 125\frac{1}{3})$ (a point P not shown in Figure 3.13) is a global maximum. Since the function f is defined only in the interval $0 \leq x \leq 10$, the function beyond these points is of no interest. The point $(1, 3\frac{5}{6})$ is a relative maximum point and $(4, -\frac{2}{3})$ is the global minimum point. If the value of the function at $x = 10$, the right-hand endpoint, were neglected, the overall maximum would not have been identified, and if this function represented a real-world phenomenon, the result of such an oversight might be very significant.

3.8 Inflection Points

One final class of points needs to be discussed, that at which the second derivative is zero because the function is changing concavity from upward to downward or the reverse. Knowing the location of such points is helpful to an analyst who wishes to understand the characteristics of a function or to sketch the graph of the function. In Figure 3.14 the function in the interval $[g; k]$ is concave upward

Figure 3.14

A Critical Point x_0 which is an Inflection Point.

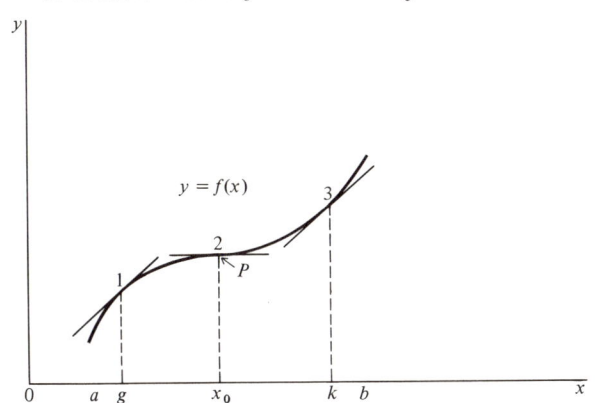

in the neighborhood of the extreme point $x = k$, and the value of the second derivative of the function is positive. In the neighborhood of the other extreme point, the relative minimum at $x = g$, the function is concave downward, and the value of the second derivative is negative in that neighborhood. At some point in the interval $g \leq x \leq k$, the function changes from concave downward to concave upward. The point at which the function changes concavity is known as an inflection point. The tangent to a curve at an *inflection point* passes through the curve at that point.

> **Definition—Inflection Point:** *Given a function f with rule $y = f(x)$ differentiable in the interval $a \leq x \leq b$, the point $(x_0, f(x_0))$ at which the function changes from concave upward to concave downward or vice versa is known as an inflection point and $f''(x_0) = 0$.*

It is stated without proof that the necessary condition for an inflection point at x_0 (the value of the second derivative function must be zero at x_0) becomes sufficient with the added condition that the *order* of the *next* derivative function of higher order whose value is nonzero at x_0 must be *odd;* that is, the order of the next nonzero derivative must be the third, or the fifth, or the seventh, etc.

Inflection points may occur at points for which the value of the first derivative function is zero; that is, at critical points, as at point *d* in Figure 3.9. They may also occur (and more generally do) at noncritical points, in the interval $b < x < c$ and $c < x < d$ in Figure 3.9, for example—at the "flop-over points" at which the function shifts the direction of its concavity. They may also occur at points at which the first derivative is infinite.

An inflection point of one type occurs at a point at which $y' = 0$, the change in the second derivative occurs when the curve has just become horizontal. In Figure 3.14 the function on the left side of the inflection point appears to have the characteristics of a maximum (this is, the part of the function approaching *P* from the left); on the right side, it appears to be a minimum (approaching *P* from the right). Inflection points of this type can be identified by the fact that they have identical roots to the first derivative at point *P*. This type of inflection point is illustrated with an example.

Example 3.11

Given the function *f*, with rule

$$y = \frac{x^3}{3} - x^2 + x$$

identify its extreme points.

Differentiating to find the rule for the derivative function gives

$$\frac{dy}{dx} = x^2 - 2x + 1.$$

Setting the first derivative equal to zero and solving for the roots gives

$$0 = x^2 - 2x + 1$$
$$0 = (x - 1)(x - 1)$$
$$\left.\begin{array}{r}+1 = x\\+1 = x\end{array}\right\} \text{ a double root at } x = 1.$$

The value of the function at $x = 1$ is

$$y = \frac{x^3}{3} - x^2 + x$$
$$y = \tfrac{1}{3} - 1 + 1 = \tfrac{1}{3}.$$

The rule for the second derivative is

$$\frac{d^2y}{dx^2} = 2x - 2.$$

The rule for the third derivative is

$$\frac{d^3y}{dx^3} = 2;$$

it is nonzero everywhere.

Substituting the value of x into the rule for the second derivative gives $2(1) - 2 = 0$; at the point $x = 1$, the second derivative is equal to zero; the third derivative is nonzero. Thus the point $(1, \frac{1}{3})$ is an inflection point. To check for other inflection points, the rule for the second derivative function is set equal to zero and solved to find all its roots.

$$0 = 2x - 2$$
$$2 = 2x$$
$$1 = x.$$

There is only the one inflection point, the one that occurs at $x = 1$.

The function is everywhere increasing as it approaches $x = 1$ from the left, reaches an inflection point at $(1, \frac{1}{3})$, and then continues to increase. To the left of $(1, \frac{1}{3})$ it becomes large negatively and to the right it becomes large positively. It has no maxima or minima, unless its domain is restricted, in which case the endpoints would be the global minimum and maximum, respectively.

When the value of the second derivative of the function is zero, that is, when $y'' = f''(x) = 0$, the second derivative test for maxima and minima breaks down. Two choices then remain: the value of the second derivative can be tested at points on both sides of the critical point to see if it does change signs at $x = x_0$, or higher order derivatives can be taken. If derivatives of all orders through the nth are equal to zero, and the derivative of order $(n + 1)$ is nonzero, then,

1. For $(n + 1)$ an even number, (a) if the value of the derivative at $x = x_0$ is < 0, the function has maximum value at $x = x_0$; (b) if the value of the derivative at $x = x_0$ is > 0, the function has a minimum value at $x = x_0$.
2. For $(n + 1)$ an odd number, the function is neither a maximum nor a minimum; it is an inflection point.

In the next chapter applications of the derivative are explored.

3.9 Summary

The derivative function f' is a function derived from an original function f; its rule is derived from the rule for f. The two functions f and f' stated in set terms are:

$$f = \{(x, y): y = f(x)\}$$
$$f' = \left\{(x, y'): y' = \lim_{\Delta x \to 0} \frac{f(x + \Delta x) - f(x)}{\Delta x}\right\}.$$

The rule for f' is derived from the *average rate of change* in the value of f:

$$\frac{f(x + \Delta x) - f(x)}{\Delta x}$$

by taking the limit of the quotient expressing the average rate of change. The equation for this limit value is the rule for f' at a given domain point

$$y' = \lim_{\Delta x \to 0} \frac{f(x + \Delta x) - f(x)}{\Delta x}$$

and is by definition the *instantaneous rate of change* of f at that point. In graphic terms the derivative of a function at a point a is the limit of the slope of the secant between the point A and another point B on the graph of the function as $A \to B$. The line passing through the point A with slope equal to the value of the derivative at A is known as the *tangent* to the function at the point A.

The specific rules for the derivative functions f' of large classes of functions f are derived by substituting the specific rules for the functions f, e.g., $y = x^n$, into the general definitional form for the rule of the derivative function

$$y' = \lim_{\Delta x \to 0} \frac{f(x + \Delta x) - f(x)}{\Delta x}$$

and then simplifying the resulting expression. For the function with rule $y = x^n$, the rule for the derivative function simplifies to $y = nx^{n-1}$. The process of finding the derivative functions for given functions is called *differentiation*.

A derivative function can itself be differentiated. Its derivative is known as the second derivative of the function f and is generally designated f''. If a function f expresses total distance traveled by a vehicle as a function of time, the first derivative function f' expresses the rate of change in distance traveled—the speed of the vehicle—as a function of time. The second derivative f'' expresses the rate of change of the rate of change—the acceleration of the vehicle as a function of time.

The derivative functions f' and f'' can be used to help identify the points at which f reaches its relative maximum and minimum values within an interval. Points at which the value of the function reaches relative maxima and minima are points at which the first derivative function equals zero. These points together with the endpoints of the interval make up the set known as *extreme points* of the function. This set plus the set of points at which f' does not exist plus inflection points make up the set known as the *critical points* of the function.

The values of the function at the *extreme* points must be compared in order to identify the largest and the smallest values the function reaches as the *global maximum* and *minimum* in a given interval. If $f(x)$ is differentiable at x_0, the condition $f'(x_0) = 0$ is necessary for an extreme point within the interval; the conditions $f'(x_0) = 0, f''(x_0) < 0$ are sufficient for a relative maximum and $f'(x_0) = 0$, $f''(x_0) < 0$ for a relative minimum within the interval. Any test for maxima and minima must include the endpoints to the interval. At the endpoints the value of the function is calculated directly and compared with values at neighboring points. The largest of the relative maxima is the global maximum; the smallest of the minima is the global minimum.

An inflection point is one at which a function changes its concavity—from concave upward (with second derivative positive) to concave downward (with second derivative negative) or vice versa. Thus a necessary condition for an inflection point is that the second derivative (if it exists) is zero. The sufficient conditions are that the second derivative is zero and the next derivative of higher order that is not zero is an odd-numbered derivative.

Restated in symbols: If $f(x)$ is differentiable twice, then

$f'(x_0) = 0, f''(x_0) > 0$ implies a relative minimum for $x = x_0$;
$f'(x_0) = 0, f''(x_0) < 0$ implies a relative maximum for $x = x_0$;
$f'(x_0) = 0, f''(x_0) = 0$ requires further analysis;
$f''(x_0) = 0$ and $f''(x_0)$ changes sign at x_0 implies an inflexion point for $x = x_0$;
$f''(x_0) = 0$ but $f''(x_0)$ does not change sign at x_0 implies no inflexion point for $x = x_0$

If $f(x)$ is three times differentiable then

$f''(x_0) = 0$ and $f'''(x_0) \neq 0$ implies that $f(x)$ has an inflexion point for $x = x_0$.

In general if $f(x)$ is n times differentiable and $f^n(x_0)$ is the *lowest order* derivative that is *different from zero* at $x = x_0$ then

if n is even and $f^n(x_0) < 0, f(x_0)$ is a maximum;
if n is even and $f^n(x_0) > 0, f(x_0)$ is a minimum;
if n is odd and $> 1, f(x_0)$ is an inflexion point.

Key Concept

Concept	Reference Section
Average Rate of Change	3.1, 3.2
Instantaneous Rate of Change	3.1, 3.2
Derivative	3.1, 3.2
Limit of a Secant	3.2
Tangent	3.2
Differentiable Function	3.2
Differentiation	3.3
The Differential	3.3
Differentiation Rules	3.4
Constant	
Constant Times a Function	
Sum of Functions	
x^n	
Product of Functions	
Chain Rule (Composite Functions)	
Quotient of Functions	
Higher Order Derivatives	3.5
Critical Points	3.6
Extreme Points	3.6
Concave Upward	3.7
Concave Downward	3.7
Convex Set	3.7
Relative Maxima, Minima	3.7
Global Maxima, Minima	3.7
Inflection Points	3.8

Appendix 3A: Proofs of Rules 3.4, 3.5, and 3.6 for Derivatives

Rule 3.4 is the rule for the derivative function of a polynomial function. The derivation of the rule requires the use of the "binomial expansion" which is presented below without proof. Given an expression in the form $(a^n - b^n)$, it can be expanded as follows:[12]

$$(a^n - b^n) = (a - b)(a^{n-1}b^0 + a^{n-2}b^1 + \cdots + a^1 b^{n-2} + a^0 b^{n-1}). \quad (3A.1)$$

Because $b^0 = 1$ and $a^0 = 1$, equation (3A.1) simplifies to

$$(a^n - b^n) = (a - b)(a^{n-1} + a^{n-2}b^1 + \cdots + a^1 b^{n-2} + b^{n-1}).$$

Rule 3.4 is formally stated.

Rule 3.4: *Given the function f with rule $y = f(x) = x^n$ differentiable in an interval, then the rule for the derivative function f' is*

$$y' = nx^{n-1}$$

for n rational.

Proof:

Given the function f with rule $y = f(x) = x^n$, the rule for the derivative function f' is given by

$$y' = \lim_{\Delta x \to 0} \frac{f(x + \Delta x) - f(x)}{\Delta x} = \lim_{\Delta x \to 0} \frac{(x + \Delta x)^n - x^n}{\Delta x},$$

by definition. The numerator is now in the form $a^n - b^n$, where $a = (x + \Delta x)$ and $b = (x)$. If the expanded form is substituted for $(a^n - b^n)$, then

$$y' = \lim_{\Delta x \to 0} \frac{[(x + \Delta x) - x][(x + \Delta x)^{n-1} + (x + \Delta x)^{n-2}x + \cdots + (x + \Delta x)x^{n-2} + x^{n-1}]}{\Delta x}.$$

The first term in the numerator reduces to Δx; thus,

$$y' = \lim_{\Delta x \to 0} \frac{[\Delta x][(x + \Delta x)^{n-1} + (x + \Delta x)^{n-2}x + \cdots + (x + \Delta x)x^{n-2} + x^{n-1}]}{\Delta x}$$

and the Δx's cancel giving:

$$y' = \lim_{\Delta x \to 0} [(x + \Delta x)^{n-1} + (x + \Delta x)^{n-2}x + \cdots + (x + \Delta x)x^{n-2} + x^{n-1}].$$

Note that there is a total of *n terms* under the limit sign. In the limit the Δx's disappear (go to zero) to give:

$$y' = x^{n-1} + (x^{n-2})x + \cdots + x(x^{n-2}) + x^{n-1}.$$

---

[12] This can be illustrated in the case of $a^3 - b^3$. If the expression is factored to the form $(a - b)(a^2 + ab + b^2)$, and then multiplied as follows,

$$\begin{array}{r} a^2 + ab + b^2 \\ \underline{a - b} \\ a^3 + a^2b + ab^2 \\ \underline{ - a^2b - ab^2 - b^3} \\ a^3 - b^3 \end{array}$$

the required result is obtained.

APPENDIX 3A

Carrying out the multiplication for each term and adding the n terms, which are now identical, gives
$$y' = n(x^{n-1}) = nx^{n-1}.$$

This establishes the rule for n, a positive integer. The extensions to establish the rule for n rational and for n irrational, are not presented.

The derivation of the rule for the derivative of x^n for n, a positive integer requires only the standard steps used in Section 3.4. In the proof in this appendix, however, the simplification step is somewhat complex.

The proof of Rule 3.5 also requires algebraic manipulation—the addition and subtraction of the same factor to an equation so that the net change is zero. The procedure is similar to that used in "completing the square," which is a familiar procedure. For example, given the equation,
$$x^2 + 4x + 1 = 0$$

it can be seen that it is not a perfect square. If it had been $x^2 + 4x + 4$ instead, it could have been factored into $(x + 2)^2$. Suppose that the factor $(3 - 3)$—which itself equals zero—were added to the original expression. The result would be
$$x^2 + 4x + 4 - 3 = 0.$$

The right-hand side is now stated as the difference of two squares $(x + 2)^2$ and $(\sqrt{3})^2$. The initial equation is left unchanged, but by "completing the square," it can then be factored and solved:
$$(x + 2)^2 - 3 = 0$$

or
$$[(x + 2) + \sqrt{3}][(x + 2) - \sqrt{3}] = 0$$
$$x = (\sqrt{3} - 2)$$
$$x = (-\sqrt{3} - 2).$$

In proving Rule 3.5 a similar approach is followed.

For Rule 3.5 to hold, the definition of a derivative function requires that
$$h' = f(x) \cdot \lim_{\Delta x \to 0} \frac{g(x + \Delta x) - g(x)}{\Delta x} + g(x) \lim_{\Delta x \to 0} \frac{f(x + \Delta x) - f(x)}{\Delta x}.$$

The proof begins with the definition of h' and must end up in the form just stated.

Rule 3.5: *Given the function f with rule $y = f(x)$, and g with rule $v = g(x)$, differentiable over an interval and the function h with rule $u = [f(x) \cdot g(x)]$, then the rule for the derivative function h' is given by*
$$u' = \frac{d}{dx}[f(x) \cdot g(x)] = f(x) \cdot g'(x) + g(x) \cdot f'(x).$$

Proof:

Given $u = h(x) = f(x) \cdot g(x)$; then the rule for the derivative function h' is
$$u' = \frac{d}{dx}[h(x)] = \lim_{\Delta x \to 0} \frac{h(x + \Delta x) - h(x)}{\Delta x},$$

by definition. Substituting $f(x) \cdot g(x)$ for $h(x)$ and $f(x + \Delta x) \cdot g(x + \Delta x)$ for $h(x + \Delta x)$ gives:
$$u' = \lim_{\Delta x \to 0} \frac{f(x + \Delta x) \cdot g(x + \Delta x) - f(x) \cdot g(x)}{\Delta x}. \tag{3A.2}$$

A special factor (which is equal to zero) must now be introduced to transform this limit to the desired form. The expression

$$\frac{f(x + \Delta x)g(x) - f(x + \Delta x)g(x)}{\Delta x}. \qquad (3A.3)$$

is inserted before the minus sign in equation (3A.2).

Since the value of this expression is zero, it can be added to the original function without changing that function or its limit. Introducing equation (3A.3) into the limit (3A.2) gives:

$$u' = \lim_{\Delta x \to 0} \frac{\overset{(1)}{f(x+\Delta x) \cdot g(x+\Delta x)} + \overset{(2)}{[f(x+\Delta x) \cdot g(x)} - \overset{(3)}{f(x+\Delta x) \cdot g(x)]} - \overset{(4)}{f(x) \cdot g(x)}}{\Delta x}.$$

The parts of the numerator of this equation are identified and numbered for convenient reference. Parts (2) and (3) are from equation (3A.3) and were just inserted between (1) and (4) of equation (3A.2). Rearranging terms, expressions (1) and (3) and (2) and (4) are placed together.

$$u' = \lim_{\Delta x \to 0} \frac{\overset{(1)}{f(x+\Delta x) \cdot g(x+\Delta x)} - \overset{(3)}{f(x+\Delta x) \cdot g(x)} + \overset{(2)}{f(x+\Delta x) \cdot g(x)} - \overset{(4)}{f(x) \cdot g(x)}}{\Delta x}.$$

The common terms are then factored out to give:

$$u' = \lim_{\Delta x \to 0} \frac{\overset{(1+3)}{f(x+\Delta x)[g(x+\Delta x) - g(x)]} + \overset{(2+4)}{g(x)[f(x+\Delta x) - f(x)]}}{\Delta x}. \qquad (3A.4)$$

By limit theorems (Theorems 2.1 to 2.3), equation (3A.4) can be rewritten as follows:

$$u' = \overset{(1+3)}{\lim_{\Delta x \to 0} f(x+\Delta x) \cdot \lim_{\Delta x \to 0} \frac{g(x+\Delta x) - g(x)}{\Delta x}} + \overset{(2+4)}{\lim_{\Delta x \to 0} g(x) \cdot \lim_{\Delta x \to 0} \frac{f(x+\Delta x) - f(x)}{\Delta x}}. \qquad (3A.5)$$

Since

$$\lim_{\Delta x \to 0} f(x+\Delta x) = f(x) \quad \text{and} \quad \lim_{\Delta x \to 0} g(x) = g(x),$$

and the other limits are in the exact form for the rule of a derivative function, equation (3A.5) reduces to:

$$u' = f(x) \cdot g'(x) + g(x) \cdot f'(x). \qquad \text{Q.E.D.}$$

Rule 3.6 The Chain Rule: *Given the function f, with rule $y = f(u)$ whose domain variable is the range variable of another function g with rule $u = g(x)$ differentiable over intervals $a \leq x \leq b$ and $c \leq x \leq d$, then the rule for the derivative function of f in the domain of g is given by*

$$\frac{dy}{dx} = \frac{dy}{du} \cdot \frac{du}{dx}.$$

Proof:

The following average rates of change can be written

$$\frac{\Delta y}{\Delta u} = \frac{f(u + \Delta u) - f(u)}{\Delta u} \qquad (3A.6)$$

APPENDIX 3A

$$\frac{\Delta u}{\Delta x} = \frac{g(x + \Delta x) - g(x)}{\Delta x} \qquad (3A.7)$$

$$\frac{\Delta y}{\Delta x} = \frac{f[g(x + \Delta x)] - f[g(x)]}{\Delta x} = \frac{f(u + \Delta u) - f(u)}{\Delta x}. \qquad (3A.8)$$

But since Δy, Δu, and Δx are small increments with finite value, the following equations can be written:

$$\frac{\Delta y}{\Delta x} = \frac{\Delta y}{\Delta u} \cdot \frac{\Delta u}{\Delta x}. \qquad (3A.9)$$

Substituting the right-hand sides of equations (3A.6) and (3A.7) into the right-hand side of equation (3A.9) gives:

$$\frac{\Delta y}{\Delta x} = \frac{f(u + \Delta u) - f(u)}{\Delta u} \cdot \frac{g(x + \Delta x) - g(x)}{\Delta x}. \qquad (3A.10)$$

From the definitions of the functions f and g and the discussion in Examples 3.5 and 3.6 it can be seen that as $\Delta x \to 0$, since $\Delta u = g(x + \Delta x) - g(x)$, $\Delta u \to 0$ also. Thus taking the limits of both sides of equation (3A.10) as the increments in their respective domain variables $\to 0$ gives:

$$\lim_{\Delta x \to 0} \frac{\Delta y}{\Delta x} = \lim_{\Delta u \to 0} \frac{f(u + \Delta u) - f(u)}{\Delta u} \cdot \lim_{\Delta x \to 0} \frac{g(x + \Delta x) - g(x)}{\Delta x}.$$

By the definition of the rule for a derivative function, this is equal to:

$$\frac{dy}{dx} = \frac{dy}{du} \cdot \frac{du}{dx}. \qquad \text{Q.E.D.}$$

This form is extremely useful for remembering the rule for the derivative of a function of a function (a composite function), but the terms dy/du and du/dx are *not* to be looked on as fractions (although $\Delta y/\Delta u$ and $\Delta u/\Delta x$ are). The statement of this rule is *not* intended to indicate that the derivative dy/dx is some sort of a fraction that cancels against another fraction du/dx. This implication is unfortunate, but the reader should be alert to the meaning of the notation.

Problems

Differentiation

Find dy/dx for the functions with the following rules.

1. $y = ax^5 - bx^3 + cx + d$

2. $y = (x^3 + 4)(x^3 - 4)$

3. $y = (x^2 + 1)^2 \cdot (x - 2)$

4. $y = (2x + 4)^3$

5. $y = (2x + 4)^{-3/2}$

6. $y = [3x^2 + (2x - 1)^2]^{1/2}$

7. $y = 3x^4 - 2(x^2 - 1)^{4/3}$

8. $y = \dfrac{x}{1 + x}$

9. $y = \dfrac{ax^2}{1/(1 + x)} + \dfrac{(1/ax^2)}{1 + x}$

10. $y = \dfrac{1 + x}{1 - x} + \dfrac{1 - x}{1 + x}$

11. $y = ax\left(bx^2 + \dfrac{x}{1 + x}\right)$

12. $y = \left[\dfrac{x}{1 - x}\right]^{1/2}$

13. $y = x \cdot \sqrt{x^2 - a^2}$

14. $y = \left[\dfrac{1}{x} + \dfrac{1}{x^2}\right]^3 \cdot \dfrac{1}{x^3}$

15. $y = \left[\dfrac{x}{2} + \dfrac{x}{2 - x}\right]^2 \cdot \dfrac{2 - x}{x}$

16. $y = \left(\dfrac{x}{x - 1}\right)(3x + 1)^2 \left(\dfrac{2x}{3x^2 + 1}\right)^{1/2}$

17.

Fill in the proper spaces below for functions with rules as indicated.

	$f(x) = \begin{cases} x^2 - 2x, & x \neq 1 \\ -2, & x = 1 \end{cases}$	$f(x) = \begin{cases} 2 - |x|, & -1 \leq x < 3 \\ |x|, & -2 < x < -1 \end{cases}$	$f(x) = \begin{cases} \pi, & x \neq 1 \\ \pi + .00001, & x = 1 \end{cases}$ (Note: π is a constant)	$f(x) = \begin{cases} 2x - x^2, & x > 0 \\ 2x, & x \leq 0 \end{cases}$
Existence of limits				
$\lim_{x \to 0} f(x) =$	Equals	Equals	Equals	Equals
$\lim_{x \to 1} f(x) =$	Does not exist	Does not exist	Does not exist	Does not exist
Continuity				
At $x = 0$, $f(x)$ is	Cont.	Cont.	Cont.	Cont.
At $x = 1$, $f(x)$ is	Not cont.	Not cont.	Not cont.	Not cont.
Differentiability				
At $x = 0$, $f'(x) =$	Equals	Equals	Equals	Equals
At $x = 1$, $f'(x) =$	Does not exist	Does not exist	Does not exist	Does not exist

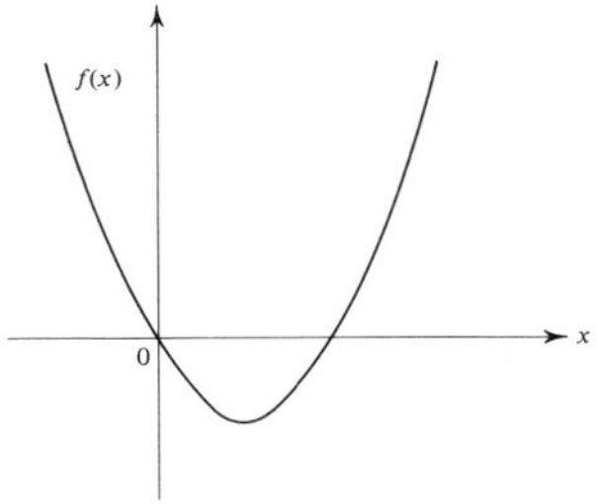

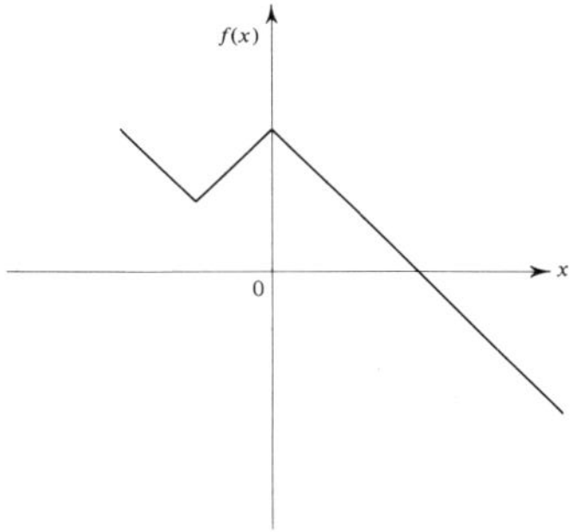

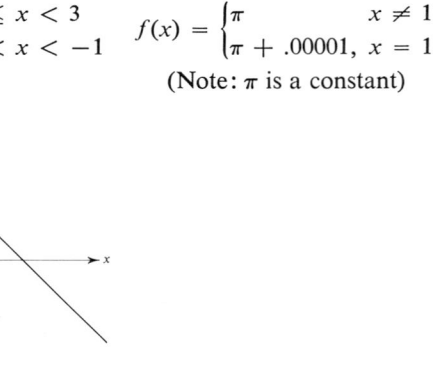

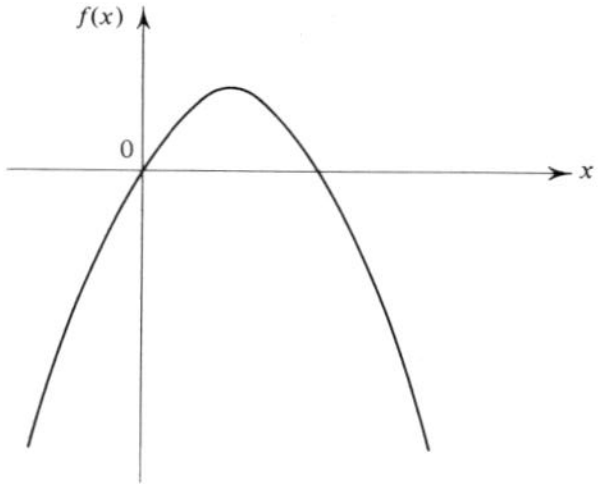

PROBLEMS 97

18. If P can be represented by the equation $2 + 50Q - 5Q^2$, what is the equation for the instantaneous rate of change of P with respect to Q?

19. In the preceding problem, what is the instantaneous rate of change in P if Q is equal to 4? Q equal to 6?

20. Find the equation for the differentials of functions with the following rules
  (a) $y = f(x) = 4 - 3x$
  (b) $y = f(x) = (4 - 3x)^2$
  (c) $y = f(x) = x(4 - 3x)^2$
  (d) $y = f(x) = (x - 2)(4 - 3x)^2$
  (e) $y = f(x) = (x - 2)^3(4 - 3x)^2$

21. Find dy and Δy for the function with the rule
  (a) $y = \dfrac{x^2}{2}$
  (b) $y = \dfrac{x^2}{2}$, for $x = 1$, $dx = .5$
  (c) $y = \dfrac{x^2}{2}$, for $x = 1$, $dx = .1$
  (d) Compare $\Delta y - dy$ for cases a, b, and c above.

22. Find Δy, dy, and $\Delta y - dy$ at the points $x = 1$ and $x = 2$ for $dx = .1$ for the functions with the following rules:
  (a) $y = f(x) = 4 - 3x$
  (b) $y = f(x) = (4 - 3x)^2$

Find the critical points for the functions with the following rules, identify them as maxima, minima, or inflection points, find the global maximum and minimum and sketch the curves.

23. $y = x^2 - 4x$

24. $y = \dfrac{x^3}{3} - 4x$

25. $y = x^3 - 6x^2 + 9x - 6$

26. $y = \tfrac{1}{2}x^{2/3}$

27. $y = x^5$

28. $y = \tfrac{1}{6}(2 + 18x + 6x^2 - 2x^3)$

29. $y = x^3 - 6x^2 + 9x + 2$
  (a) $[0; 5]$
  (b) $[0; 4]$
  (c) $(0; 4)$
  (d) $3 \le y \le 7$
  (e) How would the function change if the constant 2 was replaced by 1?

30. $y = x^3 - 6x^2 + 2$
  (a) $[-3; 3]$
  (b) $[-\infty; 2]$

31. $y = x^3 - 9x + 2$
  (a) $[2; 4]$
  (b) $[-3; 3]$

32. $y = x^2 - 4$
  (a) $[1; 3]$
  (b) $[-3; -1]$
  (c) $[-3; 2]$

33. $y = \dfrac{1}{2x}$

   (a) $[-2; 2]$
   (b) The whole real line

34. $y = x + \dfrac{1}{x}$

35. $y = \dfrac{3 + 2x}{x}$

36. The city of Itaca is considering the construction of a housing project for the elderly (or senior citizens). Two proposals are before the city fathers. The first involves separate dwelling units in clusters with a common dining and recreational facility. An alternate proposal involves one high-rise structure. The cost of either facility is a function of the number of dwelling units, construction requirements, and associated costs. The builder has estimated the *average-cost* functions under the blueprint restrictions to be of the form:

   Proposal 1: $AC_1 = x^2 - 4x + 6$

   Proposal 2: $AC_2 = \dfrac{x^3}{3} - 3x^2 + 8x + 2$

   where x is measured in hundreds of units and AC in thousands of dollars. (Thus if $x = 2$, the *average cost* per unit under proposal 1 is $(2^2 - 4(2) + 6)1000 = \$2,000$.)

   It is important that *average cost* per dwelling unit be minimized, since the necessary rent requirement will be determined in part thereby. If the city fathers insist on producing at least 400 dwelling units, but they can spend no more than \$2,800,000, (a) what is the optimal proposal, (b) how many units should be constructed, and (c) what is the total cost to the city? Use the differential calculus to obtain your solution. (*Remember* the solution involves building the number of units for which the *average cost* is minimized, subject to (1) a requirement that at least 400 units be built, and (2) a constraint on the total spending.)

   Given the composite functions with the following rules:

37. $z = 3u^2 - u$
    $u = 2x$

    Find $\dfrac{dz}{dx}$ (a) using the chain rule and (b) by direct substitution.

38. $z = 5u^2 - \dfrac{1}{u^2} + 2u$
    $u = x^3 - 2x + 1$

    Find $\dfrac{dz}{dx}$ (a) using the chain rule and (b) by direct substitution.

39. $z = 3u^2 - u + v^2$
    $u = \dfrac{v}{2}$
    $v = 2x + 1$

    Find $\dfrac{dz}{dx}$ (a) using the chain rule and (b) by direct substitution.

40. Find $\dfrac{dy}{dx}$ for the function $y = [(4x + 1)^2 + 3x]^{1/2}$.

CHAPTER 4

Applications of the Derivative

In this chapter two simple applications of differential calculus are discussed: an inventory control problem, from which the basic model for inventory control is developed under assumed conditions of certainty; and a profit-maximization problem, in which the theory of extreme points is used to determine the level of output that leads to maximum profit. Functions of more than one variable are discussed in Chapter 5 and applications of such functions in Appendix 5A, which can be viewed as a (delayed) continuation of this chapter.

4.1 Problems in Inventory Control: Optimal Reorder Quantity

Assume that a manager wants to select an optimal inventory policy for one of the basic items in his product line. Total demand Q for the product has been stable over the last several years at about 100,000 units per year, and there has been no seasonality in sales. The manager's policy has been to order 25,000 units D at a time (partly because he hates paper work). How many times does he order each year? This can be calculated from:

$$\frac{\text{Total demand}}{\text{Amount delivered each time}} = \frac{Q}{D} = \frac{100,000}{25,000} = 4 \text{ times.}$$

To decide on the number of times he should order each year, the manager must know the costs that are influenced by his inventories and the magnitudes by which these costs vary. The relevant costs are: the costs of carrying the inventory once he has it, and the costs of ordering when he runs out.

1. *Carrying costs.* If demand for an item is stable, the carrying costs can be approximated by the cost per unit multiplied by the average number of units held in the inventory. These costs consist of the warehouse expenses, the costs of funds tied up in inventories, and many incidentals. If the firm buys 25,000 units at a time and uses them up before it orders again, the average inventory would be

$$\frac{25,000 + 0}{2} = 12,500 \text{ units} = \frac{D}{2}.$$

This results in the saw-tooth diagram for inventories, as shown in Figure 4.1, a figure well known to students of inventory control.

Let k (dollars) represent the interest and other carrying costs of holding one unit of inventory for one year. Then the total carrying costs are the annual carrying

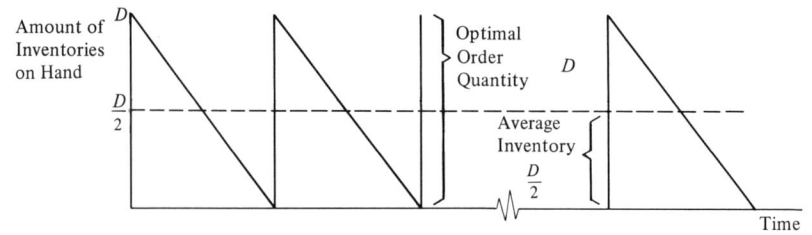

Figure 4.1

Pattern of Inventories on Hand.

cost per unit multiplied by the (average) number of units in inventory; that is,

$$\text{Total carrying cost} = k\left(\frac{D}{2}\right).$$

If k were equal to $8 per thousand units, then the carrying cost would be

$$(12.5)(\$8) = \$100.$$

2. *Order costs.* The number of orders currently placed during the year is $4 = 100/25$. More generally, if Q units are to be sold during the year and D units are delivered for each order, the required number of deliveries is Q/D. Suppose that the cost of ordering is related to the amount delivered by the expression $a + bD$, where $a = \$60$ and $b = \$3$ per thousand units. Here b may be interpreted as the shipping cost per thousand units so that the cost of sending D items is $b(D)$ dollars, that is, $\$3 \times 25 = \$75$; a represents costs such as bookkeeping and long-distance telephoning for orders—in other words, costs whose magnitude is not affected by the amount involved in the shipment. With $a = \$60$ and $b = \$3$ per thousand, the cost per delivery, $a + bD$, would total $135.

The total annual ordering cost equals the number of deliveries (Q/D) multiplied by the cost per delivery $a + bD$;[1]

$$\begin{matrix}\text{Total annual}\\ \text{ordering cost}\end{matrix} = \frac{(a + bD)Q}{D} = \frac{aQ}{D} + \frac{bQD}{D} = \frac{aQ}{D} + bQ.$$

The total cost C, which results from the manager's inventory policy is the sum of the two costs: the carrying costs and the ordering costs. It is equal to

$$C = \frac{kD}{2} + \frac{aQ}{D} + bQ. \tag{4.1}$$

4.2 Finding the Minimum Cost Inventory Policy

The only unknown in equation (4.1) for the total annual cost associated with inventory is the value of D, the number of items to be delivered per shipment. Once the number of items that leads to the minimum cost for the year is determined, all other factors are determined, because the corresponding average inventory level $D/2$ is immediately available and the number of shipments per year must be Q/D.

Equation (4.1) is the rule for the function that specifies the total cost for each value of the variable D. For example, given $Q = 100$ (thousand), $k = 8$, $a = 60$, and $b = 3$, then equation (4.1) becomes

$$C = \frac{8D}{2} + \frac{60 \times 100}{D} + 3 \times 100$$

or

$$C = 4D + \frac{6000}{D} + 300. \tag{4.2}$$

[1] For the figures given, this would equal $4 \times 135 = \$540$.

If *D* were equal to 25 (thousand), then

$$C = 100 + 240 + 300 = 640 \qquad \text{(that is, } 100 + 540\text{)}.$$

There is no assurance that $D = 25$ is the minimum cost for the year's inventory. However, with equation (4.2), the value of *D* that minimizes costs can be approximated. Thus, determining the inventory-cost equation provides the means for solving the problem. The objective is to minimize cost. The function that specifies cost is thus called the *objective function*. It shows how the firm's costs are affected by different values of the relevant variables. This is illustrated by several methods; first, by a series of trial calculations; second, graphically; and then through the use of the calculus. A formula for the least cost order quantity is derived from the calculus solution.

To find the approximate value of minimum costs one can simply take a number of alternative values of *D*, substitute them in turn into the equation, compute the corresponding values of *C*, and thus find, roughly, the value of *D* that gives the lowest cost. This is done in Table 4.1.

Table 4.1

Total Cost as a Function of Amount Ordered

D	10	20	30	40	50	60	70	80	...
C	940	680	620	610	620	640	666	695	...

From this table, it can be seen that a value of *D* of approximately 40 (thousand) units per delivery minimizes costs. In effect, then, the inventory problem has now been solved. However, some additional analysis will make it possible to extract a great deal of additional information from the equation (and perhaps clarify some of the concepts presented in the previous chapter).

The objective function, equation (4.2) and its components, can be analyzed graphically as shown in Figure 4.2. The carrying costs are represented by the straight line 4*D* in this diagram. They increase in direct proportion to the size of *D*. The order costs are made up of two components: a component that does not vary with *D*, at $300, and a continually decreasing total, variable ordering cost, indicated by the (hyperbolic) function $6000/D$. The total costs from Table 4.1 are also plotted. Note that the minimum total cost falls just slightly to the left of the point at which $D = 40$.

4.3 An Inventory Model

The objective of this analysis is to find the order quantity that minimizes the total cost of carrying inventory, without enumerating the values in a table or plotting the various values of total cost each time. Several observations can be made about the problem at the outset. First, the minimum cost occurs at the low point of the total-cost line where the slope of the curve is horizontal and thus

Sec. 4.3]   AN INVENTORY MODEL

Figure 4.2

Cost Components for Inventories: (1) *Total Inventory Costs.* (2) *Total Variable Ordering Costs.* (3) *Total Carrying Costs.* (4) *Total Fixed Ordering Costs.*

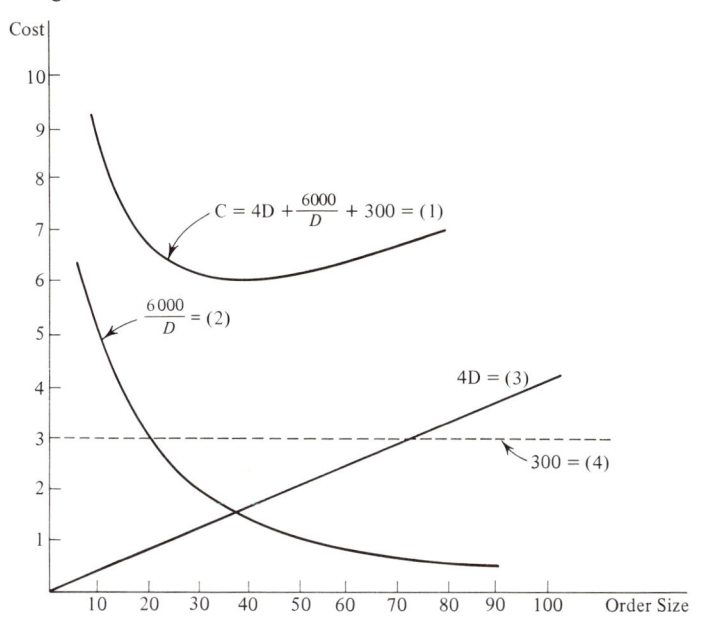

(1) Total Inventory Costs
(2) Total Variable Ordering Costs
(3) Total Carrying Costs per Year
(4) Total Fixed Ordering Costs

equal to 0. Second, both the carrying-cost line, $4D$, and the reorder line, $6000/D$ are changing continuously with D. The rate at which they change can be determined from the first derivative of the equations of the two lines. Finally, the $300 fixed-cost line does not change with D, therefore it does not affect the rate of change (the slope) of the total cost line, but it does affect its height.

The rules for extreme points can be used to find the point at which the slope of the total-cost line is equal to zero, and the second-order conditions for the function can be used to establish that the extreme point is a minimum. The point at which the rate of change of total cost with changes in D is equal to zero is found by setting the first derivative of the total-cost function dC/dD equal to zero, and solving for D. Total costs are:

$$C = \frac{kD}{2} + \frac{aQ}{D} + bQ. \qquad (4.1)$$

Differentiating the total cost equation with respect to D, and setting the derivative equal to zero, gives

$$\frac{dC}{dD} = \frac{k}{2} - \frac{aQ}{D^2} = 0. \qquad (4.3)$$

Rearranging and solving for D gives

$$\frac{k}{2} = \frac{aQ}{D^2}$$

that is,

$$D^2 = \frac{2aQ}{k},$$

or

$$D = \sqrt{\frac{2aQ}{k}} \qquad (4.4)$$

for which only the positive root is relevant.

To test for a minimum point, the second derivative must be investigated at the particular value D_0 at which the first derivative is zero. The equation for the value of the second derivative is:

$$\frac{d^2C}{dD^2} = 0 + \frac{2aQ}{D^3}. \qquad (4.5)$$

The value D_0 is then substituted into equation (4.5) to find the sign of the second derivative. All relevant values of D must be positive. All values in the numerator are positive, thus the second derivative must be positive, indicating that the extreme point is a minimum. The bQ term in equation (4.1), which corresponds to the fixed cost of $300 shown in Figure 4.2, does not involve D, therefore does not change with D, and is not found in the final equation for the optimal value of D.[2]

Equation (4.4) now represents a useful model of the problem situation as stated and makes it possible to calculate the optimal average inventory level $D/2$ and the optimal reorder quantity D corresponding to any level of sales volume Q, unit carrying cost k, and fixed reorder cost a. Equation (4.4) can be used to find the optimal value of D when any of these numbers change. It indicates, as might be expected, that the optimal inventory level $D/2$ or delivery size D should be increased when sales Q increase, or when the order (delivery) cost a increases. Similarly, inventory should be reduced if the carrying cost k goes up (because k appears in the denominator of the fraction). Given any level of D_0, such as,

$$D_0 = \sqrt{\frac{2aQ}{k}}$$

if Q increases by a factor of 4, then

$$D_1 = \sqrt{\frac{2a(4Q)}{k}} = 2\sqrt{\frac{2aQ}{k}} = 2D_0.$$

---

[2] Because b does not appear in the equation for optimal D, a change in the value of b should not lead to any change in inventory level. Intuitively, this perhaps slightly surprising result can be explained by noting that if the total amount Q to be shipped over the year is fixed, then there is nothing that can be done to save on shipping charges that are proportional to the size of the shipment. These charges will add up to bQ dollars over the year, and no change in inventory levels can reduce this number.

A similar increase in order costs a gives

$$D_2 = \sqrt{\frac{2(4a)Q}{k}} = 2\sqrt{\frac{2aQ}{k}} = 2D_0.$$

Finally, a four-fold increase in k, the carrying cost, would result in

$$D_3 = \sqrt{\frac{2aQ}{(4k)}} = \frac{1}{2}\sqrt{\frac{2aQ}{k}} = \frac{1}{2}D_0.$$

Where this formula is applicable it suggests that *inventory should increase only in proportion to the square root of sales*. When the sales of some item quadruple, the inventory should not be quadrupled; it should be only double its original amount. Many firms fix their inventory at some constant percentage of sales (a fixed number of weeks' worth of sales are kept in inventory); therefore, if one item sells five times as much as another, the firm would tend to keep five times as large an inventory of the first as of the second with no way of knowing if either level is optimal (*both* cannot be). In fact, substantial savings have often been achieved because equation (4.4) and related results have led analysts to recognize that the standard rule of thumb tends to yield excessive inventories of popular items of large sales volume and insufficient inventories of items with relatively modest sales.

The model itself is appropriate only in those cases for which the assumptions are a reasonable approximation to reality. All parameters must be known and constant. The rate of sales must be uniform over the decision time horizon. Where any one of these assumptions is not appropriate the model may not be appropriate. For example, there is a difficult problem in maintaining the right level of inventories for slow-moving items. Such items are frequently bought by just a few customers, so that demand is likely to be erratic and subject to the caprice of the few buyers. Therefore, if the firm is to be reasonably sure it has enough on hand to meet these demands, it must carry *relatively* large stocks of these items with low but erratic sales volumes. On the other hand, the demand for an item that is desired by many customers is not likely to be much affected by the whims of any particular buyer because when demands of some customers are low, those of other customers may well be high. Therefore, for large-volume items, an inventory that is not nearly as high in proportion to its sales can be expected to provide an adequate reserve against most demands. If the other conditions of the model are met, it can be used in this case and again, the optimal size of inventory would increase less than proportionately with sales volume.

From this example, it can be seen that even with a highly oversimplified inventory model, an optimality analysis, if used with sufficient caution, can produce significant results.

4.4 Problem of Profit Maximizing: Quantity to Order

The demand curve DD relates the price of goods to the quantity of goods that would be purchased at a given price, as specified in equation (4.6) where p is the

Figure 4.3

Demand Curve DD for a Given Product.

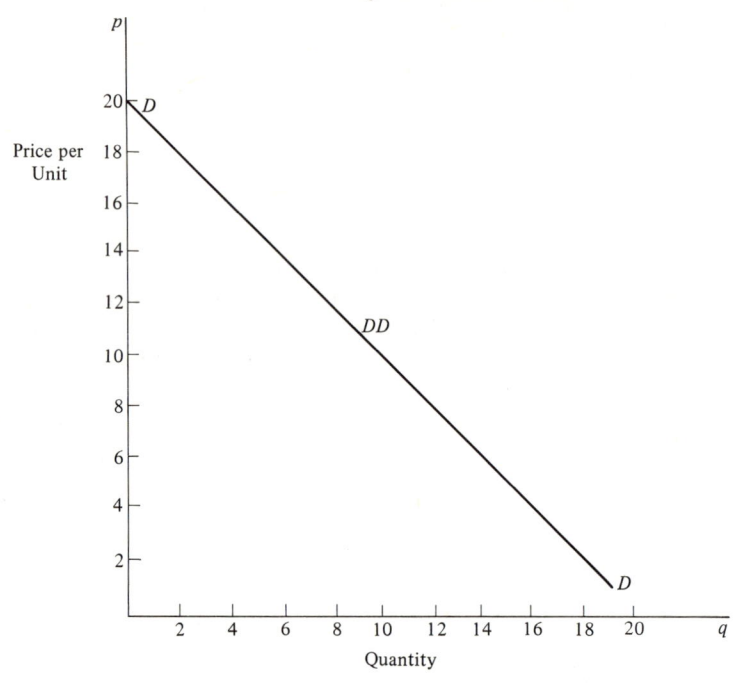

price of the product and q is the quantity demanded

$$p = 20 - \frac{q}{5}. \tag{4.6}$$

The demand curve is the line DD in Figure 4.3. It has the negative slope characteristic of a monopolist's demand curve; with a decrease in price goes an increase in the quantity demanded.[3]

Assume that the product costs $2 per unit to produce. Since profit is equal to revenue − cost, an equation for the total profit of the firm can be written as

$$\pi = r - c \tag{4.7a}$$

where π is profit, r is revenue, and c is cost. Since

$$r = p \times q \quad \text{and} \quad c = 2q$$

but

$$p = 20 - \frac{q}{5}$$

therefore

$$\pi = \left(20 - \frac{q}{5}\right)q - 2q. \tag{4.7b}$$

---

[3] In a purely competitive industry the demand curve that a firm faces for its product would be horizontal; in an oligopoly industry the demand curve would be kinked (with the apex of the kink at the existing price). Only for a monopolist (or a collusive oligopolist) would the demand curve be downsloping as illustrated in Figure 4.3.

Sec. 4.4] PROBLEM OF PROFIT MAXIMIZING

Figure 4.4

Profits in Relation to Total Quantity Sold.

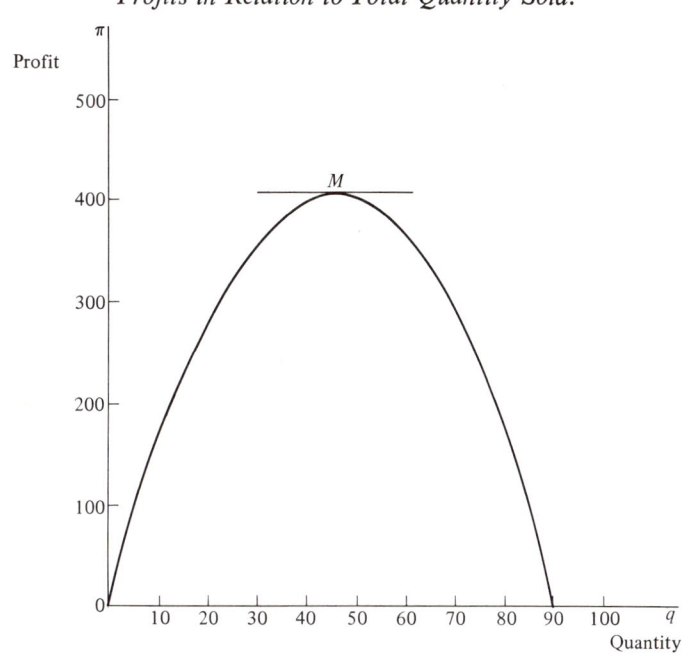

The total-profit equation (4.7b) has been developed in terms of total revenue minus total costs, with the quantity sold being a function of the price at which the product is sold. It can be simplified further to

$$\pi = 20q - \frac{q^2}{5} - 2q,$$

and

$$\pi = 18q - \frac{q^2}{5}. \tag{4.8}$$

This profit equation (4.8) can be plotted as shown in Figure 4.4. Its relative maxima and minima can be located by finding the rule for the first derivative function and setting it equal to zero. It is also necessary to check the values of the profit equation at the endpoints of any interval corresponding to the quantity of goods that can be produced. The first of the endpoints, the zero quantity point, results in zero profit. In Figure 4.4 the profit line rises to a height in excess of $400 for output (sales) quantities of approximately 40–50 units, a level greatly to be preferred to that at the zero output point. Profits decrease to the right of the point M and eventually reach zero again (at 90 units output).

The rule for the profit function has been stated in terms of the domain variable, quantity sold. The rule for the derivative function is

$$\frac{d}{dq}(\pi) = \frac{d}{dq}\left(18q - \frac{q^2}{5}\right) = 18 - \tfrac{2}{5}q. \tag{4.9}$$

This is set equal to zero:

$$18 - \tfrac{2}{5}q = 0$$
$$q = (18)(\tfrac{5}{2}) = 45.$$

There is a single value of q at which the value of the first derivative is zero; this value is $q = 45$. From Figure 4.4 the function is seen to reach a maximum value at $q = 45$, the point M. This can also be demonstrated mathematically through the use of the second derivative function. The rule for the second derivative function is

$$\frac{d}{dq}\left(\frac{d\pi}{dq}\right) = \frac{d}{dq}(18 - \tfrac{2}{5}q) = -\tfrac{2}{5}. \qquad (4.10)$$

Thus for *all* values of q, including $q = 45$, the second derivative is negative; that is, the rate of change of the slope of the function is everywhere negative. This means that the extreme point at $q = 45$ is a maximum point. The value of the profit equation (4.8) at $q = 45$ is

$$\pi = 18(45) - \frac{(45)^2}{5} = 18(45) - 9(45) = 9(45) = 405.$$

Since, from equations (4.9) and (4.10), profit must be everywhere decreasing for values of $q > 45$, then (45, 405) is the (global) maximum profit level.

Figure 4.5 shows the three functions, the profit function (Figure 4.5(a)), the first derivative function for profits (Figure 4.5(b)) and the second derivative function for profits (Figure 4.5(c)). Figures 4.5(a) and 4.5(b) show that the slope of the profit function is positive for $0 \leq q \leq 45$, becomes zero at $q = 45$, and then is everywhere negative. From Figures 4.5(b) and 4.5(c), the negative slope of the first derivative function is verified. Implicit in the foregoing discussion is the economist's widely known decision rule: produce (and sell) that output for which the marginal cost equals the marginal revenue. The origin of this rule is developed in the next paragraphs.

4.5 Marginal Cost and Marginal Revenue

In specific and general terms, equation (4.7) becomes

$$\pi = \left(20 - \frac{q}{5}\right)q - 2q \qquad (4.7)$$
$$\pi = \text{revenue} - \text{cost}.$$

The economist defines marginal revenue dr/dq to be the instantaneous rate of change of revenue with change in quantity (sold), and the marginal cost dc/dq to be the instantaneous rate of change in cost with change in quantity (produced, and presumably sold).

Using the figures in equation (4.7) the revenue is expressed as,

$$r = \left(20 - \frac{q}{5}\right)q;$$

Sec. 4.5 MARGINAL COST AND MARGINAL REVENUE 109

Figure 4.5

(a) *Profits in Relation to Total Quantity Sold.* (b) *Rate of Change of Profits with Changes in q.* (c) *Second Derivative Function.*

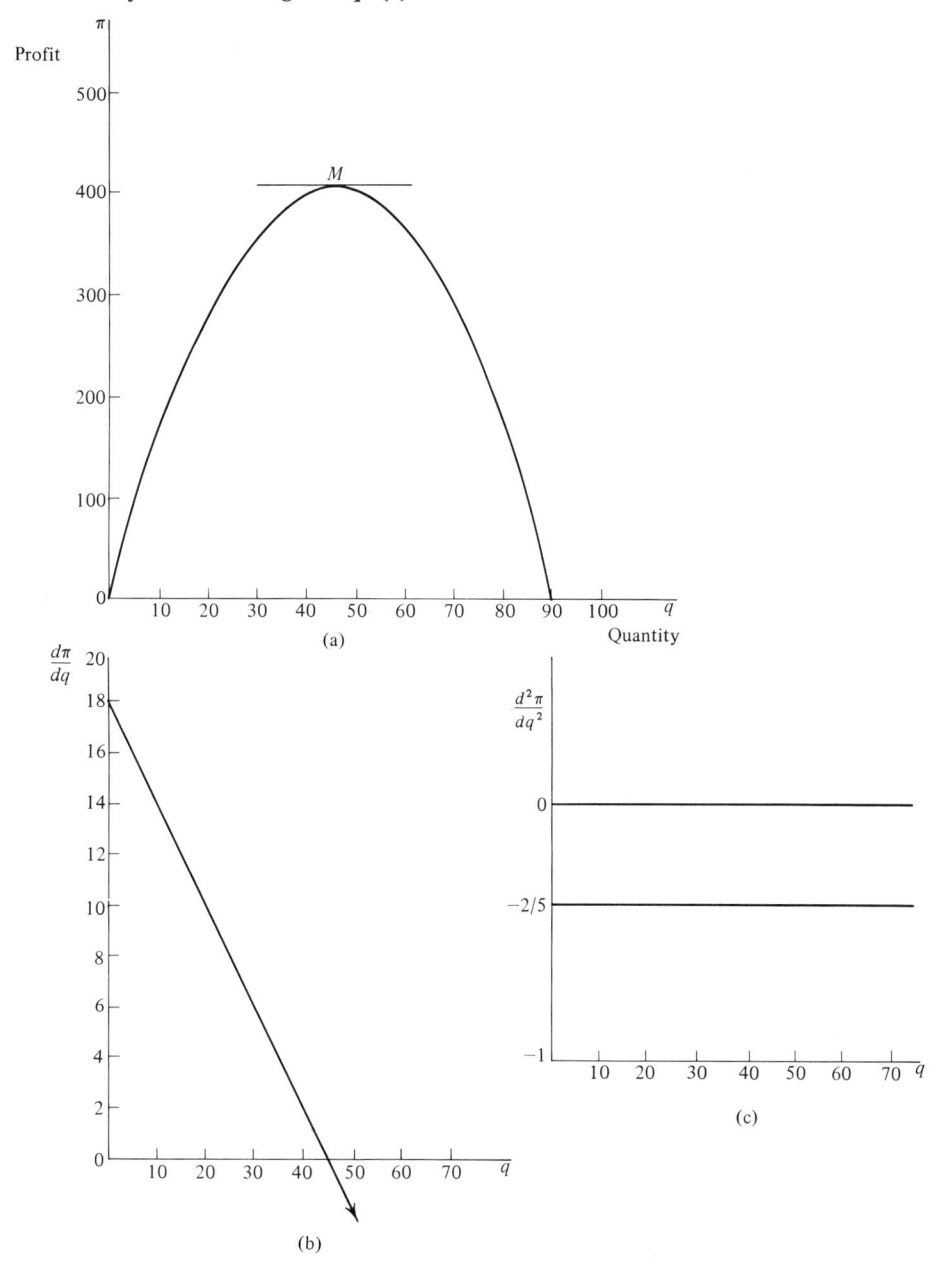

and differentiating with respect to q, gives

$$\frac{dr}{dq} = \frac{d}{dq}\left(20q - \frac{q^2}{5}\right) = (20 - \tfrac{2}{5}q). \tag{4.11}$$

Cost is given by $c = 2q$ with derivative

$$\frac{dc}{dq} = \frac{d}{dq}(2q) = 2 \qquad (4.12)$$

Note what happens when equation (4.7) is differentiated with respect to q and $d\pi/dq$ is set equal to zero:

$$\pi = \left(20 - \frac{q}{5}\right)q - 2q \qquad (4.7)$$

differentiating gives

$$\frac{d\pi}{dq} = \frac{d}{dq}\left(20q - \frac{q^2}{5}\right) - \frac{d}{dq}(2q)$$

and setting $d\pi/dq$ equal to zero gives

$$\frac{d\pi}{dq} = (20 - \tfrac{2}{5}q) - (2) = 0$$

or

$$(20 - \tfrac{2}{5}q) = 2.$$

But $20 - \tfrac{2}{5}q$ is the marginal revenue given by equation (4.11) and 2 is the marginal cost given by equation (4.12). In this example, to find that quantity of output (and sales) for which profits are at their maximum value, marginal cost is equated to marginal revenue. If the slope of the first derivative function, when stated in the form of equation (4.9) is negative, then the necessary and sufficient conditions for a maximum level of profits is established.

The analysis just given can be restated in more general terms. Given the rule for the profit function f,

$$\pi = r - c,$$

The maximum value of π is found by determining the rule for the first derivative of π with respect to q,

$$\frac{d}{dq}\pi = \frac{d}{dq}(r) - \frac{d}{dq}(c)$$

and setting it equal to zero to solve for the value of q at which π reaches an extreme point. This results in

$$\frac{d\pi}{dq} = \frac{dr}{dq} - \frac{dc}{dq} = 0,$$

or

$$\frac{dr}{dq} = \frac{dc}{dq}.$$

But dr/dq is marginal revenue (MR) and dc/dq is marginal cost (MC), therefore

$$\text{Marginal revenue} = \text{Marginal cost}$$

is a necessary condition for maximum profit. The condition

$$\frac{d^2\pi}{dq^2} < 0$$

Sec. 4.6] SUMMARY 111

at the value of q for which

$$\frac{dr}{dq} = \frac{dc}{dq}$$

when added to the necessary condition for maximum profit leads to conditions that are now sufficient for a maximum. Thus the conditions of the economist have been verified mathematically.

Further applications of the use of differential calculus are discussed in Appendix A to Chapter 5, where a production function is developed and illustrated and the normal equations of simple linear regression are derived. These applications involve functions of more than one variable, however, a topic introduced in Chapter 5.

4.6 Summary

In this chapter the theory of extreme points is applied in the solution to two simple managerial problems. A model for the optimal order quantity, that level of shipment that leads to minimum costs associated with inventories held, is developed under assumed conditions of certainty. Once the rule for the function f is expressed as an equation for the total costs of ordering and carrying inventory, the problem has essentially been solved. The rule for the first derivative function is developed, set equal to zero, and solved for D, the order quantities at the extreme points of f. The values for D are substituted into the equation for the rule of the second derivative function and the value of D_0 corresponding to the global minimum of f is the order quantity D that leads to the lowest costs of ordering and carrying inventory. The result is expressed in the illustrative problem as the model presented in equation (4.4). The model can be used for this class problem without resolving the basic equation each time.

Only those factors varying with the quantity ordered are included in the model; factors that do not vary with the quantity ordered (which are fixed) do not appear in the model. The model implies that order quantity should vary with the square root of sales or of order costs and with the reciprocal of the square root of carrying costs.

In the second application, the rule for a function representing the profits of a firm is developed from its demand and cost functions. Through the use of the first and second derivative tests that quantify is found which when produced and sold, maximizes profit. The maximum profit is then determined by substituting this amount into the profit equation and solving. The economists' decision rule that maximum profit is found at that output for which marginal revenue equals marginal cost is then shown to be implicit in the procedure just described for determining maximum profit through use of the calculus. Since profit π equals revenue r minus costs c, when the profit equation

$$\pi(q) = r(q) - c(q)$$
$$\pi = r - c$$

is differentiated with respect to q, the result is

$$\frac{d\pi}{dq} = \frac{dr}{dq} - \frac{dc}{dq}.$$

But dr/dq is marginal revenue (the rate of change of revenue r with minute change in output q) and dc/dq is marginal cost (the rate of change of cost c with minute change in output q). When $d\pi/dq$ is set equal to zero

$$\frac{d\pi}{dq} = \frac{dr}{dq} - \frac{dc}{dq} = 0,$$

or

$$\frac{dr}{dq} = \frac{dc}{dq}.$$

That is, marginal revenue equals marginal cost. Given that the second order conditions indicate a maximum, this condition implies that a maximum profit level has been attained.

Key Concept

Concept	Reference Section
Objective Function	4.2
Inventory Model	4.3
Optimal Order Quantity	4.3
Profit Maximizing Price	4.4
Marginal Cost	4.5
Marginal Revenue	4.5

Problems

1. Assume that predicted sales volume for product A is 10,000 units annually.
    (a) What is the optimal reorder quantity if the fixed reorder cost is $20 and the carrying cost per unit is $0.10?
    (b) How many times will the company order the product each year?
    (c) What is the average inventory?
    (d) What are the total carrying costs per year?
2. Using the figures from the previous problem:
    (a) What is the optimal reorder quantity if the fixed reorder cost is increased to $180? When fixed reorder cost increases to nine times the earlier level, how much does the optimal reorder quantity change?
    (b) If sales increase to 90,000 units, what is the optimal reorder quantity? What is the average inventory? When sales increase to nine times the earlier level, how much does the average inventory increase?
    (c) What happens to the optimal reorder quantity and average inventories when sales and carrying costs per unit each double?
3. Study the following case.
    Given the following definitions:
    1. Marginal cost (at a point) equals the rate of change of total cost (at the point). $MC = d/dq(c)$.

PROBLEMS

2. Marginal revenue (at a point) equals the rate of change of total revenue (at the point). $MR = d/dq(r)$.
3. Profit equals total revenue minus total cost. $\pi = (r - c)$.
4. The profit maximizing price for a given function is that price at which marginal revenue equals marginal cost. π_m at $MC = MR$.
5. Variable cost is that portion of cost which varies with output. And given: The demand equation for a good

$$q = 500 - p$$

where q equals quantity sold and p equals price per unit, whose cost equation is

$$c = 3000 + \left(10 - \frac{q}{10}\right)q.$$

Find:

(a) The equation for total revenue stated as a function of p only.
(b) The equation for total cost expressed as a function of p only.
(c) The equation for marginal revenue.
(d) The equation for marginal cost.
(e) The profit maximizing output by using the equations for marginal cost and revenue only. Show that this output corresponds to a maximum not a minimum.
(f) Write an expression for total variable cost.
(g) Write an expression for average variable cost.
(h) Find the maximum profit point by using an equation for total profit stated as a function of p alone. Compare your figures (and variables).

4. The demand curve for a certain product is given by $p = 12 - 2q$ where p denotes price and q quantity.

(a) What is the derivative of p with respect to q?
(b) What is the derivative of q with respect to p?
(c) What price will lead to no demand?
(d) What price will lead to a demand of 5 units?
(e) How many units will be demanded for a price of $p = 8$?
(f) Form the expression for total revenue.

5. The total revenue, r, for a certain product is given by the equation $r = p \cdot q = 12q - 2q^2$.

(a) Derive the expression for the demand curve.
(b) What would total revenue be for a price of $p = 8$?
(c) Find the expression for marginal revenue.
(d) What price would maximize r?
(e) Show that that price is a maximum and not a minimum.

6. Total revenue, r, and total cost, c, for a certain product are given by the equations:

$$r = p \cdot q = 12q - 2q^2, \quad c = 4q.$$

(a) Find the quantity at which total revenue is a maximum.
(b) Find the quantity at which total cost is a minimum.
(c) Form the expression for net profit, π.
(d) Find the quantity at which profit is a maximum.
(e) What price will lead to maximum profit?
(f) Show that marginal revenue equals marginal cost at that price.
(g) Would the optimum price change if there had been a fixed cost of 10, i.e., $c = 4q + 10$?
(h) Would a decision to produce or not be affected by the fixed cost?
(i) Graph the situation and show that the graphical solution agrees with the mathematical one.

7. The demand curve for a certain product is given by $p = a - bq$, where a and b are positive constants
   (a) Form the expression for total revenue.
   (b) Show that the marginal revenue curve has a slope equal to twice that of the demand curve. (This is true for all linear demand curves only.)

8. A company produces a daily output of Q tons of product A that has a variable cost of $(Q^3/3) - 3Q^2 + 5Q$ and a fixed cost of \$100. The forecasted sales are represented by the equation $Q = 20 - .5p$ where p is the price at which the product is sold.
   (a.1) Determine the level of output at which profits are maximized.
   (a.2) Determine the profit maximizing price level.
   (a.3) Determine the maximum profits.
   (b) How are the answers to (a) affected when fixed costs are doubled to \$200?

9. The demand curve facing a monopolist is represented by $q = 6 - p/3$ where q is the amount of product sold and p is the price. If the company's total-cost curve is given by $25 - 6q + q^2$, determine
   (a) the profit maximizing level of output.
   (b) the profit maximizing price.
   (c) the maximum profits.

10. Metal Company has scrap pieces of metal sheeting left over at the end of its production line. Recently, the marketing department discovered that there was a market for lidless metal boxes made of this material. The company has no other use for the scrap and it can manufacture the new boxes on present under-utilized machinery. The market is willing to pay .54 cent per cubic inch of such boxes so Metal Company wishes to maximize the volume that can be made by cutting equal squares out of the corners of the scrap pieces that measure 3" × 8". The cost of manufacturing and selling the boxes is 2.5 cents per box. The production department states that the metal costs .08 cent per square inch.
    (a) What is the volume of the largest box that can be made from the scrap? Show that the box has maximum, not minimum volume.
    (b) Should the company produce the box?

11. A company found that the total cost of its product could be represented by the equation $500 + 300q - 45q^2 + 2q^3$ where q is the quantity of units produced (in thousands). The company desires to maximize its profits after four years of losses.
    (a) If the company is in a purely competitive industry, at what level should the company produce? (State the general rule and a function representing the general case.)
    (b) If the company is under a "cost plus" contract and cannot sell more than 11,000 units, at what level might the company wish to produce?
    (c) What does the answer to (b) imply about "cost plus" contracts?

CHAPTER 5

Partial Differentiation

Differential calculus and its applications were discussed in Chapters 3 and 4 in finding maximum and minimum values of functions with a single domain variable. But the manager lives in a world in which there are many factors influencing the attainment of his goals, not just one. The earlier techniques are now extended to more realistic situations in which the maximum or minimum values of a function are influenced by many variables.[1] Sections 5.1 and 5.2 present the exposition in terms of common physical objects in three dimensions—apples and bananas—to facilitate intuitive interpretation. Formal procedures for differentiation of functions of more than one variable are then developed, implicit differentiation discussed, and then the topic of finding extreme points for variables subject to constraints is developed. Appendix 5A demonstrates the relationship between these topics and the economist's production function.

The essential difference between the more complex functions presented in this chapter and the simpler ones discussed in earlier chapters is that, in place of the domain with single variable x, the domain is now an ordered pair (x, y) (n-tuple). To carry out the process of partial differentiation, the complex case is merely condensed to the simple one; that is, one of the variables in the ordered pair is held constant, while differentiation is carried on in relation to the other by procedures exactly analogous to those of the earlier chapters (for an ordered n-tuple, all but one domain variable is held constant). Thus there is little that is completely new in the first half of this chapter except that the earlier rules and procedures are modified to fit the more complex conditions. In the second half of the chapter a new application, the use of Lagrange multipliers, for dealing with particular restrictions on domain variables is presented.

5.1 Partial Differentiation

Assume that an apple has been sliced perpendicular to its core, and one half placed on its flat side on a table. Let the table be an xy plane in which coordinates are established with the $(0, 0)$ point directly under the center of the apple. Next, erect an axis perpendicular to the table and passing through the stem of the apple. Call this third axis the z axis. There is now an axis with coordinates for each of the apple's three dimensions.[2]

To determine the rate at which the height of the apple (skin) above the table changes for changes in each respective variable in the domain, one can use derivatives with respect to each of the domain variables. To begin with, consider the piece of apple that is bordered by planes containing the three axes, that shown in Figure 5.1. Different parts of the apple are at different heights above the xy plane. The height z of a given point on the surface of the apple corresponds to a particular ordered pair (x_0, y_0) in the xy plane.

The function f with rule $z = f(x, y)$ represents the various heights of the apple above the xy plane. Stated in set notation, this function is

$$f = \{((x, y), z): z \in R, x \in R, y \in R; z = f(x, y) = \text{height of the apple above the point } (x, y)\}$$

---

[1] Later in the chapter it will be shown that a manager can use the procedures of this chapter to help him find the highest profit point when profits depend on more than one variable, on both price and advertising expenditure, for example. The exposition is generally given in three dimensions, though it can be generalized to n-dimensions in most instances.

[2] It is possible to use many different coordinate systems, but here cartesian coordinates in three dimensions, coordinates that are mutually perpendicular in three-space, are used. This is the most familiar coordinate system and in many ways the most useful.

Sec. 5.1]   PARTIAL DIFFERENTIATION   117

Figure 5.1

One Eighth of an Apple Cut by a Plane at $y = y_0$.

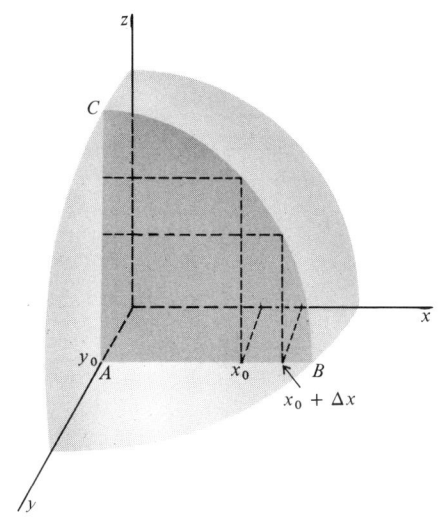

The domain consists of ordered pairs (x, y) of the variables x and y, and the range of the variable z. But if y is held constant at $y = y_0$, the function is reduced back to one with a single variable over the domain, and the rules for differentiation presented in Chapters 3 and 4 can be applied.

When the value of y is fixed, a single plane passing through the apple is established. The intersection ABC of this plane with the apple, in Figure 5.1 is indictated by the shaded area (one side of the slice) in the diagram. This slice can be shown in only two dimensions as shown in Figure 5.2. With y fixed at y_0, any change in z is a result of changes in x alone. The average rate at which z changes with x in the interval $[x_0; x_0 + \Delta x]$ is given by the value $\Delta z/\Delta x$; that is,

$$\frac{\Delta z}{\Delta x} = \frac{f(x_0 + \Delta x, y_0) - f(x_0, y_0)}{(x_0 + \Delta x) - x_0} = \frac{f(x_0 + \Delta x, y_0) - f(x_0, y_0)}{\Delta x}.$$

The limit of this ratio as $\Delta x \to 0$ is equal to the slope of the tangent to the curve at the point $(x_0, f(x_0, y_0))$, just as it was in the two-dimensional analysis in Chapter 3. This limit is the *partial derivative* of z with respect to x at $x = x_0$ with y held constant at y_0.

Definition—Partial Derivative of f: *Given the continuous function f with rule $z = f(x, y)$ and y held constant at $y = y_0$, then the*

$$\lim_{\Delta x \to 0} \frac{f(x_0 + \Delta x, y_0) - f(x_0, y_0)}{\Delta x} \tag{5.1}$$

if it exists, is the value of the partial derivative of f with respect to x at $x = x_0$. The symbol for the partial derivative of f with respect to x is $\partial f/\partial x$.

A plane can be passed through the sample piece of apple in the x direction if x is held constant at some value, say $x = x_{00}$, to obtain the rate of change in z

Figure 5.2

Slice of Apple Cut by the Plane $y = y_0$.

$z_0 = f(x_0, y_0)$
$z_2 = f(x_0 + \Delta x_2, y_0)$
$z_1 = f(x_0 + \Delta x_1, y_0)$

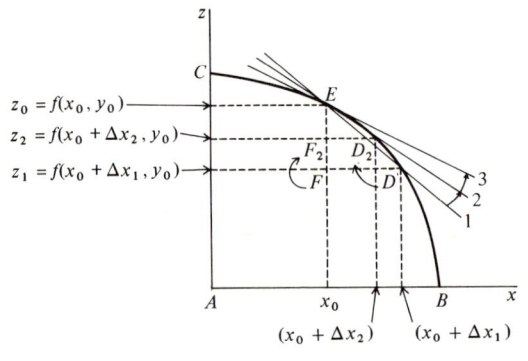

Figure 5.3

One Eighth of an Apple Cut by a Plane at $x = x_{00}$.

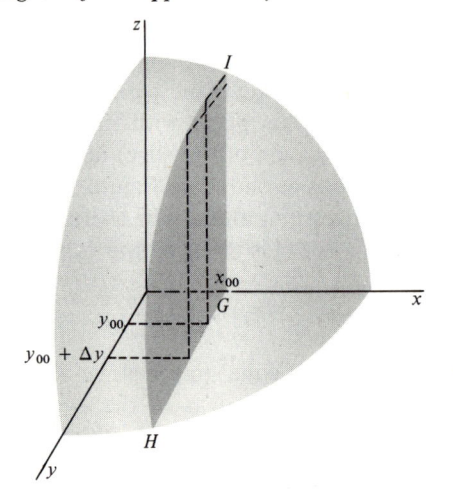

with respect to small changes in y (with fixed x), as shown in Figure 5.3. With x fixed, the plane *GHI* is established; as the apple is sliced in a direction at right angles to the first. If the *GHI* plane is rotated 90° to the right, a figure similar to Figure 5.2 is generated, as shown in Figure 5.4, a two-dimensional view of the slice. The rate of change of z with respect to y at y_{00} is equal to the slope of the tangent to *HLI* at *L*. This is the partial derivative of f with respect to y. The change in z

Sec. 5.1]   PARTIAL DIFFERENTIATION   119

Figure 5.4

The Slice of Apple Cut by Plane $x = x_{00}$.

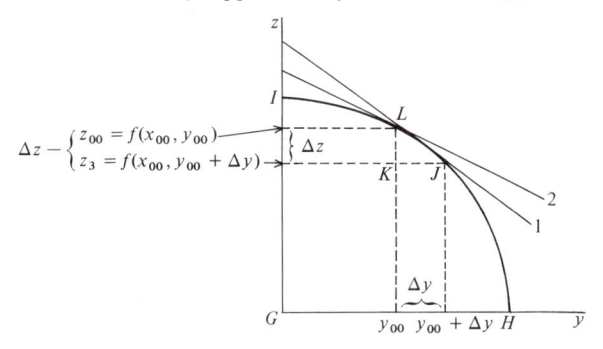

$\Delta z - \begin{cases} z_{00} = f(x_{00}, y_{00}) \\ z_3 = f(x_{00}, y_{00} + \Delta y) \end{cases}$

with a small change in y is again a negative amount, given by the equation

$$\frac{\Delta z}{\Delta y} = \frac{f(x_{00}, y_{00} + \Delta y) - f(x_{00}, y_{00})}{\Delta y}.$$

The limit of this value as $\Delta y \to 0$ is the value of the partial derivative of z with respect to y at y_{00}, with x held constant at x_{00}. The partial derivative at y_{00} is equal to:

$$\frac{\partial z}{\partial y} = \lim_{\Delta y \to 0} \frac{f(x_{00}, y_{00} + \Delta y) - f(x_{00}, y_{00})}{\Delta y}.$$

This is a statement of the definition of the rule for a partial derivative function of f, this time with the rule stated with respect to y.

The process of taking partial derivatives with respect to each of the domain variables has been illustrated in the three-dimensional case to facilitate a graphic interpretation of the process. However, the basic mathematics, as well as much of the "three-dimensional intuition" can be extended to as many dimensions as desired. This discussion is the basis for the formal rules for partial differentiation of a function whose domain is a point specified by an ordered n-tuple. These rules are essentially the same as those for differentiation of a function with a single domain variable. For each partial derivative, only one of the variables in the domain n-tuple is allowed to vary—all others are held constant. For example, when taking a partial derivative of f with respect to x to get $\partial f/\partial x$, all other variables are treated as constants. This is necessary to reduce the problem to one in which the change in f depends only on the change in a single variable.

> **Rule 5.1—Procedure for Partial Differentiation:** Given the function f with rule $z = f(x, y)$ differentiable in a neighborhood around (x_0, y_0), when a partial derivative is taken with respect to one variable in the domain, all other domain variables are treated as constants, and the normal rules of differentiation for a function with a single domain variable are applied.

There are several symbols commonly used for the rule for a partial derivative of f, a function with rule $z = f(x, y)$. In addition to $\partial f/\partial x$, f_x or $f_x(x, y)$, or $\partial z/\partial x$ are used also.

Example 5.1

Given the function f with rule $z = f(x, y) = 2xy^3 + x^2y^2 + 4y^5$, find $\partial z/\partial x$ and $\partial z/\partial y$:

$$\frac{\partial z}{\partial x} = 2y^3 + 2xy^2 + 0$$

$$\frac{\partial z}{\partial y} = 6xy^2 + 2x^2y + 20y^4.$$

In the next example the procedure is illustrated for a function f with three variables in its domain n-tuple.

Example 5.2

Given the function f with rule $z = f(w, x, y) = 4w^2xy^3 + 7wx^4 + x^6y + 3y^7$, find $\partial z/\partial w$, $\partial z/\partial x$, and $\partial z/\partial y$:

$$\frac{\partial z}{\partial w} = 8wxy^3 + 7x^4 + 0 + 0$$

$$\frac{\partial z}{\partial x} = 4w^2y^3 + 28wx^3 + 6x^5y + 0$$

$$\frac{\partial z}{\partial y} = 12w^2xy^2 + 0 + x^6 + 21y^6.$$

It is also possible to take partial derivatives of higher order for a function f with rule $z = f(x, y)$ by following the same procedure as for functions of a single variable, but applied to the rules for the first partial derivative functions; that is, $\partial z/\partial x$ may be differentiated again with respect to x to obtain:

$$\frac{\partial}{\partial x}\left(\frac{\partial z}{\partial x}\right) = \frac{\partial^2 z}{\partial x^2}$$

the second partial derivative of f with respect to x. Again there are many symbols used in the literature to indicate the second partial derivative of f with respect to x; for example,

$$f_{xx} \quad \text{or} \quad f_{xx}(x, y) \quad \text{or} \quad \frac{\partial^2 f}{\partial x^2} \quad \text{or} \quad z_{xx}.$$

With partial derivatives further complication in higher order derivatives is possible. Those just discussed are called "straight" partial derivatives, but it is possible, for example, to take "cross partial" derivatives as well as "straight" second (or third or fourth, and so on) partial derivatives. A cross-partial derivative, this time with respect to x, may be taken for first partial derivative taken originally with respect to y as follows:

$$\frac{\partial}{\partial x}\left(\frac{\partial z}{\partial y}\right) = \frac{\partial^2 z}{\partial x\, \partial y} = f_{yx} = f_{yx}(x, y) = \frac{\partial^2 f}{\partial x\, \partial y} = z_{yx}.$$

Sec. 5.1] PARTIAL DIFFERENTIATION 121

This is called the second partial derivative of z with respect to x and y. It should be noted that the sequence of differentiation is indicated by the symbols used (and that this is indicated differently with different symbols). The cross partial of z with respect to x and then y is:

$$\frac{\partial}{\partial y}\left(\frac{\partial z}{\partial x}\right) = \frac{\partial^2 z}{\partial y\, \partial x} = f_{xy} = f_{xy}(x, y) = \frac{\partial^2 f}{\partial y\, \partial x} = z_{xy}.$$

This procedure is now illustrated in the next two examples for the partial derivatives developed in Examples 5.1 and 5.2.

Example 5.3

Use the partial derivatives found in Example 5.1, and find $\partial^2 z/\partial x^2$, $\partial^2 z/\partial y^2$, $\partial^2 z/\partial y\, \partial x$, and $\partial^2 z/\partial x\, \partial y$:

$$\frac{\partial z}{\partial x} = 2y^3 + 2xy^2$$

$$\frac{\partial z}{\partial y} = 6xy^2 + 2x^2 y + 20y^4$$

$$\frac{\partial^2 z}{\partial x^2} = 2y^2$$

$$\frac{\partial^2 z}{\partial y^2} = 12xy + 2x^2 + 80y^3$$

$$\frac{\partial^2 z}{\partial y\, \partial x} = 6y^2 + 4xy$$

$$\frac{\partial^2 z}{\partial x\, \partial y} = 6y^2 + 4xy.$$

Example 5.4

Use the partial derivatives from Example 5.2 and find ① $\dfrac{\partial^2 z}{\partial w^2}$, ② $\dfrac{\partial^2 z}{\partial x^2}$, ③ $\dfrac{\partial^2 z}{\partial y^2}$, ④ $\dfrac{\partial^2 z}{\partial x\, \partial w}$, ⑤ $\dfrac{\partial^2 z}{\partial y\, \partial w}$, ⑥ $\dfrac{\partial^2 z}{\partial w\, \partial x}$, ⑦ $\dfrac{\partial^2 z}{\partial y\, \partial x}$, ⑧ $\dfrac{\partial^2 z}{\partial w\, \partial y}$, ⑨ $\dfrac{\partial^2 z}{\partial x\, \partial y}$

$$\frac{\partial z}{\partial w} = 8wxy^3 + 7x^4$$

$$\frac{\partial z}{\partial x} = 4w^2 y^3 + 28wx^3 + 6x^5 y$$

$$\frac{\partial z}{\partial y} = 12w^2 xy^2 + x^6 + 21y^6$$

① $\dfrac{\partial^2 z}{\partial w^2} = 8xy^3$

② $\dfrac{\partial^2 z}{\partial x^2} = 84wx^2 + 30x^4 y$

③ $\dfrac{\partial^2 z}{\partial y^2} = 24w^2 xy + 126y^5$

④ $\dfrac{\partial^2 z}{\partial x\, \partial w} = 8wy^3 + 28x^3$

⑤ $\dfrac{\partial^2 z}{\partial y\, \partial w} = 24wxy^2$

⑥ $\dfrac{\partial^2 z}{\partial w\, \partial x} = 8wy^3 + 28x^3$

⑦ $\dfrac{\partial^2 z}{\partial y\, \partial x} = 12w^2 y^2 + 6x^5$ ⁣ ⑨ $\dfrac{\partial^2 z}{\partial x\, \partial y} = 12w^2 y^2 + 6x^5$

⑧ $\dfrac{\partial^2 z}{\partial w\, \partial y} = 24wxy^2$

A few general observations can now be made about these examples. In Example 5.3,

$$\frac{\partial^2 z}{\partial x\, \partial y} = 6y^2 + 4xy = \frac{\partial^2 z}{\partial y\, \partial x}.$$

In Example 5.4 there are pairs of values which are the same

④ $\dfrac{\partial^2 z}{\partial x\, \partial w} = 8wy^3 + 28x^3 = \dfrac{\partial^2 z}{\partial w\, \partial x}$ ⑥

⑤ $\dfrac{\partial^2 z}{\partial y\, \partial w} = 24wxy^2 = \dfrac{\partial^2 z}{\partial w - y}$ ⑧

⑦ $\dfrac{\partial^2 z}{\partial y\, \partial x} = 12w^2 y^2 + 6x^5 = \dfrac{\partial^2 z}{\partial x\, \partial y}$ ⑨.

This leads to a general rule:

> **Rule 5.2:** *Given f a function with rule $z = f(x, y)$, differentiable partially to any order, then the rules for the second cross-partial derivatives $\partial^2 f/\partial y\, \partial x$ and $\partial^2 f/\partial x\, \partial y$ are equal; or, in general terms, if f is differentiated i times with respect to x and j times with respect to y, the resulting rule will be the same no matter in what order the differentiation is carried out.*[3]

Rule 5.2 is tested for a third-order partial derivative in Example 5.5.

Example 5.5

Beginning with the results of differentiation from Example 5.4 find:

$$\frac{\partial^3 z}{\partial y\, \partial x\, \partial w},\ \frac{\partial^3 z}{\partial x\, \partial y\, \partial w},\ \frac{\partial^3 z}{\partial w\, \partial y\, \partial x};$$

$$\frac{\partial^3 z}{\partial y\, \partial x\, \partial w} = \frac{\partial}{\partial y} \frac{\partial^2 z}{\partial x\, \partial w} = \frac{\partial}{\partial y}(④) = \frac{\partial}{\partial y}(8wy^3 + 28x^3) = 24wy^2$$

$$\frac{\partial^3 z}{\partial x\, \partial y\, \partial w} = \frac{\partial}{\partial x} \frac{\partial^2 z}{\partial w\, \partial x} = \frac{\partial}{\partial x}(⑤) = \frac{\partial}{\partial x}(24wxy^2) = 24wy^2$$

$$\frac{\partial^3 z}{\partial w\, \partial y\, \partial x} = \frac{\partial}{\partial w} \frac{\partial^2 z}{\partial y\, \partial x} = \frac{\partial}{\partial w}(⑥) = \frac{\partial}{\partial w}(12w^2 y^2 + 6x^5) = 24wy^2.$$

Thus the three-way cross-partial derivatives fulfill the requirement that they all be equal, no matter what the sequence of their differentiation. Since all three of

---

[3] This general rule has already been illustrated, but it is not proved. It may be extended to n domain variables.

these are the same, it would make no difference which one was used in further differentiation with respect to one variable say, w; the result would be the same. The result would be derivatives taken twice with respect to w, and once each with respect to x and y in several different sequences, but with the same result—$24wy^2$.

5.2 Maxima and Minima, First Order Conditions

The theory of extreme points in three dimensions is similar to that in two dimensions, but again it is modified for the additional complications of the third dimension. In Figures 5.2 and 5.4, a partial derivative with respect to each independent variable is a value equal to the slope of a curve determined by the intersection between a plane and the three-dimensional figure, in these cases the piece of apple. By setting the partial derivative in one direction equal to zero—say the partial derivative of f with respect to x, holding y constant at y_0 as shown in Figure 5.2, it is possible to find an extreme point (a maximum) for this curve in this direction. But the plane that determines the curve was arbitrarily selected as $y = y_0$ in Figure 5.1. This plane does not necessarily lead to the extreme point for the entire figure, the highest point on the apple. In order to arrive at the extreme point for the entire figure, the highest point in the x direction and the highest in the y direction would have to occur simultaneously. Thus the conditions for an extreme point in the direction of each independent variable—that is with respect to each independent variable—must be simultaneously satisfied.

The apple half that has been used thus far in illustrations was placed such that the z axis passed through its stem and only one part of it was considered in the discussions and Figures 5.1 to 5.4. Now it is convenient to move the apple in a southeasterly direction away from the intersection of the three axes, as shown in Figure 5.5.

To find the highest point on the apple, it must be cut with planes in the x and y directions, in such a way that an extreme point in each direction occurs at the same point. Intuitively, it seems reasonable that this point will be the point of tangency of the apple with a plane parallel to the xy plane.

If the apple is cut with a yz plane to the right or left of the center of the apple, it would be possible to find the optimum for the resulting two-dimensional figure. However, only if the yz plane passes through the center of the apple would the two-dimensional high point be at the high point of the whole fruit. In that case a horizontal plane including the line tangent to the two-dimensional curve would be tangent to the apple. If the xz plane does not pass through the center of the apple, a horizontal plane including the tangent to the curve at its maximum point in the yz plane would slice off the top of the apple. Only when the high point on the xy plane is simultaneously the high point on the yz plane is an overall maximum reached.

Several steps were suggested in Chapter 3 to find the maxima or minima for functions with one variable over the domain. Essentially the same procedure can be used to find the maxima and minima for a function whose domain is an ordered pair. The main difference is that the maximum in one direction in the

Figure 5.5

One-Half Apple on the xy Plane with Maximum Value of z Occurring at the Stem.

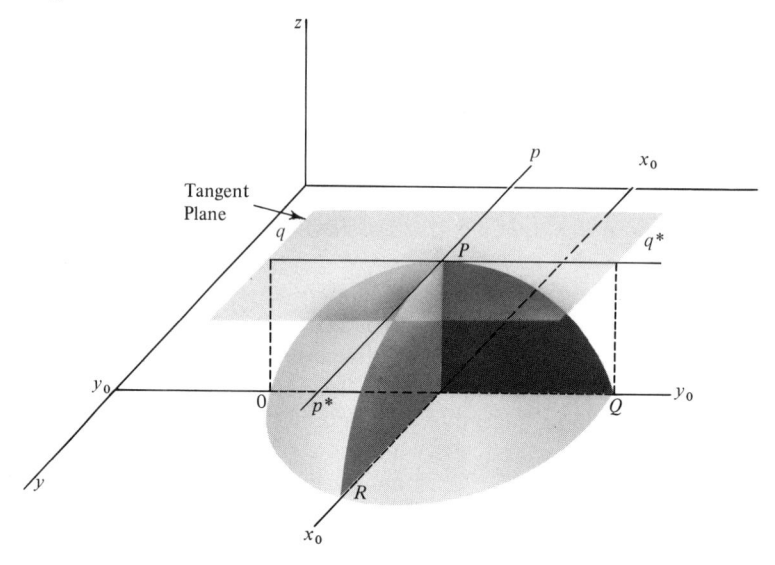

xz plane must simultaneously be a maximum in all others, e.g. in the yz plane. The steps to be followed are:

1. Find the rule for the first partial derivative of the function f with respect to each domain variable, and set it equal to zero.
2. Solve the resulting system of equations simultaneously for the values of the domain variables at which the derivatives are simultaneously zero.
3. Substitute the values of the ordered pairs (n-tuples) of domain variables into the rule for f to find the value z corresponding to each extreme point identified in step 2.

These steps constitute the first-order conditions for maxima and minima within an interval. But there are problems in identifying which values of z represent which type of extreme point; therefore, it is again necessary to investigate the second-order conditions. The three additional steps 4, 5, and 6 that are required are presented in Section 5.3.

The simultaneous solution of the equations developed in step 2 ensures that a single point, an ordered pair, is located in the domain for which an extreme point exists in the range. Given that the tangents (the partial derivatives) to the fruit at this point are parallel to the x and y axes (each partial equal to zero), it is now stated without proof that these tangents define a tangent plane that is parallel to the xy plane as shown in Figure 5.5. In Figure 5.5 the tangent lines, pp^* and qq^*, determine the tangent to the apple at its high point.

For a point under study (point P in Figure 5.5) to be established as a maximum, the slope of the curve must be decreasing for increasing values of each domain

Figure 5.6

The Apple Slice Cut by the Plane $y = y_0$ with Tangent Line at P.

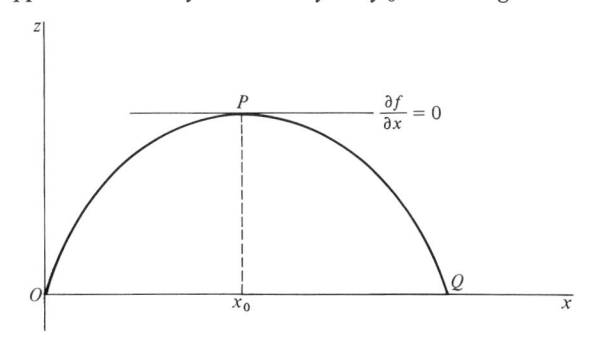

variable (for *each* of the partial derivatives) at that point. This can be tested analytically by the second *straight* partial derivative with respect to the domain variables, x and then y. If both are negative, the rate of change of the slope of the function in both directions is negative. This condition is also necessary (but not sufficient) for the value of the function at the extreme point to be a maximum. Both second partial derivatives must be positive at the point; that is, the rate of change of the slope must be positive in both directions, in order for the value of the function to be a minimum at that point (again a necessary, but not sufficient condition).

5.3 Sufficient Conditions

The apple example can be used to provide an intuitive justification for the second-order conditions, (asserted without proof) which along with steps 1 through 3 of section 5.2 are sufficient for maxima and minima. Figure 5.6 is a diagram derived from passing the $y = y_0$ plane through the apple shown in Figure 5.5. The slope of the curve OPQ traced out by the intersection of the plane with the skin of the apple is positive from O to P, zero at P, and negative from P to Q. This means that the value of the function increases, reaches a maximum at $x = x_0$, and then begins to decrease. The fact that the rate of change of the slope is negative as x increases from $x_0 - \Delta x$ to $x_0 + \Delta x$ indicates that the function reaches a maximum with respect to the x variable at x_0. The second (partial) derivative evaluated at $x = x_0$ provides a measure of the rate of change of the slope at $x = x_0$; thus $\partial^2 f / \partial x^2 < 0$.

Now the yz plane is passed through the apple at $x = x_0$, and the resulting figure rotated 90° to the right. The result is Figure 5.7. The second derivative of f with respect to y must show that the rate of change of the slope of the curve in this plane is also negative at $y = y_0$: $\partial^2 f / \partial y^2 < 0$, as indicated in Figure 5.7. These second-order conditions alone are not sufficient for a maximum value of f, but they are now formally stated and with added conditions (5.2) and (5.3) together become sufficient conditions.

Figure 5.7

Apple Slice Cut by Plane $x = x_0$ with Tangent at P.

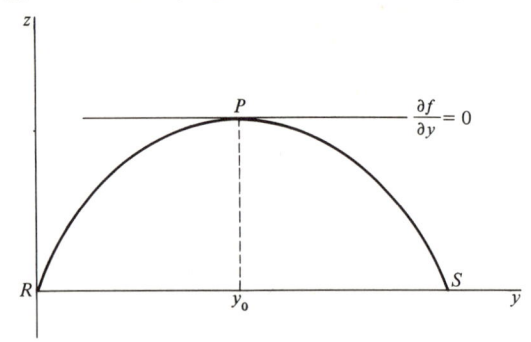

$$\frac{\partial f}{\partial y} = 0$$

Rule 5.3—A maximum (minimum) point for a function f, with rule $z = f(x, y)$: Given a function f with rule $z = f(x, y)$, differentiable partially in a neighborhood of (x_0, y_0), a relative maximum for the function occurs at a point (x_0, y_0) if the following conditions are met. The conditions (5.1) must hold simultaneously

Then if
$$\frac{\partial f}{\partial x} = 0 \quad \text{and} \quad \frac{\partial f}{\partial y} = 0. \tag{5.1}$$

$$\left[\frac{\partial^2 f}{\partial x^2}\right]\left[\frac{\partial^2 f}{\partial y^2}\right] - \left[\frac{\partial^2 f}{\partial x \, \partial y}\right]^2 > 0 \quad \text{at } (x_0, y_0) \tag{5.2}$$

the function is either a maximum or a minimum at the point. If

$$\frac{\partial^2 f}{\partial x^2} < 0, \quad \frac{\partial^2 f}{\partial y^2} < 0 \quad \text{at } (x_0, y_0) \tag{5.3a}$$

the value of the function at the point is a maximum. If

$$\frac{\partial^2 f}{\partial x^2} > 0, \quad \frac{\partial^2 f}{\partial y^2} > 0 \quad \text{at } (x_0, y_0) \tag{5.3b}$$

the value of the function at the point is a minimum.

These requirements can be generalized into n-space, but their statement is beyond the scope of the current text.

When the first partial derivatives are both zero at (x_0, y_0), the function is at a critical point. Condition (5.2) must be fulfilled in order for the critical point to be *either* a maximum or a minimum. However, if

$$\left[\frac{\partial^2 f}{\partial x^2}\right]\left[\frac{\partial^2 f}{\partial y^2}\right] - \left[\frac{\partial^2 f}{\partial x \, \partial y}\right]^2 < 0,$$

then the critical point is *neither* a maximum nor a minimum. (It is what is called a saddle point.) If

$$\left[\frac{\partial^2 f}{\partial x^2}\right]\left[\frac{\partial^2 f}{\partial y^2}\right] - \left[\frac{\partial^2 f}{\partial x \, \partial y}\right]^2 = 0,$$

Sec. 5.3] SUFFICIENT CONDITIONS   127

Figure 5.8

The Function f with Rule z = f(x, y) with a Saddle Point at Point P.

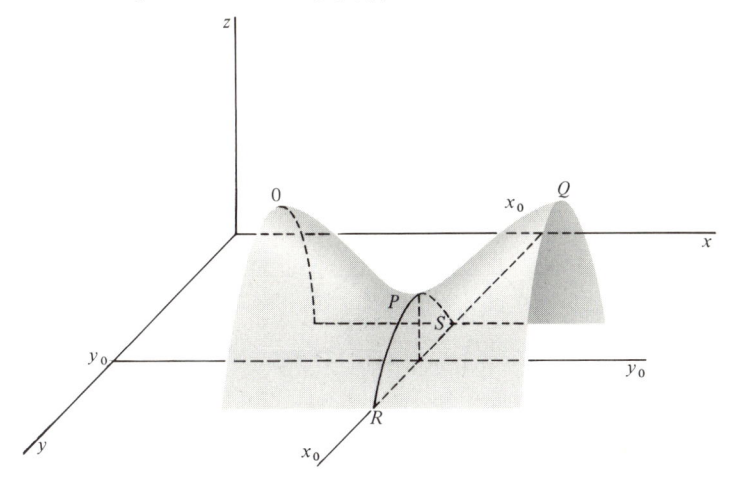

the test fails; there is no information about the function at (x_0, y_0), but the function can be investigated directly.[4]

When conditions (5.1) and (5.2) have both been met, then through conditions (5.3a) and (5.3b), the maximum or minimum can be identified. If

$$\frac{\partial^2 f}{\partial x^2} < 0, \quad \frac{\partial^2 f}{\partial y^2} < 0 \quad \text{at } (x_0, y_0)$$

the function is at a maximum point. If

$$\frac{\partial^2 f}{\partial x^2} > 0, \quad \frac{\partial^2 f}{\partial y^2} > 0,$$

the function is at a minimum point. As in the case of functions of a single variable, it is also necessary to investigate the value of the function at its boundaries. This is a separate topic that is discussed in Section 5.5 for functions of more than one variable.

The reader may wonder about the conditions that lead to no maximum or minimum, despite the first partial derivatives being equal to zero. One such case is that referred to as a "saddle point" such as that shown in Figure 5.8, which is the three-dimensional analog to the inflection point.

Given the function f with rule $z = f(x, y)$, as graphed in Figure 5.8, if a partial derivative of the function is taken with respect to x and then with respect to y,

---

[4] Actually when

$$\left[\frac{\partial^2 z}{\partial x^2}\right] \cdot \left[\frac{\partial^2 z}{\partial y^2}\right] = \left[\frac{\partial^2 z}{\partial x \, \partial y}\right]^2,$$

(x_0, y_0) could correspond to a maximum, a minimum, or neither. For this case, it is necessary to take further partial derivatives or investigate the function directly. A discussion of this situation is given in I. S. Sokolnikoff and R. M. Redheffer, *Mathematics of Physics and Modern Engineering*, (New York; McGraw-Hill, 1958), Ch. 3 and in many other texts.

Figure 5.9

A Saddle with Longitudinal Axis Running Diagonally Across the xy Plane.

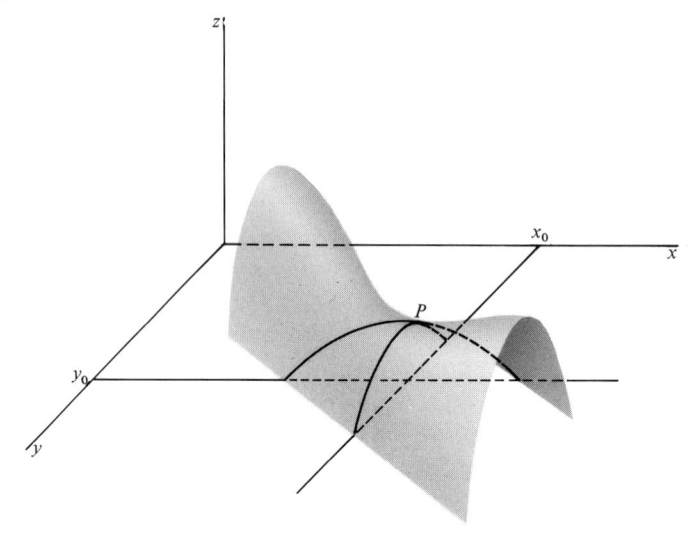

each partial derivative set equal to zero and the resulting equations solved simultaneously for x and y, as in

$$\frac{\partial f}{\partial x} = 0, \qquad \frac{\partial f}{\partial y} = 0$$

then the values x_0 and y_0 that represent the solution to the simultaneous equations determine neither a maximum nor a minimum for the function! In the plane parallel to the xz plane at y_0, the function f has a minimum value. In the plane parallel to the yz plane at x_0, the function f has a maximum value. Together the pair of values (xy) determines a location corresponding to a very significant point on the saddle.

Given the characteristics of Figure 5.8, the two straight second partial derivatives should have opposite signs:

$$\frac{\partial^2 f}{\partial x^2} > 0, \qquad \frac{\partial^2 f}{\partial y^2} < 0$$

and therefore,

$$\frac{\partial^2 f}{\partial x^2} \cdot \frac{\partial^2 f}{\partial y^2} < \left(\frac{\partial^2 f}{\partial x\, \partial y}\right)^2. \tag{5.4}$$

The direction of the inequality in equation (5.4) must hold, since the second cross-partial derivative is squared (and thus greater than 0), while one straight second partial derivative is positive and one negative and thus their product is necessarily less than zero. The value of f is neither a maximum nor a minimum.

It is also possible, however, for the second straight partial derivative in each domain direction to be negative (positive) at a saddle point. Such a situation is

Sec. 5.3] SUFFICIENT CONDITIONS   129

illustrated in Figure 5.9 in which the longitudinal axis of the saddle is along a diagonal in the xy plane (rather than parallel to the x axis as in Figure 5.8). In this case the straight second partial derivatives are both negative, but the point is a saddle point for the function.[5] Condition (5.2) identifies this situation in which both other conditions for a maximum or minimum value are met; expression (5.2) would be < 0.

In practice, then, several steps must be taken to assure that second-order conditions for maxima and minima are met in the three-dimensional case:

4. Take the second straight partial derivative with respect to each domain variable.
5. Take the second cross partial derivative $\partial^2 f / \partial x\, \partial y$ (for the three-dimensional case).
6. Make the three second-derivative tests and identify the extreme points as maxima, minima, or saddle points for the ordered pairs of values for which the first partial derivatives are simultaneously zero.

The use of partial differentiation is now illustrated in an economic example.

Example 5.6

Assume that in the absence of advertising for its product a company faces a demand function given by

$$q = 200 - 4p$$

where q = quantity sold and p = price per unit, and additional net revenues r can be generated through advertising expenditures A. Assume that the (unlikely) equation

$$r = (5pA - 3A^2 + 50A)$$

expresses the net change in revenues resulting from changes in the advertising budget A. The negative term $-3A^2$ reflects the cost implications of the advertising expenditures. The cost directly associated with production (and sale) of the product exclusive of advertising costs is given by

$$c = 300 + 3q$$

then

Total (basic) revenue $r = p \cdot q = p(200 - 4p)$
Total cost $c = (300 + 3q)q$
$c = 300q + 3q^2.$

Since $q = 200 - 4p$, substituting for q gives

$$c = 60{,}000 - 1200p + 120{,}000 - 4800p + 48p^2$$
$$c = 180{,}000 - 6000p + 48p^2.$$

---
[5] This situation can be illustrated in a home experiment by using a banana as a sample saddle. Cuts at 45° and 135° to the longitudinal axis of the banana will produce results similar to those shown in Figure 5.9.

The profits of the firm π can be stated as the difference between revenues and costs, including net revenues associated with advertising; that is,

π = revenue — cost + advertising
$\pi = (200p - 4p^2) + (-180{,}000 + 6000p - 48p^2) + 5pA - 3A^2 + 50A.$

To find the maximum level of π, the first partial derivatives must be taken and set equal to zero:

$$\frac{\partial \pi}{\partial p} = 200 - 8p + 6000 - 96p + 5A = 0$$

$$\frac{\partial \pi}{\partial A} = 5p - 6A + 50 = 0.$$

Simplifying gives

$$6200 - 104p + 5A = 0$$
$$50 + 5p - 6A = 0.$$

Multiplying the first equation above by 6 and the second equation by 5 and adding gives

$$37{,}200 - 624p + 30A = 0$$
$$\underline{250 + 25p - 30A = 0}$$
$$37{,}450 - 599p \qquad\quad = 0$$

then

$$p = \frac{37{,}450}{599}$$
$$p = \$62.50$$
$$A = \frac{50 + 312.50}{6} = \frac{362.50}{6} = \$60.42.$$

Thus \$62.50 appears to be the profit-maximizing price—when used in conjunction with an expenditure of approximately \$60 on advertising. The two straight second partial derivatives are:

$$\frac{\partial^2 \pi}{\partial p^2} = -104, \qquad \frac{\partial^2 \pi}{\partial A^2} = -6$$

and both are negative. Condition (5.3a) is met. The second cross-partial derivative is

$$\frac{\partial^2 \pi}{\partial p\, \partial A} = 5.$$

Condition (5.2) thus is also met:

$$\left[\frac{\partial^2 \pi}{\partial p^2}\right]\left[\frac{\partial^2 \pi}{\partial A^2}\right] > \left[\frac{\partial^2 \pi}{\partial p\, \partial A}\right]^2$$
$$(-104)(-6) > (5)^2$$
$$624 > 25.$$

The second-order conditions show the figures to be a maximum and not a minimum or a saddle point.

5.4 Implicit Differentiation

Another use of the partial derivative is for implicit differentiation. Two methods of implicit differentiation are described, the second being the use of partial derivatives.

Given the function f with rule $x^2 + 3xy + 2y = 10$, which is stated implicitly rather than solved explicitly for x or y, implicit differentiation can be used to find the derivative function f'. This requires that one of the variables be treated as the domain variable and the other as the range variable.

First, with x as the domain variable and y as the range variable, the rule is assumed to be of the form $y = f(x)$. Thus the rules for differentiation of a function of a function, the chain rule, must be used for differentiating the terms in which y appears. The term $3y^2$ would be treated as $u = f(x)$, for example. The equation for the chain rule appropriate for this case is $du/dx = du/dy \cdot dy/dx$. Differentiating with respect to y and multiplying by dy/dx gives $du/dx = du/dy \cdot dy/d = 6y \cdot dy/dx$. This procedure is used in the equation for the rule of the function given, and then solved for dy/dx.

Example 5.7

Given the function f with rule stated implicitly, $x^2 + 3xy + 2y - 10$, find dy/dx:

$$\frac{d}{dx}(x^2 + 3xy + 2y - 10) = 0.$$

The second term is treated as a product of two functions of x and the third as a function of x:

$$2x + \left(3x\frac{dy}{dx} + 3y\right) + 2\frac{dy}{dx} - 0 = 0$$

$$2x + 3y + \frac{dy}{dx}(3x + 2) = 0$$

$$\frac{dy}{dx}(3x + 2) = -(2x + 3y)$$

$$\frac{dy}{dx} = -\frac{(2x + 3y)}{3x + 2}.$$

Note that the differentiation of the factor $3xy$ is treated as the differentiation of a product of two functions of x, the second requiring the use of the chain rule. The process of implicit differentiation is repeated, treating y as the domain variable and x as the range variable using the chain rule, with x considered to be a function of y.[6] The appropriate form of the chain rule is now

$$\frac{du}{dy} = \frac{du}{dx} \cdot \frac{dx}{dy}$$

---

[6] The reader is reminded that when an equation is referred to as a "function," the implication is that the equation is the rule for mapping from the domain of the real numbers into the range specified by the equation as the rule for mapping. To fit this definition of a function, the mapping in this case must be restricted to, say, the positive reals. (Why?)

then

$$\frac{d}{dy}(x^2 + 3xy + 2y - 10) = 0$$

$$2x\frac{dx}{dy} + \left(3y\frac{dx}{dy} + 3x\right) + 2 = 0$$

$$\frac{dx}{dy}(2x + 3y) = -(3x + 2)$$

$$\frac{dx}{dy} = -\left(\frac{3x + 2}{2x + 3y}\right).$$

It can be observed that

$$\frac{dy}{dx} = \frac{1}{dx/dy} = \frac{1}{-(3x+2)/(2x+3y)} = -\left(\frac{2x+3y}{3x+2}\right).$$

This relationship is presented formally as the rule 6.1 in Chapter 6.

Partial differentiation can also be used to differentiate this function implicitly. There is a rule stating that given a function f with rule stated as an implicit function of both x and y, then dy/dx, the derivative of y with respect to x, can be determined from the partial derivatives of f with respect to both x and y. This rule is included without proof as Rule 5.4.

Rule 5.4: *Given the function f with rule stated as an implicit function of both x and y, then dy/dx the derivative of f with respect to x, is given by the following:*[7]

$$\frac{dy}{dx} = \frac{-\partial f/\partial x}{\partial f/\partial y} \qquad \frac{\partial f}{\partial y} \neq 0. \tag{5.5}$$

Example 5.8

Given the function f with rule stated in implicit form $x^2 + 3xy + 2y = 10$, find the rule for the derivative function f' with rule stated $y' = f'(x) = dy/dx$. To do this define a new function g for which $z = g(x, y) = x^2 + 3xy + 2y - 10$ is the rule. Then,

$$\frac{\partial g}{\partial x} = 2x + 3y \qquad \text{and} \qquad \frac{\partial g}{\partial y} = 3x + 2$$

so

$$\frac{dy}{dx} = \frac{-\partial g/\partial x}{\partial g/\partial y} = -\left(\frac{2x + 3y}{3x + 2}\right)$$

also

$$\frac{dx}{dy} = \frac{-\partial g/\partial y}{\partial g/\partial x} = -\left(\frac{3x + 2}{2x + 3y}\right).$$

The result achieved by means of partial derivatives is identical to that achieved by implicit differentiation by use of the chain rule (although this has been merely illustrated, not proved).

---

[7] The condition $\partial f/\partial y \neq 0$ justifies solving an expression containing x and y for y in terms of x as implied by implicit differentiation, although this is not proved here.

5.5 Constrained Maxima and Minima

The manager is continually required to choose paths that achieve maximum value, minimum cost, and so on, but more important, he must frequently achieve such goals under particular limiting conditions, often referred to as "constraints." The manager could minimize total costs by producing nothing at all, but then this would profit him nothing. He would be better served to minimize cost while at the same time producing enough output to meet the requirements of his customers. The manager may be able to maximize revenue by producing an infinite output of goods, but he may have only enough resources of labor, equipment, and cash to produce at most 15,000 units of product.

Techniques for solving the unconstrained minimum or maximum problems have already been dealt with. Fortunately there are, in addition, certain techniques that make possible the solution of the problem of maximization or minimization subject to constraints. One of the most useful of these, the Lagrange multiplier technique for maximizing or minimizing a function subject to one or more constraints is considered next as illustrated by an extended example. The illustrative function used in Example 5.9 was chosen for its simplicity, but it is one that can also be used as the basis for developing a production function, as discussed in Appendix 5A. In the example, the implications of the procedure for finding an unconstrained maximum are illustrated as a basis against which to explore the significance of the introduction of simple constraints, and then a formal procedure for finding maximum (minimum) values of functions subject to constraints is introduced and illustrated for the function under discussion.

The function f is graphed three dimensionally in Figure 5.10. It should be useful in visualizing the discussion that follows.

Example 5.9

Given a function f with rule $z = f(x, y)$ or specifically

$$z = 4 - (x - 4)^2 - (y - 5)^2,$$

investigate the function f to locate its extreme points.

The first partial derivatives are taken and set equal to zero:

$$\frac{\partial f}{\partial x} = -2(x - 4) = 0$$

$$\frac{\partial f}{\partial y} = -2(y - 5) = 0.$$

Solving these equations for x and y gives

$$x_0 = 4$$
$$y_0 = 5.$$

The value z_0 at this point is

$$z_0 = 4 - (0)^2 - (0)^2 = 4.$$

Figure 5.10

The Function f with Rule $z = 4 - (x - 4)^2 - (y - 5)^2$.

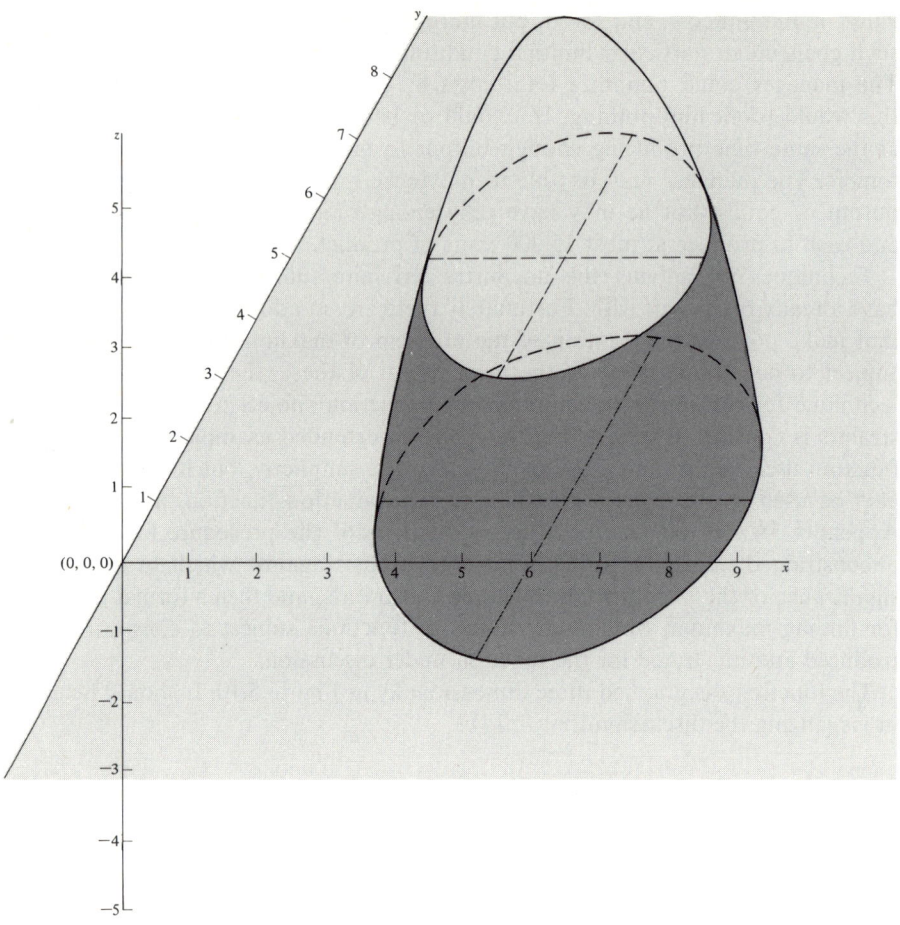

Investigating the second-order conditions gives

$$\frac{\partial^2 f}{\partial x^2} = -2, \quad \frac{\partial^2 f}{\partial y^2} = -2$$

$$\frac{\partial^2 f}{\partial x \, \partial y} = 0, \quad \frac{\partial^2 f}{\partial y \, \partial x} = 0$$

so

$$\frac{\partial^2 f}{\partial x^2} \cdot \frac{\partial^2 f}{\partial x^2} - \left(\frac{\partial^2 f}{\partial x \, \partial y}\right)^2 > 0.$$

The point $((4, 5), 4)$ is a maximum; conditions (5.1) through (5.3a) are met.

The function has a maximum value $z_0 = 4$ at the point $x_0 = 4$, $y_0 = 5$. The intersection of f with the plane $y_0 = 5$ is graphed in Figure 5.11 and with the plane $x_0 = 4$ in Figure 5.12. The functions representing these intersections are

Figure 5.11

Function f_1 with Rule $f(x, 5) = 4 - (4 - x)^2$ for the Intersection of the Function f with Plane $y_0 = 5$.

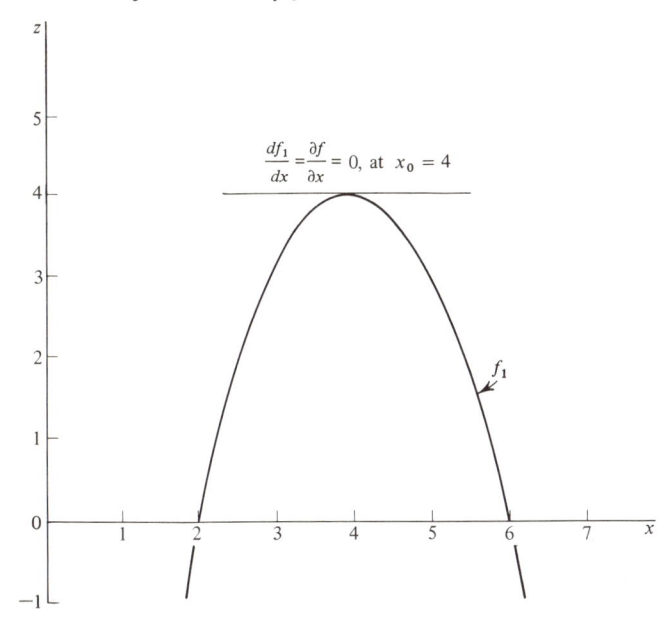

now developed and analyzed both algebraically and graphically. This discussion forms the basis for understanding the significance of the constraints that are introduced next.

The unconstrained maximum point occurs when the first partial derivative of z with respect to x is zero, with y constant at the value $y_0 = 5$. The rule for the function f_1 representing the intersection of the function f with the plane $y_0 = 5$ is found from the rule of f with y set at $y_0 = 5$:

$$f_1(x) = f(x, 5) = 4 - (x - 4)^2 - (5 - 5)^2 = 4 - (x - 4)^2. \qquad (5.6)$$

This is the equation for the intersection of the function f with the plane $y_0 = 5$, a plane parallel to the xz plane. The plane has been chosen to pass through the maximum value of z as determined earlier in the example.

The maximum value of f_1 and f should occur at the same value of x. To test this, take the first and second derivatives of f_1 to get[8]

$$\frac{df_1}{dx} = -2(x - 4)$$

$$\frac{d^2 f_1}{dx^2} = -2.$$

By setting the first derivative equal to zero, it is possible to find the value of x

---

[8] Note that df_1/dx is the same as $\partial f/\partial x$ at $y_0 = 5$. In this illustration, the emphasis is on the planes determined by the simultaneous solution of the equations of the first partial derivatives of f set equal to zero.

Figure 5.12

Function f_2 with Rule $f(4, y) = 4 - (y - 5)^2$ for the Intersection of Function f with Plane $x_0 = 4$.

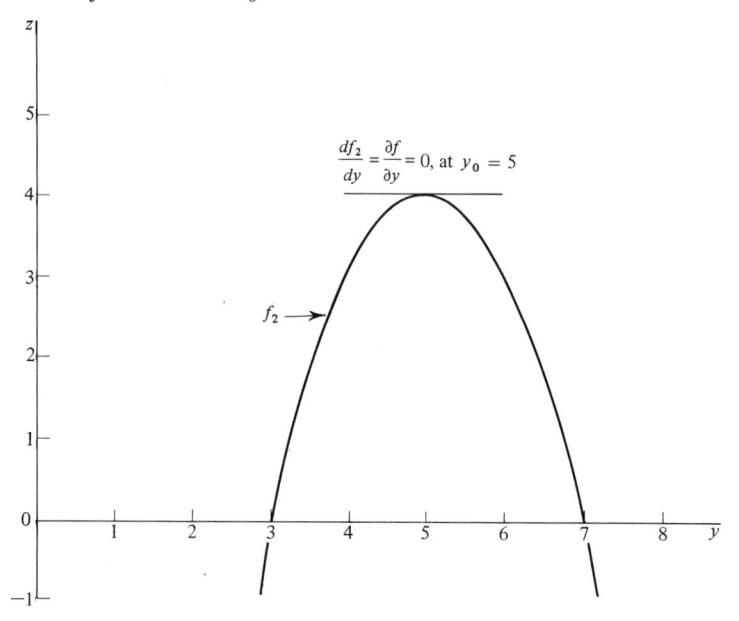

at which the tangent to the curve f_1 is horizontal. This value is $x_0 = 4$ at which $z_0 = 4$, which is a maximum since the second derivative is (everywhere) negative. Thus the curve f_1 has a maximum value of z at $x_0 = 4$. The curve, with x intercepts at $x = 2$ and $x = 6$, is graphed in Figure 5.11.

Now exploring the intersection of the yz plane at $x_0 = 4$ with the function f; that is, in the plane cutting through the (original) Figure 5.10 at the value for x at which f reaches its maximum, results in the new function f_2:

$$f_2 = f(4, y) = 4 - (4 - 4)^2 - (y - 5)^2 = 4 - (y - 5)^2. \qquad (5.7)$$

Again the functions f and f_2 should have maximum values at the same values of y. Taking the first and second derivatives of f_2 gives[9]

$$\frac{df_2}{dy} = -2(y - 5)$$

$$\frac{d^2 f_2}{dy^2} = -2.$$

Setting the first derivative equal to zero gives $y_0 = 5$, $z_0 = 4$, which is a maximum since the second derivative again is (everywhere) negative. The curve is graphed in Figure 5.12.[10] The significance of introducing a constraint into the problem is now investigated.

[9] Note that df_2/dy is the same as $\partial f/\partial y$ for $x_0 = 4$.
[10] Both cross sections are parabolas; therefore, the function itself is a paraboloid (actually a paraboloid of revolution since it is symmetric in x and y). Its axis is parallel to the z axis and passes through the point $x = 4$, $y = 5$.

Sec. 5.5] CONSTRAINED MAXIMA AND MINIMA 137

Figure 5.13

Function f_1 Subject to the Constraint $x \leq 3$.

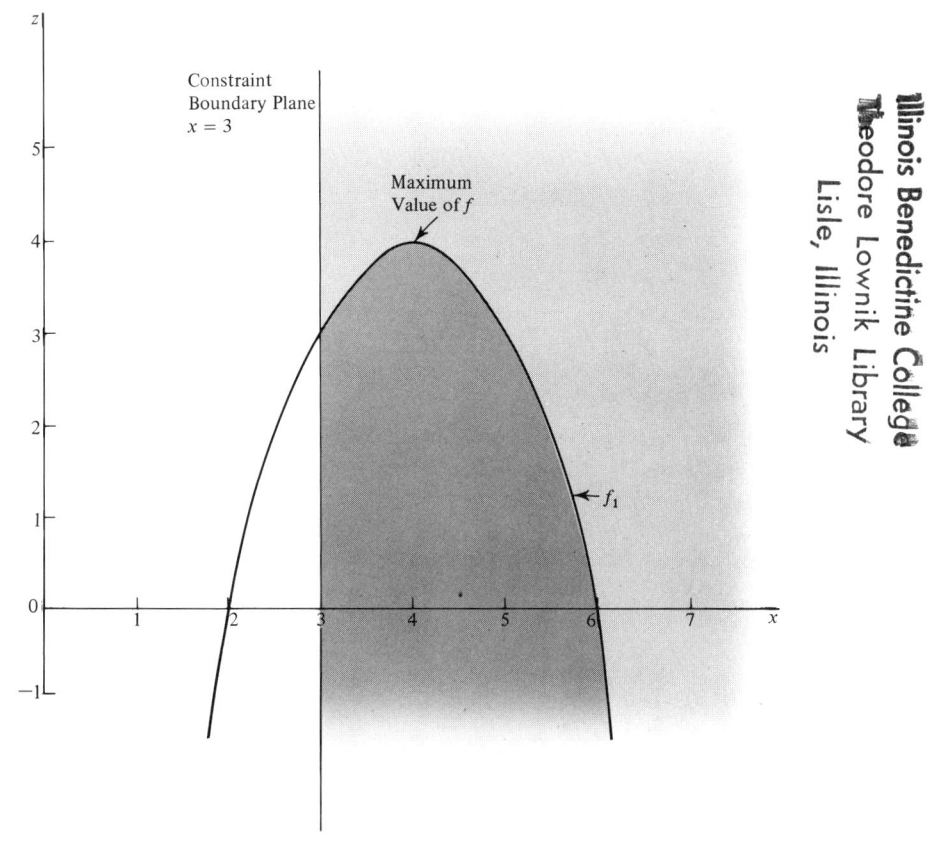

Now assume that the constraint $x \leq 3$ is introduced and the objective is to maximize f subject to this constraint. The values of x are now restricted to the interval $-\infty < x \leq 3$ and the value corresponding to the (global) maximum for f is not in the permitted interval. This is shown in Figure 5.13 where the shaded area is the area in which the constraint is exceeded. The boundary of the constraint is the plane $x = 3$ parallel to the yz plane shown in Figure 5.13 as the line at $x = 3$. The highest value of z occurs at the boundary of the constraint, as can be observed from the graph.

The change brought about in Figure 5.12 by introducing the constraint is shown in Figure 5.14. The upper curve is the graph of f_2, the intersection of the plane $x_0 = 4$ with the function f. But, because $x_0 = 4$ is not in the region permitted by the constraint, it is necessary to look at the intersection of the plane $x_1 = 3$, with f. This intersection is shown in Figure 5.14 as the graph of the new function f_3. The rule for f_3 is found from the rule for f with x set equal to 3; that is,

$$f_3 = f(3, y) = 4 - (3 - 4)^2 - (y - 5)^2 = 3 - (y - 5)^2. \tag{5.8}$$

138 PARTIAL DIFFERENTIATION [Ch. 5

Figure 5.14

Functions f_2 and f_3 for the Intersections of the Function f with the Planes $x = 4$ and $x = 3$, Respectively.

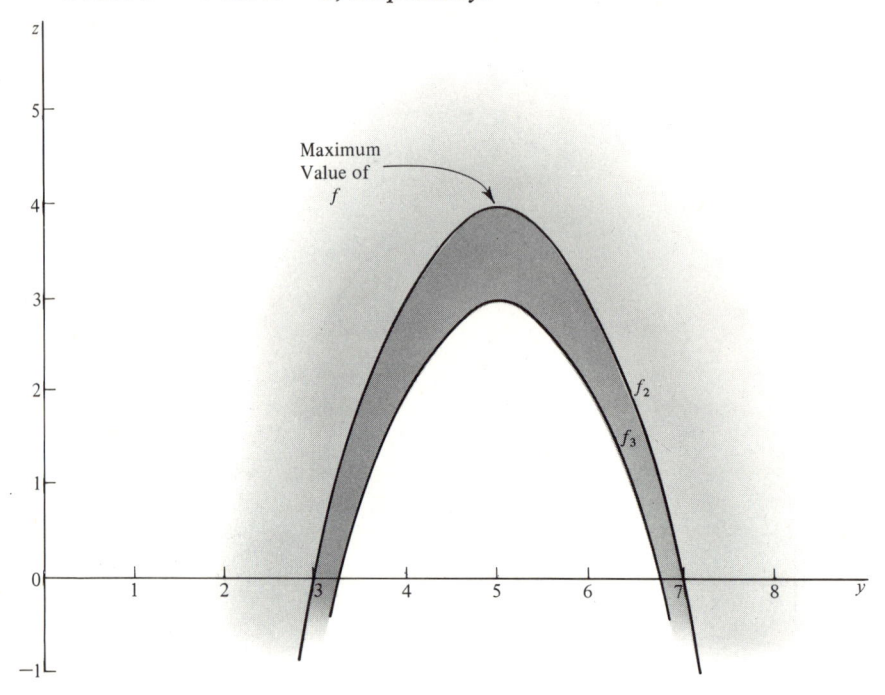

The maximum value of f_3 can be found. Taking the first and second derivatives of f_3 gives:

$$\frac{df_3}{dy} = -2(y - 5)$$

$$\frac{d^2 f_3}{dy^2} = -2.$$

Setting the first derivative equal to zero gives $y = 5$ at which point $z = 3$, a maximum since the second derivative is everywhere negative. No matter at what value of x the constraint boundary occurs, the maximum point on the curve of its intersection with f always falls at $y_0 = 5$ as can be seen in Figure 5.14 (and just confirmed for f_3). The maximum value of f under the constraint $x \leq 3$ can be seen to be at $x_1 = 3$, $y_1 = 5$, for which $z_1 = 3$.

The intersections with f of planes parallel to the yz plane for smaller values of x will have successively lower maxima. (They fall further and further to the left in Figure 5.13; in Figure 5.14 the maximum for f_3 is lower than for f_2.)

Another case should be noted. Given the constraint $x \leq 5$, the intersection of the function f with the boundary plane of the constraint results in a function f_4 with rule:

$$f_4 = f(5, y) = 4 - (5 - 4)^2 - (y - 5)^2 = 3 - (y - 5)^2. \tag{5.9}$$

This constraint is graphed in Figures 5.15 and 5.16.

Figure 5.15

Function f_1 Subject to Constraint $x \leq 5$.

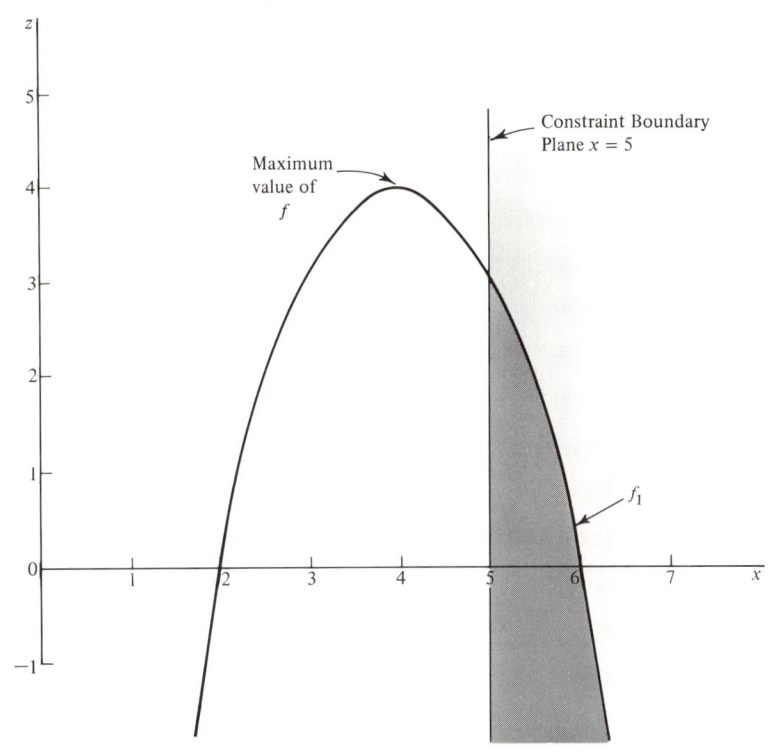

The constraint does not eliminate the value of x at which the function f reaches its maximum value, and in Figure 5.16 the function f_4 falls behind (or below) f_2. The maximum for the constrained function is identical with that for the unconstrained. Thus investigating the intersection(s) of the original function f with the constraint plane(s) is not a sure way of identifying the constrained maximum or minimum; the constraint on the domain variable(s) may not constrain the achievement of the maximum (range) value of the function! An analytic procedure for determining maxima and minima subject to constraints must distinguish cases in which the constraint is and is not binding.

5.6 More than One Constraint[11]

Suppose now that a constraint $y \leq 3$ is introduced into the problem in addition to the constraint $x \leq 3$. The rule for the function f_5 representing the intersection of f with the boundary plane $y = 3$ (alone) is

$$f_5 = f(x, 3) = 4 - (x - 4)^2 - (3 - 5)^2 = -(x - 4)^2. \quad (5.10)$$

---

[11] This section may be skipped without loss of continuity. It represents merely a complication of the discussion of the single-constraint situation.

Figure 5.16

Functions f_2 and f_4, the Intersections of Function f with Planes $x = 4$ and $x = 5$, Respectively.

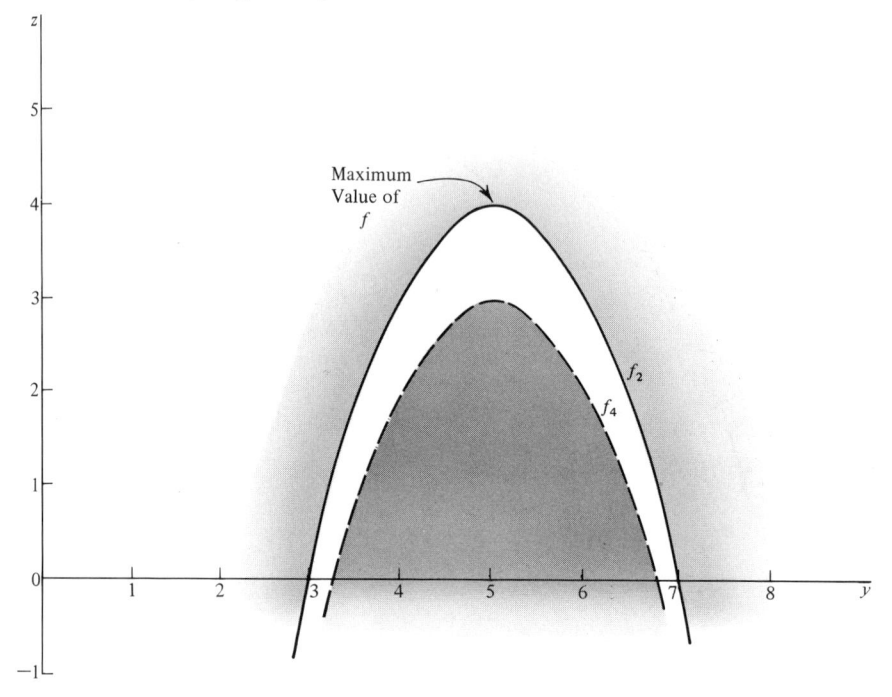

The graph of the function f and the two constraints are shown in Figures 5.17 and 5.18. The upper curve in Figure 5.17 is the intersection of f with the plane $y_0 = 5$ and the lower curve is the intersection of f with the plane $y_2 = 3$. Also shown is the influence of the plane $x = 3$. The area *permitted* under the two constraints is the unshaded area. (Figure 5.19 shows this three dimensionally.)

The upper curve in Figure 5.18 is the intersection of the plane $x_0 = 4$ with the function f, the lower curve is the intersection of the plane $x_3 = 3$. The area permitted under the two constraints (the area in which the constraints are not violated) is unshaded. It appears from Figures 5.17 and 5.18 that the maximum of $f = 4 - (x - 4)^2 - (y - 5)^2$ under the constraints $x \leq 3$, $y \leq 3$ occurs for $x = 3$, $y = 3$, which gives $z = -1$. Let this point be denoted $((x_2, y_2), z_2) = ((3, 3), -1)$.

The highest point attainable with both inequalities satisfied is the point that is usually identified in an analytic approach as the solution to the constrained maximization problem. In the discussion of the particular example here this intersection point, the point $((3, 3), -1)$ is the maximum point in the intersection of the point sets for planes $x = 3$, $y = 3$, and the function f, and is the maximum point in the permitted region. However, if the direction of one of the inequalities is reversed, if $x \geq 3$ for example, then the point of intersection of all the boundary functions need not be the maximum point on the *overall boundary* of the

Sec. 5.6]　MORE THAN ONE CONSTRAINT　　　　　　　　　　　　　　　141

Figure 5.17

Functions f_1 and f_5, the Intersections of the Function f with Planes $y_0 = 5$, $y_2 = 3$, Respectively. Unshaded Area Represents Portion Permitted under Constraints $x \leq 3$, $y \leq 3$.

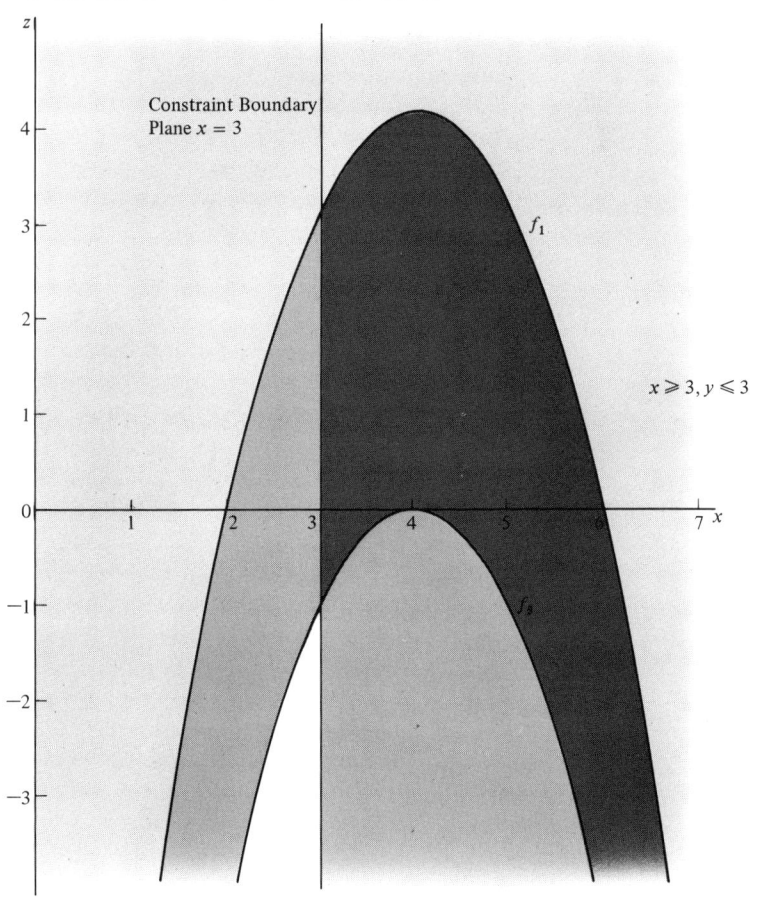

point sets. This is shown in Figure 5.17(a). (The corresponding figure to Figure 5.18 is not shown.) The intersection of the two constraint planes and the surface of the function f occur at $((3, 3), -1)$ as before, but the maximum point in the permitted region *now* falls at the point $((4, 3), 0)$, the maximum point of function f_5. Figure 5.19 shows the situation in three dimensions. The planes $x = 3$ and $y = 3$ intersect f at the point $((3, 3), -1)$; this is the point reached in the solution to the full Lagrangian problem, the solution appropriate to the constraints in the form $x - 3 = 0$, $y - 3 = 0$ for constraints $x \leq 3$, $y \leq 3$. (The Lagrangian Function is discussed in Section 5.7.) The solution to the problem for constraints in the form $x \geq 3$, $y \leq 3$ lies on the plane $y = 3$ at the high point $((4, 3), 0)$ on the boundary of the intersection of that plane with f; that is, on the function f_5.

Figure 5.17(a)

Functions f_1 and f_5, the Intersections of the Function f with Planes $y_0 = 5$, $y_2 = 3$, Respectively. Unshaded Area Represents Portion Permitted Under Constraints $x \geq 3$, $y \leq 3$.

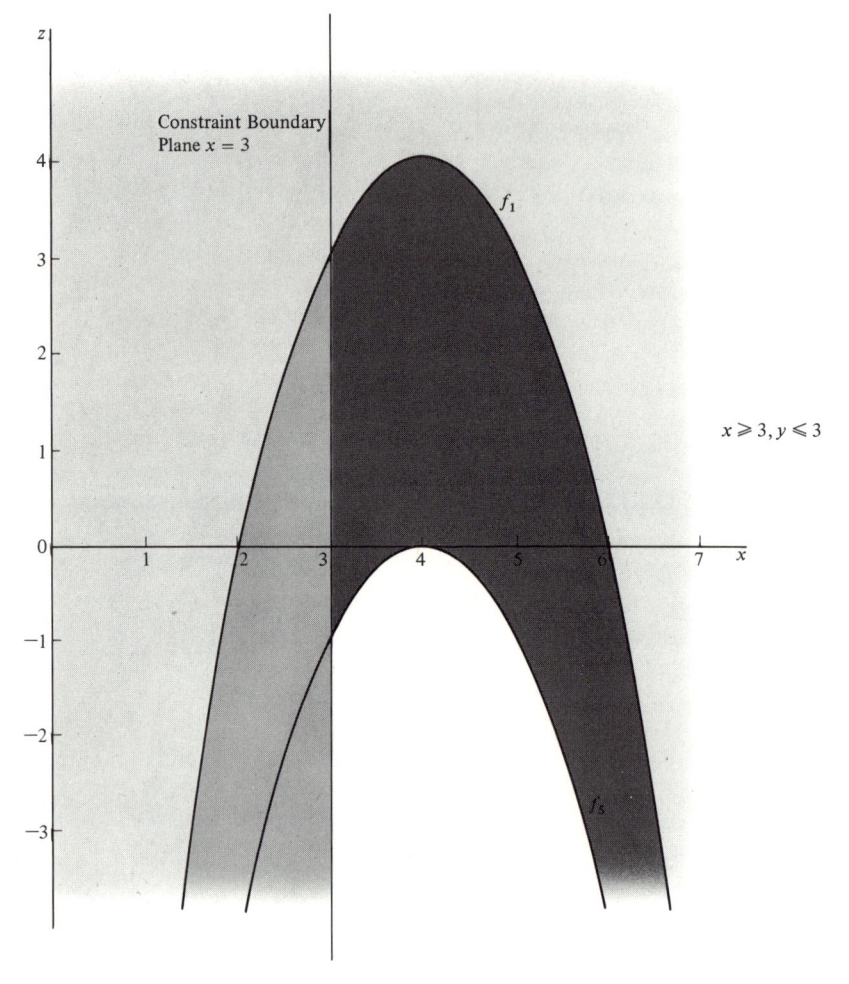

The procedure that allows the proper situation to be identified is as follows:

1. The values of the domain variables for the maximum (minimum) value of the unconstrained function f are found.
2. These values for the domain variables are substituted into all constraints to see if any constraint is violated.
3. If any constraint is violated, then the function f must be tested for maxima (minima) at its intersection with all possible combinations of constraints. The maxima (minima) not violating any constraints are compared and the largest (smallest) is the global maximum (minimum). It is wise to begin the comparison among the maxima (minima) for those constraints that do bind at the optimal point.

Sec. 5.6]   MORE THAN ONE CONSTRAINT   143

Figure 5.18

Functions f_2 and f_3, the Intersections of the Function f with Planes $x_0 = 4$, $x_1 = 3$, Respectively. Unshaded Area Represents Portion Permitted under $x \leq 3$, $y \leq 3$.

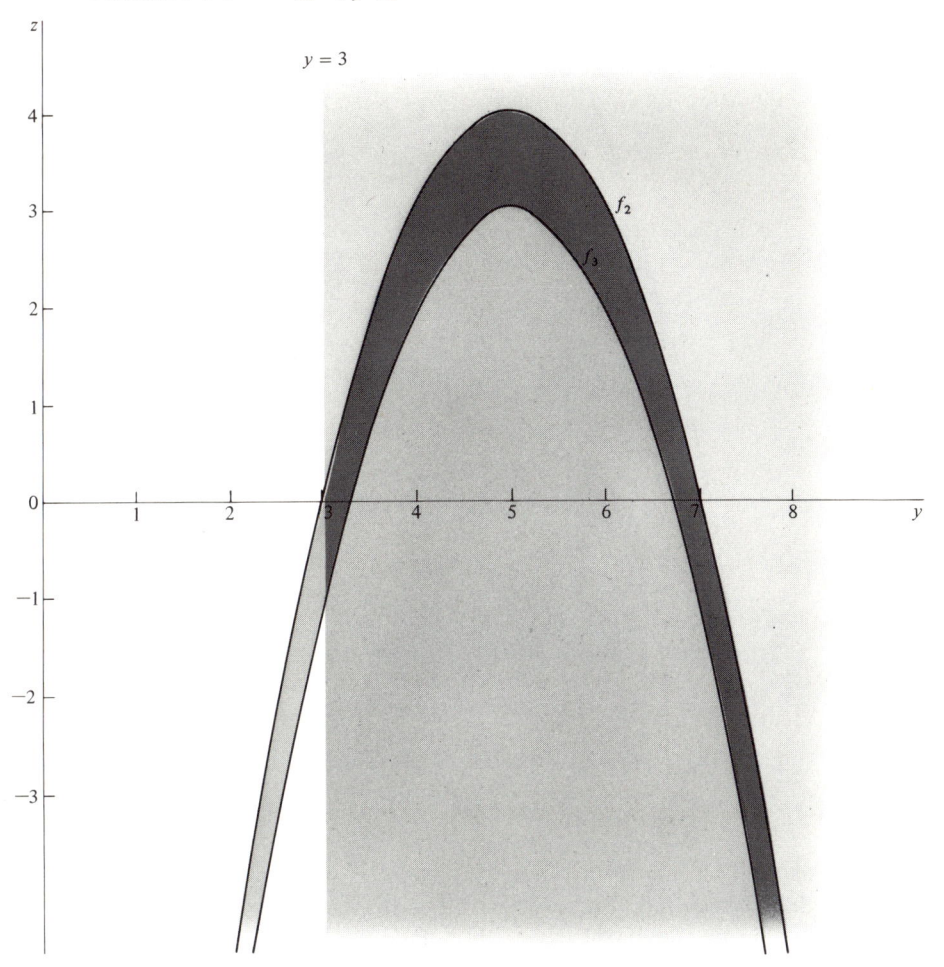

If this procedure were followed in the current two constraint case, the overall maximum for f unconstrained would be found to be the point $((4, 5), 4)$. At this point the constraint $y \leq 3$ is violated; thus it is necessary to test the function f along with all possible combinations of constraints for maxima. The combinations are: f alone (done), f plus constraints $x \geq 3$, $y \leq 3$; f plus $y \leq 3$; and f plus $x \geq 3$.

The value at the intersection of f plus $x \geq 3$, $y \leq 3$ is the point $((3, 3), -1)$ as just shown. The maximum value on the boundary of f plus $y \leq 3$ is the point $((4, 3), 0)$. That on the boundary of f plus $x \geq 3$ is $((3, 5), 3)$ but the constraint on y is violated. The maximum value of z for which no constraint is violated is 0; the maximum occurs at $((4, 3), 0)$.

Figure 5.19

The Function f Cut by Constraint Planes $x = 3$ and $y = 3$.

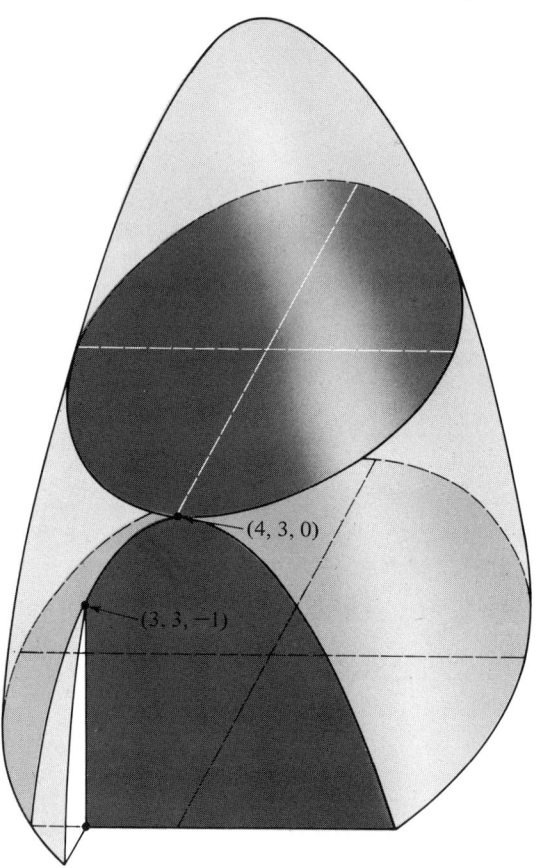

(4, 3, 0)

(3, 3, −1)

For constraints in the form of strict inequalities such as $x < 3$, no change in the procedure is required. The solution differs from that achieved for $x \leq 3$ only in that one point is eliminated from consideration, the point $x = 3$. The solution for a binding strict inequality constraint falls one (dimensionless) point over from that for a binding inequality of the form $x \leq 3$ (or $x \geq 3$). The method of Lagrange multipliers is now applied to this example with a single constraint to show how this type of result can be reached algebraically.

5.7 The Method of Lagrange Multipliers

Constraints faced by the manager are often stated as inequalities; for example, the amount of labor used must be less than or equal to the number of man hours available to the organization. The discussion in Sections 5.5 and 5.6 deals with cases in which functions are maximized (or minimized) in relation to inequalities bounded by the intersection of a plane or planes at one or more values of x, y, or $(x + y)$ with a three-dimensional surface of f, for example. The graphic results can be reproduced algebraically through the use of Lagrange multipliers for constraints in both the inequality and equality forms. The technique is applied as if the equality form held; then by interpretation of the values of the Lagrange multiplier, the analyst can determine whether the inequality is satisfied at its limit (the equality in fact held) or the straight inequality form was satisfied. Again the discussion is in terms of a three-dimensional example, but the procedure is generally applicable and can be extended to n dimensions.

Given the function f with rule

$$z = f(x, y) = 4 - (x - 4)^2 - (y - 5)^2$$

subject to the constraints that

$$x \leq 3 \tag{5.11}$$
$$y \leq 3 \tag{5.12}$$

find the maximum value of z. (This could be a production function, for example, for which the maximum amounts of labor x and capital y available are each 3, and z is the maximum possible output.) The procedure for solution of the problem is to restate the constraints as equalities in implicit form and then define a new function G_λ, whose rule is the rule for f plus terms made up of a multiplicative factor (a multiplier λ) of the right-hand sides of each constraint equation subtracted from f[12]; that is,

$$g_1(x) = x - 3 = 0 \tag{5.11a}$$
$$g_2(y) = y - 3 = 0 \tag{5.12a}$$

and

$$G_\lambda = G(x, y, \lambda_1, \lambda_2)$$
$$G_\lambda = 4 - (x - 4)^2 - (y - 5)^2 - \lambda_1(x - 3) - \lambda_2(y - 3). \tag{5.13}$$

The factors λ_1 and λ_2 are called Lagrange multipliers and their values are unknown; their values are determined in the solution to the system of equations such as that shown in equations (5.14)–(5.17). Both constraints are *equal to zero;* therefore, multiplication of them by *any* factor will not change their value, nor will their subtraction from f change its value.

By the usual rules for maxima and minima, the maximum value of z associated with the constrained value of f is determined. First, the first partial derivative of each function with respect to each of the (four) domain variables must be zero. Differentiating G_λ with respect to x, y, λ_1, and λ_2, respectively, and setting them equal to zero gives

$$\frac{\partial G_\lambda}{\partial x} = -2(x - 4) - \lambda_1 = 0 \tag{5.14}$$

$$\frac{\partial G_\lambda}{\partial y} = -2(y - 5) - \lambda_2 = 0 \tag{5.15}$$

$$\frac{\partial G_\lambda}{\partial \lambda_1} = -(x - 3) = 0 \tag{5.16}$$

$$\frac{\partial G_\lambda}{\partial \lambda_2} = -(y - 3) = 0. \tag{5.17}$$

---

[12] Constraint equations can be added or subtracted in forming the Lagrangian expression. They are subtracted here and the interpretation of the sign of λ is dependent on this assumption. It is also dependent on the practice of writing $x = 3$ as $x - 3 = 0$ rather than as $3 - x = 0$.

Equations (5.16) and (5.17) are merely a restatement of the equality form of the two constraints.[13] By solving them, a critical point is identified at $x = 3$, $y = 3$. Substituting these values into equations (5.14) and (5.15) gives

$$\lambda_1 = 2 \quad \text{and} \quad \lambda_2 = 4$$

thus

$$G_\lambda = -1.$$

The maximum value of the constrained function is -1 at the point $((3, 3), -1)$. The graph of the function in each domain direction appears in Figures 5.17 and 5.18.

The significance of the procedure just described is that the constraints themselves in implicit form have been added to the set of first partial derivatives of the original function f, *and* the first partial derivatives of f have been modified by a factor $(-\lambda_1)\partial g_1/\partial x$ or $(-\lambda_2)\partial g_2/\partial y$. The implications of this are now considered.

5.8 The Meaning of λ

For the unconstrained problem, the maximum occurs at $x = 4$, $y = 5$, at which point, tangents to f in each domain direction (in the xz and the yz planes) are horizontal; that is, both first partial derivatives are zero. For the constrained problem, the maximum occurs at $x = 3$, $y = 3$, at which point, however, neither tangent to the curves of intersection is horizontal. Thus neither partial derivative is zero *at that point on the function f*. Instead, however, the partial derivative of the function G_λ is *forced* to zero for each such partial derivative; that is,

$$\frac{\partial f}{\partial x} - \lambda_1 \frac{\partial g_1}{\partial x} = 0$$

$$\frac{\partial f}{\partial y} - \lambda_2 \frac{\partial g_2}{\partial y} = 0.$$

Thus, the value λ_1 times the partial derivative of the constraint g_1 with respect to x is just large enough to offset $\partial f/\partial x$ from its value at (3, 3) to zero; and λ_2 times the partial of g_2 with respect to y is just large enough to offset $\partial f/\partial y$ to zero.[14] This means that

$$-2(x - 4) - (2)(1) = 0 \quad \text{for } x = 3$$
$$-2(y - 5) - (4)(1) = 0 \quad \text{for } y = 3.$$

---

[13] Because this is the case, the problem can be approached in an alternative way dealing only with the partial derivatives of G_λ with respect to x and y (each domain variable) and then the constraints stated implicitly in equality form. Mathematically this latter approach may be preferable, since it does not require differentiation with respect to λ, but the approach used in the text is pedagogically more appealing; the equation system is the same in either case.

[14] If the constraints g_1 and g_2 had been functions of two variables, that is, $g_1(x, y)$, $g_2(x, y)$, the expressions given would have been, instead,

$$\frac{\partial f}{\partial x} - \lambda_1 \frac{\partial g_1}{\partial x} - \lambda_2 \frac{\partial g_2}{\partial x} = 0$$

$$\frac{\partial f}{\partial y} - \lambda_1 \frac{\partial g_1}{\partial y} - \lambda_2 \frac{\partial g_2}{\partial y} = 0.$$

This would be the normal case and the preceding case, a special case, in which $\partial g_2/\partial x$ and $\partial g_1/\partial y$ are zero. The special case has been used to simplify the exposition.

Sec. 5.8] THE MEANING OF λ 147

Figure 5.20

Tangent at the Point $x = 3$ to Curve Representing Intersection of f with Plane $y = 3$. $\partial f/\partial x = 2$ at $x = 3$; $\partial g_1/\partial x = 1$, and $\lambda_1 = 2$; thus $\partial f/\partial x - \lambda_1(\partial g_1/\partial x) = 0$.

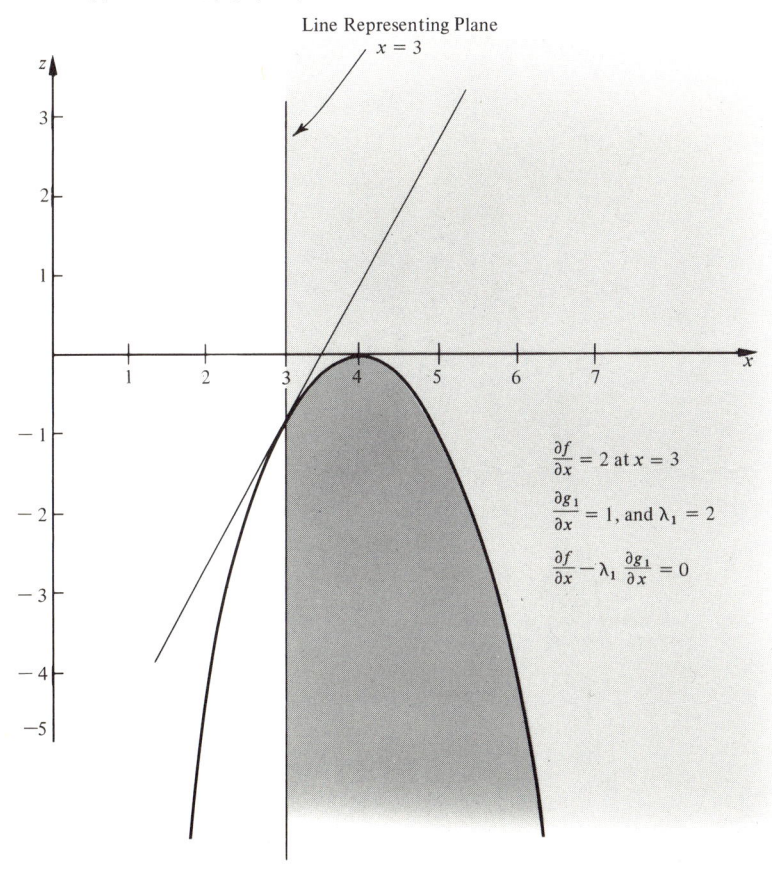

$\dfrac{\partial f}{\partial x} = 2$ at $x = 3$

$\dfrac{\partial g_1}{\partial x} = 1$, and $\lambda_1 = 2$

$\dfrac{\partial f}{\partial x} - \lambda_1 \dfrac{\partial g_1}{\partial x} = 0$

The slopes of the tangents to f are not horizontal at (3, 3); they are $+2$ and $+4$ as shown in Figures 5.20 and 5.21; but the partial derivatives to G_λ *must* be zero; they have been forced to zero by the λ factors. The multiplier λ for a constraint that is binding has an application to economics that is very important. Suppose that a small constant ϵ is added to the first constraint. How will this affect the value of the objective function f (or the constrained function G_λ)? This can be seen by adding $\epsilon > 0$ to the right hand side of the rule for g_1 to get $x = 3 + \epsilon$ and including this in G_λ. Call the new function G_{λ_1}, then

$$\begin{aligned}G_{\lambda_1} &= 4 - (x_1 - 4)^2 - (y_1 - 5)^2 - \lambda_3(x_1 - 3 - \epsilon) - \lambda_4(y_1 - 3) \\ &= 4 - (x_1 - 4)^2 - (y_1 - 5)^2 - \lambda_3(x_1 - 3) - \lambda_4(y_1 - 3) + \lambda_3\epsilon \\ &= G_\lambda + \lambda_3\epsilon.\end{aligned}$$

Thus relaxing the constraint by a small amount ϵ leads to an increase in the ob-

Figure 5.21

Tangent at the Point $y = 3$ to Curve Representing Intersection of f with Plane $x = 3$. $\partial f/\partial y = 4$ at $y = 3$; $\partial g_2/\partial y = 1$, and $\lambda_2 = 4$; thus $\partial f/\partial y - \lambda_2(\partial g_2/\partial y) = 0$.

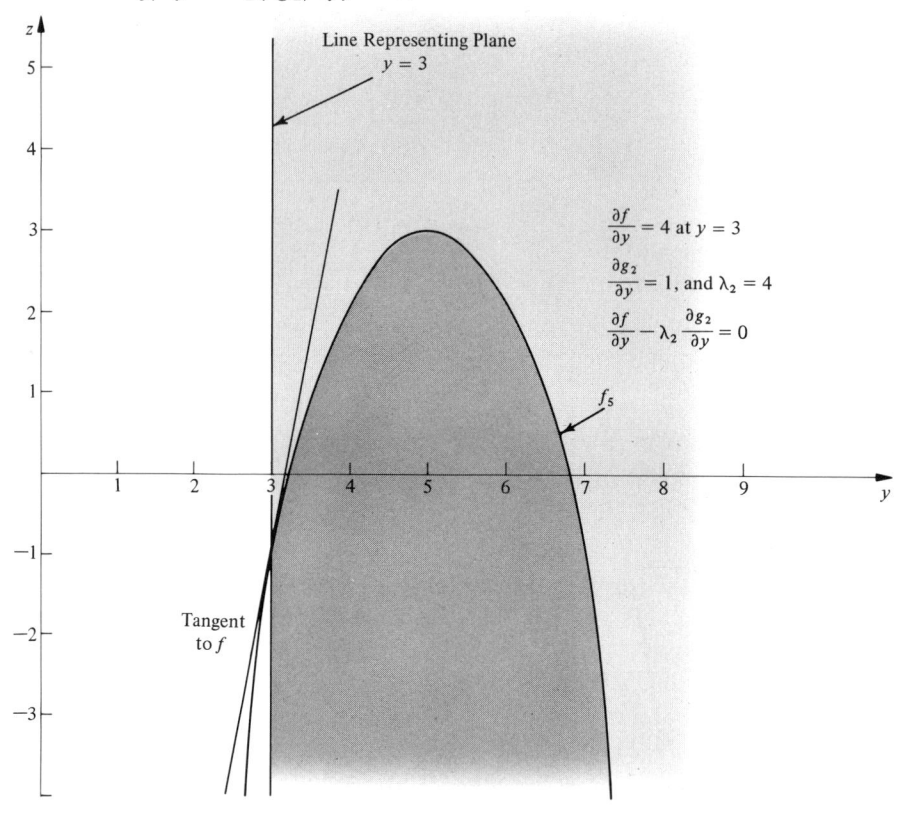

jective function by λ_3 times ϵ (as long as another constraint does not prevent the increase).[15]

The value λ_3 is the change in f for an infinitesimal change in the (binding) constraint plane. It applies only in the neighborhood around $x = 3$ (unless, of course, the intersection between f and the constraint is a straight line). The value λ_3 is the *marginal change in f associated with a change in this* (binding) *constraint*. If ϵ were .1 in the example, the change in f would be about $2(.1) = .2$. Since λ_3 is a marginal value, it applies only approximately to any finite relaxation of the constraint; that is, to the relaxation by $\epsilon = .1$. However, the smaller ϵ is, the better the approximation. The value $\lambda\epsilon$ is the differential of z, dz; it is a good approximation to Δz for small changes in the constraint plane.

---

[15] The constraint itself has been relaxed by ϵ. In the case illustrated this is equivalent to allowing the domain variable x to increase by ϵ, but this is not always so. If the constraint had been $2x = 3$ as stated in equality form, then relaxing it by ϵ to $2x = 3 + \epsilon$ would allow the domain variable x to increase by $\epsilon/2$. In either case λ measures the marginal impact on the objective function from relaxing the constraint by ϵ.

Sec. 5.8]   THE MEANING OF λ

At the beginning of Section 5.7, the statement is made that the analyst can identify those conditions under which the constraint is binding and those under which it is not. This is now demonstrated for the single constraint case for functions that have a single extreme point within the permitted region. In the illustration just given, the value of λ for the maximum point is positive and the constraint binding. The significance of the positive λ is that the maximum point occurs at the intersection of the constraint plane and the surface of the function f at a point where the first partial derivatives were not zero and have to be forced to that value; that is,

$$\frac{\partial G_\lambda}{\partial x} = \frac{\partial f}{\partial x} - \lambda_1 \frac{\partial g_1}{\partial x} = 0$$

where

$$\frac{\partial f}{\partial x} = -2(x - 4) \quad \text{and} \quad \frac{\partial g_1}{\partial x} = 1.$$

The positive value $\lambda_1 = 2$ is required for $-(\lambda_1)(\partial g_1/\partial x)$ to be negative, that is, $(\lambda_1)(\partial g_1/\partial x)$ is positive, as must be the case if the slope of the function at $x = 3$ is to be forced from a value of 2 to a value of zero:

$$\frac{\partial G_\lambda}{\partial x} = -2(3 - 4) - (2)(1)$$
$$= (2) - (2) = 0.$$

It is the partial derivative of the constrained function that is zero.

If the constraint plane were at $x = 6$ (for the moment ignoring the constraining value of y), the new constraint g_3 stated implicitly as an equality would be $g_3 = x - 6 = 0$; the new Lagrangian function would be

$$G_\lambda = 4 - (x - 4)^2 - (y - 5)^2 - \lambda_5(x - 6)$$

and its solution would be

$$\frac{\partial G_\lambda}{\partial x} = -2(x - 4) - \lambda_5 = 0$$

$$\frac{\partial G_\lambda}{\partial y} = -2(y - 5) = 0, \quad \left(\text{unchanged from } \frac{\partial f}{\partial y}\right)$$

$$\frac{\partial G_\lambda}{\partial \lambda_5} = x - 6 = 0$$

thus

$$x = 6$$
$$y = 5$$
$$\lambda_5 = -2(x - 4) = -2(6 - 4) = -4.$$

The value of λ_5 is negative; the value $-\lambda_5(\partial g_3/\partial x)$ is thus positive. The height of z at the point (6, 5) in the domain is

$$z = 4 - (6 - 4)^2 - (5 - 5)^2 = 0.$$

But the constraint $x \leq 6$ would not be violated at the domain point (4, 5), at which

Figure 5.22

The Intersection of the Function f with Rule $z = f(x, y)$ with the Plane $y = y_0$ Subject to (a) the Binding Constraint $x \leq a$ and (b) the Nonbinding Constraint $x \leq c$.

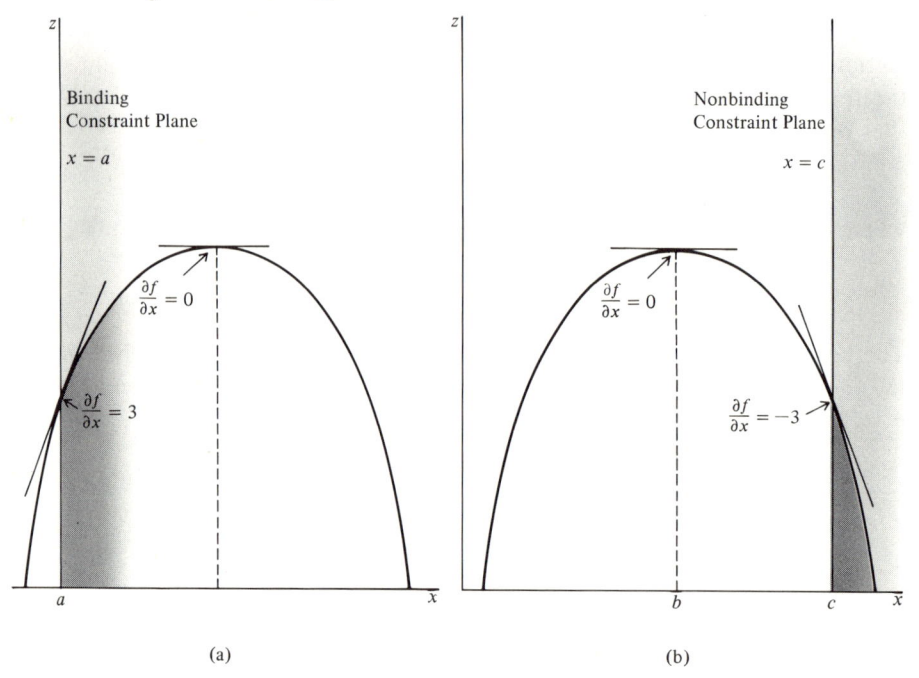

(a) Binding Constraint Plane $x = a$

(b) Nonbinding Constraint Plane $x = c$

value z is 4, its (global) maximum. The constraint *does not* restrict the attainment of the maximum value. The slope of f is negative at (6, 5) and thus

$$-(\lambda_5) \frac{\partial g_3}{\partial x} = (-4)(1)$$

must be positive (and λ_5 negative) for $\partial G_\lambda / \partial x$ to be equal to zero. The two situations are compared in Figure 5.22 for a function subject to general constraints $x \leq a, x \leq c$.

Thus for a constraint stated as $g(x, y) \leq 0$ in the neighborhood of a maximum point for f, the sign of λ indicates if the constraint is binding. For $\lambda > 0$, the constraint is binding; for $\lambda \leq 0$ the constraint is not binding. (If $\lambda = 0$, the constraint passes through the maximum point.)

If it is established that a constraint, $g(x, y) \leq 0$, falls in the neighborhood of a *minimum* value of f, a similar rationale applies. The value of λ is negative for *binding* values of the constraint. The values of λ are ≥ 0 where the constraint does not restrict attainment of the minimum value. Both cases are illustrated in Figure 5.23.

For the binding constraint plane at $x = a$, the value of λ must be negative to offset the negative slope of $-\frac{3}{2}$ for $\partial f / \partial x$ (where $\partial g_j / \partial x$ is subtracted from $-\frac{3}{2}$). In the nonbinding case for plane $x = c$ for the constraint $x \leq c$, the minimum is attained and the value of λ is ≥ 0 in order for the $\partial G_\lambda / \partial x$ to be zero. A similar

Sec. 5.8]   THE MEANING OF λ   151

Figure 5.23

The Intersection of the Function f with Rule $z = f(x, y)$ with the Plane $y = y_0$ Subject to (a) the Binding Constraint $x \leq a$ and (b) the Nonbinding Constraint $x \leq c$.

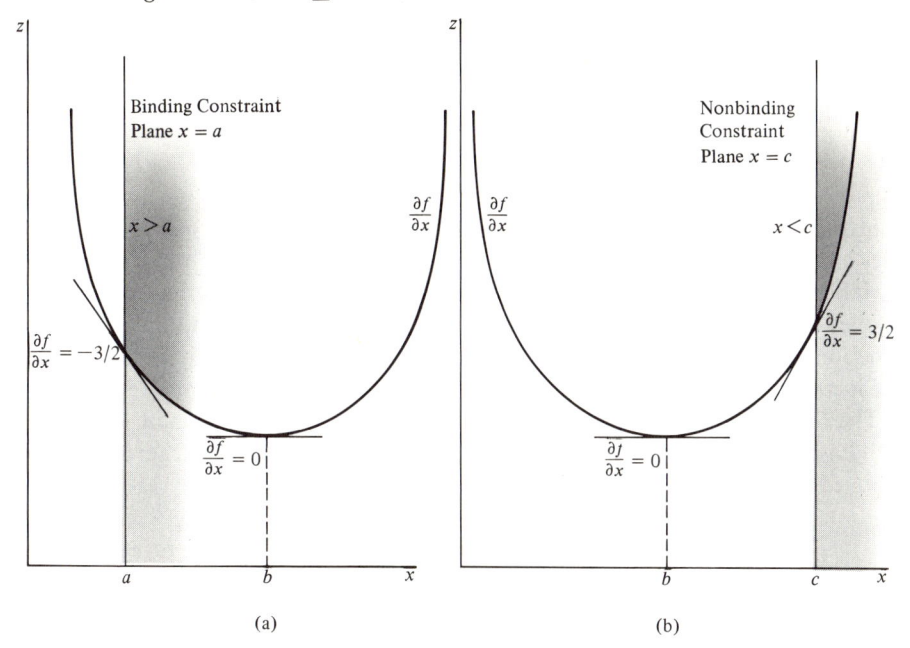

(a)   (b)

diagram and discussion would apply for the $\partial f/\partial y$ at $x = x_0$, if the constraint plane were expressed in terms of y.

In the neighborhood of a maximum if $\lambda \leq 0$ for the Lagrangian expression, then the *unconstrained* value of the function f is the local maximum. For the values of λ that are greater than or equal to zero in the neighborhood of a minimum, the unconstrained value of the function is the local minimum.

Table 5.1 summarizes the preceding discussion and Theorem 5.1 is the statement of the general theorem. (Table 5.1 is further extended as Table 5.2 in the Summary of this chapter.)

Table 5.1

Values of λ in Relation to the Various Conditions for the Single Constraint Case with ≤ Inequality.

	Constraint Binding	Constraint Nonbinding
Maximum	$\lambda > 0$	$\lambda \leq 0$
Minimum	$\lambda < 0$	$\lambda \geq 0$

Theorem 5.1: Given a function f with rule $z = f(x, y)$, the point (x, y) is a local maximum (minimum) of f subject to $g(x, y) \leq 0$ only if there exists a non-negative (nonpositive) λ that satisfies the following conditions:

$$\frac{\partial f}{\partial x} - \lambda \frac{\partial g}{\partial x} = 0 \tag{5.18}$$

$$\frac{\partial f}{\partial y} - \lambda \frac{\partial g}{\partial y} = 0 \tag{5.19}$$

$$\lambda g(x, y) = 0 \tag{5.20}$$

$$g(x, y) \leq 0. \tag{5.21}$$

These conditions are known as the Kuhn-Tucker conditions. They apply for one constraint, involving both domain variables in the three-dimensional case, and they may be generalized to n dimensions in relation to a constraint involving n variables and inequality constraints. Strictly speaking the Lagrangian expression applies only to constraints expressed in equality form (but when modified in conjunction with Table 5.1 or the procedure suggested in Section 5.6 it may be used for inequality constraints also).

The formal test for a maximum or minimum at the critical point (x_0, y_0) for the Lagrangian expression is modified only slightly from that for an unconstrained function f.

Rule 5.5 A maximum (minimum) for a function f subject to a constraint: Given a function f with rule $z = f(x, y)$, differentiable in a neighborhood of (x_0, y_0) and subject to the constraint $g(x, y) = 0$, where $\lambda g(x, y) = 0$, a relative maximum (minimum) occurs for the Lagrangian expression,

$$G_\lambda = f(x, y) - \lambda g(x, y)$$

at the point (x_0, y_0) if several conditions are met:

$$\frac{\partial G_\lambda}{\partial x} = 0, \quad \frac{\partial G_\lambda}{\partial y} = 0 \quad (at\ x_0, y_0) \tag{5.22}$$

$$\left[\frac{\partial^2 G_\lambda}{\partial x^2}\right]\left[\frac{\partial^2 G_\lambda}{\partial y^2}\right] - \left[\frac{\partial^2 G_\lambda}{\partial x\ \partial y}\right]^2 > 0 \quad (at\ x_0, y_0). \tag{5.23}$$

Where these conditions are met, then

$$G_\lambda\ is\ a\ maximum\ when\ \frac{\partial^2 G_\lambda}{\partial x^2} < 0, \quad \frac{\partial^2 G_\lambda}{\partial y^2} < 0 \quad at\ (x_0, y_0) \tag{5.24}$$

$$G_\lambda\ is\ a\ minimum\ when\ \frac{\partial^2 G_\lambda}{\partial x^2} > 0, \quad \frac{\partial^2 G_\lambda}{\partial y^2} > 0 \quad at\ (x_0, y_0). \tag{5.25}$$

When the expression in the second condition (5.23) is equal to or less than zero, the test fails, and it is necessary to investigate the function directly near (x_0, y_0).

The method of Lagrange multipliers can be extended to n dimensions for constraints stated as equalities; that is, $f(x_1, x_2, \ldots, x_n)$ subject to k constraints $g_i(x_1, x_2, \ldots, x_n) = 0$, $i = 1, 2, \ldots, k$ where $k \leq n$. In such cases

$$G_\lambda(x_1, \ldots, x_n; \lambda_1, \ldots, \lambda_k) = f(x_1, \ldots, x_n) - \sum_{i=1}^{k} \lambda_i g_i(x_1, \ldots, x_n). \tag{5.26}$$

Sec. 5.9]   SUMMARY   153

The solution to the equations for partial derivatives involves $n + k$ equations in $n + k$ unknowns. It is also possible to extend the test for maxima and minima to further dimensions, but this is beyond the scope of this volume.

The method of Lagrange multipliers is very useful in applications in management and economics as suggested in Appendix 5A.

5.9 Summary

Functions with more than one domain variable can be differentiated by procedures very similar to those for the differentiation of functions with a single domain variable. For functions of more than one variable differentiation is carried out with respect to each domain variable by treating all other domain variables as constants and proceeding by the rules of Chapter 3. When each partial derivative is set equal to zero, the result is a system of equations for *partial derivatives* with respect to each variable.

Partial derivatives of higher order may be taken also. If the first and later partial derivatives are all taken with respect to the same variable, the derivatives are known as straight partial derivatives; if the derivatives are taken with respect to more than one domain variable, they are known as cross-partial derivatives. For any given function the cross-partial derivatives of the same order with respect to each domain variable, that is, i times with respect to the first domain variable and j times with respect to the second (and so on for n variables) are identical no matter what the sequence in which the partial derivatives are taken. For example,

$$\frac{\partial^3 f}{\partial x \, \partial y \, \partial x} = \frac{\partial^3 f}{\partial x \, \partial x \, \partial y} = \frac{\partial^3 f}{\partial y \, \partial x \, \partial x}.$$

To find the interior extreme points for differentiable functions of two domain variables (extreme points that do not occur at the boundary of the region of definition), the following procedure is followed:

1. Find the rule for the first partial derivative of the function f with respect to each domain variable and set it equal to zero.
2. Solve the resulting system of equations simultaneously for the values of the domain variables at which the derivatives are simultaneously zero.
3. Substitute the values of the ordered pairs of domain variables into the rule for f to find the value z corresponding to each point identified in step 2.
4. Take the second straight partial derivatives with respect to each domain variable.
5. Take the second cross-partial derivative $\dfrac{\partial^2 f}{\partial x \, \partial y}$ for the function.
6. Make the three second-derivative tests and identify the critical points as maxima, minima, or saddle points for the ordered pairs of values for which the first partial derivatives are simultaneously zero. If

$$\left[\frac{\partial^2 f}{\partial x^2}\right]\left[\frac{\partial^2 f}{\partial y^2}\right] - \left[\frac{\partial^2 f}{\partial x \, \partial y}\right]^2 > 0 \qquad \text{at } (x_0, y_0) \qquad (5.2)$$

the function is either a maximum or a minimum at the point. If

$$\frac{\partial^2 f}{\partial x^2} < 0, \qquad \frac{\partial^2 f}{\partial y^2} < 0 \qquad \text{at } (x_0, y_0) \qquad (5.3\text{a})$$

the value of the function at the point is a maximum. If

$$\frac{\partial^2 f}{\partial x^2} > 0, \qquad \frac{\partial^2 f}{\partial y^2} > 0 \qquad \text{at } (x_0, y_0) \qquad (5.3\text{b})$$

the value of the function at the point is a minimum.

When the first partial derivatives are both zero at (x_0, y_0), the function is at a critical point. Condition (5.2) must be fulfilled in order for the critical point to be *either* a maximum or a minimum. However, if

$$\left[\frac{\partial^2 f}{\partial x^2}\right]\left[\frac{\partial^2 f}{\partial y^2}\right] - \left[\frac{\partial^2 f}{\partial x\, \partial y}\right]^2 < 0,$$

then the point is neither a maximum nor a minimum, it is a saddle point. If

$$\left[\frac{\partial^2 f}{\partial x^2}\right]\left[\frac{\partial^2 f}{\partial y^2}\right] - \left[\frac{\partial^2 f}{\partial x\, \partial y}\right]^2 = 0,$$

the test fails, there is no information about the function at (x_0, y_0), and the function must be investigated directly. These requirements can be generalized into n-space, but their statement is beyond the scope of this volume.

Partial differentiation can be used to differentiate a function implicitly. The rule for the derivative of an implicit function f can be stated

$$\frac{dy}{dx} = -\frac{\partial f/\partial x}{\partial f/\partial y}. \qquad (5.5)$$

The maximum and minimum values of a function can be found subject to constraints involving its domain variables. The following procedure is followed:

1. The maximum (minimum) value of the unconstrained function f is found by steps 1–6 discussed above.
2. The values of the domain variables at the maxima (minima) are substituted into all constraints to see if any constraint is violated.
3. If any constraint is violated, then the intersection of the function f with all possible combinations of constraints must be tested for maxima (minima). The maxima (minima) so obtained and for which no constraint is violated are then compared, and the largest (smallest) is the global maximum (minimum). It is wise to begin the comparison among the maxima (minima) for those constraints that do bind at step two.

A procedure using Lagrange multipliers for determining maxima and minima of a (unimodal) function subject to a single constraint is summarized as follows: The analyst must

1. Form the Lagrange expression subtracting the λ term multiplied by the implicit statement of the constraint (stated in the form $x - a = 0$).

APPENDIX 5A   155

2. Apply Rule 5.5.
3. Check Table 5.2 below to see if the constraint is binding.
4. If the constraint is binding, the maximum or minimum for the constrained function has been found. If the constraint is not binding, disregard it and apply Rule 5.3. (Steps 3 and 4 work for unimodal functions even if Rule 5.5 fails to determine whether the extreme point is a relative maximum or minimum.)

Table 5.2

Significance of the Values of λ for Maxima and Minima for a Unimodal Function Subject to a Single Constraint.[a]

	Constraint Type	Constraint Binding	Constraint Not Binding
Maximization problem	$x \leq a$	$\lambda > 0$	$\lambda \leq 0$
	$x \geq a$	$\lambda < 0$	$\lambda \geq 0$
Minimization problem	$x \leq a$	$\lambda < 0$	$\lambda > 0$
	$x \geq a$	$\lambda > 0$	$\lambda < 0$

[a] When $\lambda = 0$, the constraint passes through the optimal point itself.

Key Concept

Concept	Reference Section
Partial Derivative	5.1
Partial Differentiation	5.1
Straight Partial Derivatives	5.1
Cross-Partial Derivatives	5.1
Maxima and Minima	5.2
First Order Conditions	5.2
Sufficient Conditions	5.3
Saddle Points	5.3
Implicit Differentiation	5.4
Constrained Maxima and Minima	5.5
More than One Constraint	5.6
Lagrange Multipliers	5.7
Interpretation of λ	5.8
Kuhn-Tucker Conditions	5.8

Appendix 5A: Applications to the Production Function and to Simple Linear Regression

5A.1 The Production Function

The preceding discussion of optimization of functions of more than one variable can be transferred almost in its entirety to a discussion of the economist's production function

Figure 5A.1

(a) *Side-View Production Opportunity Surface with Contour Lines at Each Unit Level of Output in the Rational Output Region.* (b) *Plan View.*

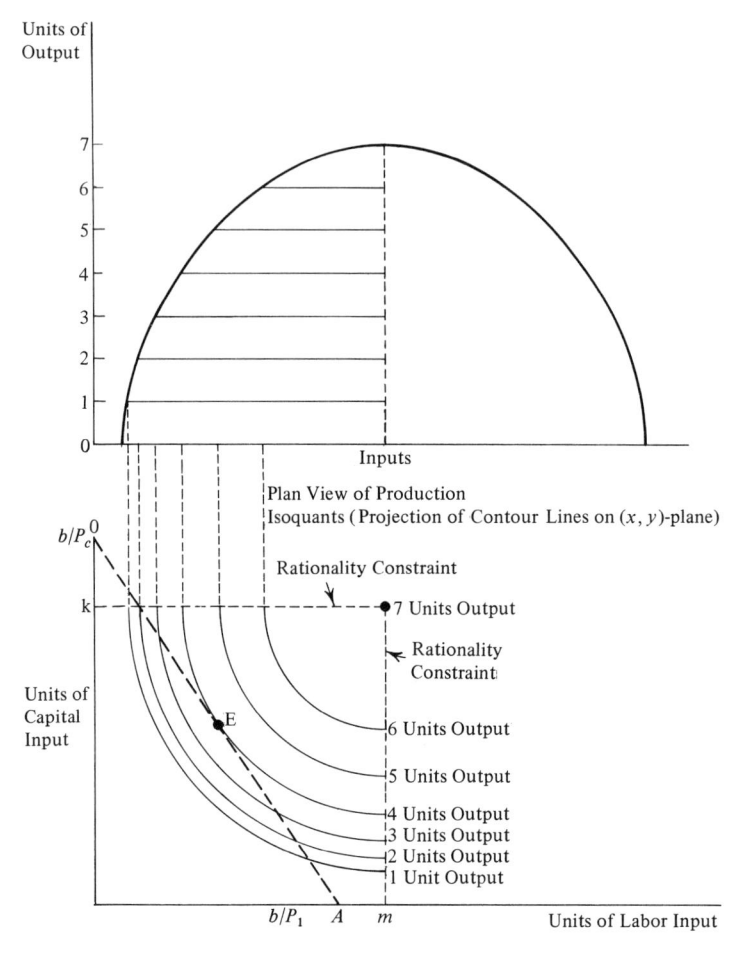

as suggested in the chapter. Given the function f, which expresses the output (O) of a firm in relation to its inputs, labor (ℓ) and capital (c), with rule

$$O = f(\ell, c) \tag{5A.1}$$

it is possible to determine the maximum level of unconstrained output, and output subject to particular constraints.

A function similar to the illustrative paraboloid (of Sections 5.5–5.8) with rule $z = f(x, y)$ can be readily transformed into a production function. The height in the z direction can represent the level of output O and the two domain variables x and y the inputs labor and capital, respectively. The first modification necessary is the introduction of constraints that might be termed "rationality constraints." It is assumed that the production surface—equivalent to the skin of an apple—is smooth (and, for this discussion, symmetrical).

The rationality constraints require that inefficient input combinations for a given level of output be eliminated from consideration in the production function. If it is possible to produce u units of output with $3\ell + 2c$ units of input, then it is irrational to pro-

APPENDIX 5A

duce u units with $3\ell + 3c$ or $4\ell + 2c$. Input combinations of this sort (portions of semicircles left out of Figure 5A.1) are eliminated from consideration, by the two lines representing the rationality constraints at $\ell = m$ and $c = k$. Thus, approximately (though exactly in the case shown) one fourth of the function remains to be considered further. The value of f is optimized subject to the rationality constraints; effectively their impact is noted through the elimination of many values of the basic inputs from further consideration. Another constraint that is crucial for analysis is the budget constraint, line OA. Also, in Figure 5A.1, contour lines representing different levels of output have been indicated.

Next, if a budget constraint is introduced into the problem, output can be maximized subject to the constraint of the amount of money available for the organization. The plan view of the function (Figure 5A.1b) showing the tracings of the contour lines for different levels of production from 1 to 7 units is the usual form of the function presented in textbooks on economics.

It is possible to define a Lagrangian expression for the constrained problem. The objective is to maximize output subject to the constraint of a limited budget. Given the production function f with rule,

$$O = f(\ell, c) \tag{5A.2}$$

maximize output subject to

$$b = p_\ell \ell + p_c c \tag{5A.3}$$

where p_ℓ is the price of labor and p_c the price of capital. Restating equation (5A.3) in its implicit form gives

$$p_\ell \ell + p_c c - b = 0 \tag{5A.4}$$

and including it in equation (5A.2) along with a multiplier λ gives:

$$G_\lambda = G_\lambda(\ell, c, \lambda) = f(\ell, c) - \lambda(p_\ell \ell + p_c c - b). \tag{5A.5}$$

Taking partial derivatives with respect to each variable and setting them equal to zero gives:

$$\frac{\partial G_\lambda}{\partial \ell} = \frac{\partial f}{\partial \ell} - \lambda p_\ell = 0 \tag{5A.6}$$

$$\frac{\partial G_\lambda}{\partial c} = \frac{\partial f}{\partial c} - \lambda p_c = 0 \tag{5A.7}$$

$$\frac{\partial G_\lambda}{\partial \lambda} = (p_\ell \ell + p_c c - b) = 0. \tag{5A.8}$$

Simplifying gives

$$\frac{\partial f}{\partial \ell} = \lambda p_\ell$$

$$\frac{\partial f}{\partial c} = \lambda p_c$$

$$0 = p_\ell \ell + p_c c - b.$$

The last equation is equation (5A.4) restated. Solving for λ gives

$$\frac{\partial f}{\partial \ell} \cdot \frac{1}{p_\ell} = \lambda$$

$$\frac{\partial f}{\partial c} \cdot \frac{1}{p_c} = \lambda.$$

Thus

$$\frac{\partial f}{\partial \ell} \cdot \frac{1}{p_\ell} = \frac{\partial f}{\partial c} \cdot \frac{1}{p_c}$$

or

$$\frac{\partial f / \partial \ell}{\partial f / \partial c} = \frac{p_\ell}{p_c}. \tag{5A.9}$$

When equation (5A.9) is solved simultaneously with equation (5A.4), the optimal value of output is found (subject to the verification of the second-order conditions). But $\partial f/\partial \ell$ is the marginal physical product (MPP) of labor MPP_ℓ (the rate of change of output from a small change in labor input), and $\partial f/\partial c$ is the marginal physical product of capital MPP_c. Thus the familiar relationship $\text{MPP}_\ell/\text{MPP}_c = p_\ell/p_c$, the ratio of the marginal physical products of labor to capital, must be equal to the ratio of their respective prices at an optimal point as derived from the necessary conditions for an optimum.

The maximum level of output occurs at the highest point on the intersection of the plane representing a budget of b dollars with the production surface. This is the point identified in the paragraph following equation (5A.9). It is also possible to find the maximum output point in a diagram such as that in the lower portion of Figure 5A.1. Here lines of equal-output quantities, *isoquants*, are shown for various combinations of inputs. The original three-dimensional relationships have been captured in this two-dimensional diagram.

For given prices of labor p_ℓ and of capital p_c, a fixed dollar budget b will purchase b/p_ℓ units of labor or b/p_c units of capital, or some (proportional) combination of the two. A budget plane is also shown in the plan view of Figure 5A.1 as a line for a budget of b dollars with intercepts on the ℓ axis of b/p_ℓ and on the c axis of b/p_c. The slope of this budget line is thus

$$-\frac{b/p_c}{b/p_\ell} = -\frac{p_\ell}{p_c}.$$

Given these prices for the inputs, all budget lines have the same slope and differ only in that lines representing smaller budget expenditures lie closer to the origin; those representing larger expenditures lie above and to the right of those of lesser amount. The highest output level achievable with the given budget b is that represented by the isoquant to which the budget line is just tangent. (If the line passed *through* a given isoquant, it would automatically provide access to the next higher isoquant.) Thus the slope of the isoquant is equal to the slope of the budget line at the point of highest output attainable under the given budget; in addition, the budget must be fully utilized at that point.

It is known from the intercepts of the budget line with the respective axes that the slope of that line is $-p_\ell/p_c$. The slope of the isoquants can be determined by implicit differentiation. The slope of the isoquants is $dc/d\ell$ the derivative of c with respect to ℓ, but the rule for f is stated as $O = f(c, \ell)$. By Rule 5.4, however,

$$\frac{dc}{d\ell} = -\frac{\partial f/\partial \ell}{\partial f/\partial c}.$$

Thus,

$$-\frac{p_\ell}{p_c} = -\frac{\partial f/\partial \ell}{\partial f/\partial c} \tag{5A.10}$$

and when equation (5A.10) holds at the same time that

$$p_c c + p_\ell \ell - b = 0,$$

the value of output is at a maximum. But it should be noted that these simultaneous equations are identical with the necessary conditions for an optimal solution in the three-dimensional case. Thus either approach may be used in the determination of the optimal level of output.

5A.2 Simple Linear Regression

As an application of partial differentiation, the mathematics required for fitting a straight line to a number of data points is discussed next. The procedure is known as

APPENDIX 5A

Figure 5A.2

Simple Regression with Line $\hat{y} = a + bx$ Fit to n Data Points.

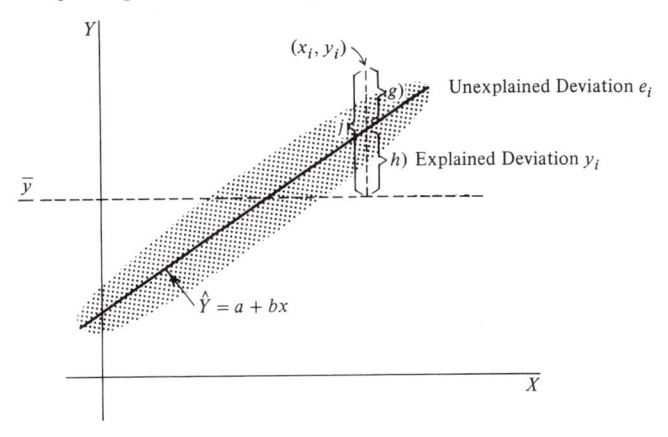

[Figure 5A.2: scatter plot with regression line $\hat{Y} = a + bx$, showing point (x_i, y_i), mean $\bar{y}$, with labeled deviations g) Unexplained Deviation e_i and h) Explained Deviation y_i]

simple linear regression, a technique that is explored in depth in a number of statistical texts.[15] The topic is introduced here only sufficiently to allow the mathematics to be understood.

Assume that n pairs of observed data are being used to estimate the relationship of a dependent variable Y to an independent variable X. The true relationship of Y to X is given by

$$y_i = \alpha + \beta x_i + \epsilon_i \ (i = 1, 2, \ldots, n) \tag{5A.11}$$

where ϵ_i is an "error term" that accounts for the deviation of individual observations y_i from the direct linear relationship expected from $\alpha + \beta x_i$. The criterion used in fitting a line to the data is that the line should be such that it minimizes the sum of the squares of the error terms remaining after the line has been fitted. The line is fitted by estimating the intercept and the slope of the true line. The equation for the line fitted is

$$\hat{y}_i = a + bx_i \tag{5A.12}$$

where a and b are the estimates of the true factors α and β. A graph of the fitting of the regression line may be helpful to an understanding of the subsequent discussion and is shown in Figure 5A.2. The total vertical deviation of a given point (x_i, y_i) from the mean $\bar{y}$ of the observed values of the dependent variable, a distance designated by j and equal to $(\bar{y} - y_i)$ consists of two parts as seen in Figure 5A.2:

g = that portion of the deviation not explained by the regression line; that is, $(y_i - \hat{y}_i)$.
h = that portion of the deviation explained by the regression line, that is, $(\hat{y}_i - \bar{y})$.

Because the sum of the y deviations around $\bar{y}$ must, by definition, equal zero, the sum of the deviations themselves cannot be used as a measure of dispersion; the dispersion around any line would sum to zero. However, by squaring the deviations the problems of working with the actual deviations are avoided. The method of least squares amounts to choosing a and b in the equation $\hat{y}_i = a + bx_i$ in such a way that the sum of the squares of the residuals $e_i = (y_i - \hat{y})$ are minimized.

[15] For example, see: Taro Yamane, *Statistics, an Introductory Analysis* (2nd ed.), New York: Harper & Row, 1967, Chapter 14; or Samuel B. Richmond, *Statistical Analysis* (2nd ed.), New York: Ronald Press, 1964, Chapter 19.

To achieve this, the minimum point for the function f must be found. This is done by choosing a and b appropriately for f, a function of a and b:

$$f(a, b) = \sum_{i=1}^{n} e_i^2 = \sum_{i=1}^{n} (y_i - \hat{y}_i)^2 = \sum_{i=1}^{n} [y_i - (a + bx_i)]^2. \quad (5A.13)$$

The necessary conditions for a minimum are obtained by setting the first partial derivatives of f with respect to a and b equal to zero as follows.[16]

$$\frac{\partial f}{\partial a} = \sum (-2[y_i - (a + bx_i)]) = 0 \quad (5A.14)$$

$$\frac{\partial f}{\partial b} = \sum (-2x_i[y_i - (a + bx_i)]) = 0 \quad (5A.15)$$

Simplifying equation (5A.14) gives

$$2\sum (-y_i + a + bx_i) = 0$$

or

$$-\sum y_i + \sum a + b\sum x_i = 0.$$

Carrying out the summations and solving for a gives

$$-n\bar{y} + na + bn\bar{x} = 0$$
$$a = \bar{y} - b\bar{x}. \quad (5A.16)$$

Simplifying equation (5A.15) gives

$$-\sum x_i y_i + \sum a x_i + b\sum x_i^2 = 0.$$

Substituting for a and solving explicitly for b:

$$-\sum x_i y_i + \sum x_i (\bar{y} - b\bar{x}) + b\sum x_i^2 = 0$$
$$-\sum x_i y_i + n\overline{xy} - bn\bar{x}^2 + b\sum x_i^2 = 0$$
$$-\sum x_i y_i + n\overline{xy} - b(n\bar{x}^2 - \sum x_i^2) = 0$$
$$\sum x_i y_i - n\overline{xy} - b(\sum x_i^2 - n\bar{x}^2) = 0$$

$$b = \frac{\sum x_i y_i - n\overline{xy}}{\sum x_i^2 - n\bar{x}^2}. \quad (5A.17)$$

Equations (5A.16) and (5A.17) are known as the normal equations for simple linear regression.

Problems

1. Find all *first* and *second* partial derivatives.

$$f(x, y, z) = ax + by + z$$

2. Find all *first* and *second* partial derivatives.

$$f(x, y, z) = ax^3 + \frac{1}{y} + x^2 z^4$$

---

[16] Assume that the sample consists of n pairs of values. The following are intended as review statements for reference:

1. $\sum a = na$
2. $\bar{y} = \frac{\sum y_i}{n}$
3. $n\bar{y} = \sum y_i$
4. $\sum (y_i - \bar{y}) = 0$,
5. $\sum c(y_i - \bar{y}) = 0$ (where c is a constant), but
6. $\sum y_i(y_i - \bar{y}) \neq 0$

PROBLEMS

3. Let $f(x, y, z) = 4x^2y + 3yz + xyz^2$. Find all *first* and *second* partial derivatives at the point $(x, y, z) = (1, 2, 3)$.

4. (a) Find the *first*, *second*, and *third* straight partial derivatives of
$$f(x, y) = (a^2x^2 - 2abx + b^2)(y^2 + 1), \quad a, b > 0,$$
(b) Does $f(x, y)$ have an inflexion point at $(x, y) = (0, 0)$?

5. Find all first- and second-order partial derivatives: $f(x, y) = \left(\dfrac{x}{y} - \dfrac{y}{x}\right) = 0$.

6. Find $\dfrac{\partial^2 f}{\partial x\, \partial y}$ and $\dfrac{\partial^2 f}{\partial y\, \partial x}$ at the point $(x, y) = (1, 1)$ of the function
$$f(x, y) = x^a y^b + x^c y^d = 0.$$

7. Given the function f with rule stated implicitly, $\dfrac{1}{w} = \dfrac{1}{x} + \dfrac{1}{y^2}$, solve explicitly for w, find $\partial w/\partial x$, $\partial w/\partial y$, and also dy/dx.

8. Find $\partial^2 w/\partial x^2$ of the function $w^4 = xy$ at the point $(x, y) = (1, 3)$.

9. Find $\partial w/\partial x$ and $\partial w/\partial y$ for the function $wx + x^2 + y^2 = 4$. Find dy/dx by implicit differentiation, using the method of partial derivatives.

10. For Problems 5 and 6 $\dfrac{dy}{dx}$ by implicit differentiation.

11. Find $\dfrac{dy}{dx}$ for the function f with rule $x^3 + 3x^2y + 3xy^2 + y^3 = 0$ using implicit differentiation.

12. Find $\dfrac{dy}{dx}$ for the function f with rule $x^4 + x^3y + xy^2 + y^3 + e^b = 0$ using implicit differentiation, where e and b are constants.

13. Use the implicit function rule for partial derivatives to find dy/dx, given $4x^2 + 5xy - 6y^2 - 3 = 0$.

14. Given the function with rule $x^3 + 3xy + y^2 - 3 = 0$
    (a) Find dy/dx by the differentiation method of partial derivatives.
    (b) Find dy/dx by the implicit differentiation method for ordinary (total) derivatives.

15. Given the function with rule $2x^3y + 5xy^2 + 7y^2 + 1 = 0$
    (a) Find $\dfrac{dy}{dx}$ by the differentiation method of partial derivatives.
    (b) Find $\dfrac{dy}{dx}$ by the implicit differentiation method for ordinary (total) derivatives.

16. Find maxima and minima of the function f with rule $z = (x - 2)^2 + (y - 3)^2 + 2$.

17. Find maxima and minima of the function f with rule $z = 5 - (x - 1)^2 - (y - 2)^2$.

18. Find maxima and minima of the function f with rule $z = 5 - (x - 1)^2 + (y - 2)^2$.

19. Find maxima and minima of the function f with rule $z = 5 - (x - 1)^2 - (y - 2)^2$
    (a) in the plane $y = 2$, and (b) in the plane $x = 1$.

20. Find maxima and minima of the function f with rule $z = 5 - (x - 1)^2 + (y - 2)^2$
    (a) in the plane $y = 2$, and (b) in the plane $x = 1$.

21. For f with rule $z = 5 - (x - 1)^2 + (y - 2)^2$, find $\dfrac{\partial z}{\partial x}$ and $\dfrac{\partial z}{\partial y}$ at the following points
    (a) $(x, y) = (0, 2)$
    (b) $(x, y) = (1, 1)$
    (c) $(x, y) = (2, 0)$.

22. Find maxima and minima of the function f with rule $z = 5 - (x - 1)^2 - (y - 2)^2 - xy$.

23. Find maxima and minima of the function f with rule $z = 5 - (x - 1)^2 - (y - 2)^2 - xy$
    (a) in the plane $y = 2$
    (b) in the plane $x = 0$
    (c) in the plane $x = 1$.

24. Given $f(x, y) = 3x^2 - 3x + y^2$, find the values of x and y at which $f(x, y)$ is an extremum, and state whether these are maxima or minima.

25. Let $f(x, y) = x^3 y - x^2 - xy^2 + 2$.
    (a) Find the expression for the slope of the function in the plane $y = 2$.
    (b) Does $f(x, y)$ have a maximum at the point $(x, y) = (1, 2)$?

26. Given the function with rule $z = xy$, find its maxima and minima if they exist.

27. The following is a production function for a company:

    $$Q = f(C, H, L, T) = 20C + 50H + 30L + 5T - \tfrac{1}{4}C^2 - H^2 - \tfrac{1}{3}L^2 - \tfrac{1}{10}T^2$$

    where Q represents the total output and C, H, L, and T are factor inputs.
    (a) Find the marginal productivity function for each of the factors of production.
    (b) Are there any diminishing marginal returns?
    (c) How do you know?
    (d) Compute the marginal products for each of the factors for $C = 20$, $H = 2.5$, $L = 45$, and $T = 40$.
    (e) If the company wishes to maximize output, should the number of units of any of the factor inputs be increased or decreased?
    (f) If the company wished to maximize output, what would be the optimal level of each factor input? What is the maximum output?

28. The two major inputs to the product made by a firm are capital (C) and Labor (L). The equation for the total cost of the output produced is:

    $$T = C^2 - 2C - CL + 2L^2 + 2 \quad \text{(in millions of dollars)}$$

    Find the amount of each input which leads to the lowest cost point for the firm and show that it is lowest.

29. An eighty-five year old farmer from Strawberry Point, Iowa, recently found that his profits per acre of corn planted on his 30-acre plot could be represented by the following production function:

    $$\text{Profit} = 10S + 50L + 30F - 10SF - S^2 - 5L^2 - 5F^2$$

    where S is the cost of the seed per acre, L is the labor cost, and F is the fertilizer expenditure per acre.
    (a) Determine the optimal level of each factor input.
    (b) What will his maximum profits per acre be?

30. The demand functions facing a monopolist for products x and y are:

    $$Q_x = 4 - P_x \quad \text{and} \quad Q_y = 6 - P_y$$

    and the joint cost function is $TC = Q_x^2 + Q_y^2$.
    (a) What are the profit maximizing price and production levels of x and y?
    (b) What are the maximum profits of x and y?

PROBLEMS

31. The following function represents the demand curve for product A:

$$Q = 268 - .1Y + .2P_b - 2P_a^2$$

where Q is the quantity demanded, Y is the consumer personal disposable income, P_b is the price of a related product, and P_a is the price of A.

(a) Compute the price elasticity of demand

$$\left[\frac{\partial Q}{\partial P_a} \cdot \frac{P_a}{Q} \right]$$

when $P_a = 5$, $P_b = 60$, and $Y = 300$.

(b) Compute the cross elasticity of demand for A with respect to P_b

$$\left[\frac{\partial Q}{\partial P_b} \cdot \frac{P_b}{Q} \right]$$

when $P_a = 5$, $P_b = 60$ and $Y = 300$.

(c) Compute the income elasticity of demand for A

$$\left[\frac{\partial Q}{\partial Y} \cdot \frac{Y}{Q} \right]$$

when $P_a = 5$, $P_b = 60$ and $Y = 300$.

(d) What is your intuitive feeling about the characteristics of product A when its income elasticity is considered? (That is, what happens to the demand for P when income changes? What type of product might be affected this way?)

(e) How is the demand for A affected by the prices of A and B? Would you say that A and B are complimentary or substitute products? Give an example of what product A might be.

32. Suppose a bank produces a single output (bank services) from four different input factors—deposits, equity funds, borrowed funds, and labor.

Let R = gross revenue in dollars
Q = quantity of output in physical units
P = price of output in dollars per unit
D = average amount of deposits in dollars
E = amount of equity capital
B = amount of borrowed funds
L = amount of labor in physical units

Let the bank's production function be $Q = f(D, E, B, L)$. Let the demand for the bank's services be $P = g(Q)$.

(a) Form the expression for gross revenue.
(b) Find $\partial R/\partial D$.
(c) State in words what $\partial R/\partial D$ represents.
(d) How would the expression for $\partial R/\partial D$ change if the market for the bank's services was perfect?
(e) Abstracting from handling costs, and so on, how much should the bank be willing to pay for deposits?

33. The demand function for snow shovels is

$$q_d = ap^{-b}J^c$$

where q_d is the rate of consumption, p is the price per unit, J is per capita income, and a, b, and c are positive constants. The supply function is

$$q_s = dp^k$$

where q_s is the rate of production, and d and k are positive constants. The market is in equilibrium when $q_s = q_d$, and p_e is the value of p in that condition. The above relations define p_e as a function of a, b, c, d, k, and J. Find dp/dJ.

34. Find the maximum of the function f with rule $z = 5 - (x - 1)^2 - (y - 2)^2$ subject to the constraint $x \leq -1$
    (a) by graphing the function and the constraint.
    (b) by the method of Lagrange multipliers.

35. Find the maximum of the function f with rule $z = 5 - (x - 1)^2 - (y - 2)^2$ subject to $x \geq 2$
    (a) by graphing the function and the constraint.
    (b) by the method of Lagrange multipliers.

36. Find the minimum of the function f with rule $z = 1 + (x - 2)^2 + y^2$ subject to the constraint $x \leq a$.
    (a) Show graphically where the minimum will occur for different values of a.
    (b) Verify the results by the method of Lagrange multipliers.

37. Find the maximum of the function f with rule $y = x^2 - 2x$ subject to $x - y \geq 0$
    (a) by the method of Lagrange multipliers.
    (b) by graphing the function and the constraint.

38. Find the maximum of $u = 2x^2 - 6y^2$ under the condition (or constraint) that $x + 2y = 4$. What is the value of u at the maximum?

39. Using the method of Lagrange multipliers, find the extreme values of $x^2 + 2xy + y^2$, subject to $x - y = 3$.

40. Optimize $z = 6x^2 - 2xy$, subject to the constraint $5x - 2y - 4 = 0$. Use Lagrangian multipliers.

41. If the cost equation in Problem 28 were subject to the following constraint

    $$C - 2L + 1 = 0$$

    what would be the amount of each input leading to the lowest cost? Show that your answer corresponds to a minimum.

42. Optimize $2x^2 + 6y - 4$ subject to the constraint $x + 3y - 5 = 0$. Use the method of Lagrange multipliers, and show that you found a maximum.

43. Given $z = x^2 - xy + y$ subject to the constraint that $x + y = 2$, find the extreme points of the function.

44. The production function for product A has been found to be represented by

    $$Q_A = 10M + 16N + 2MN - 2M^2 - N^2$$

    where Q_A is the production level of A, and M and N are production inputs. The labor inputs are twice as expensive for N as they are for M. In addition, the firm can spend only $55 on labor inputs. If M costs $1, find the maximum quantity that can be produced (use Lagrange multipliers).

45. (a) Maximize the function f with rule $z = 5x + 4y$ subject to $x \leq 2$, $y \leq 4$ using the method of Lagrange multipliers.
    (b) State the values of λ_1, λ_2, and z at the maximum point.
    (c) Relax the second constraint by one unit making it $y \leq 5$ and find the new maximum point with the rest of the problem unchanged. Compare the value of z from part (b) above with that just obtained. How does their difference compare with the value of λ_2 in part (b)?
    (d) Repeat part (c), this time tightening the second constraint by one unit making it $y \leq 3$. How does $z_b - z_d$ compare with λ_2?

46. Let $f(x, y) = (x - 3)(y - 3)$ be a person's indifference map for the goods x and y. Suppose his budget is $48 and the prices of x and y are $P_x = \$4$ and $P_y = \$4$. Find the optimal "basket" of x and y.

CHAPTER 6

Exponential and Logarithmic Functions

Exponential and logarithmic functions have many applications to problems in business and economics. Exponential functions are encountered in consumer demand studies, in studies of cost relationships, in projections of sales, and in computing present values at continuous interest rates. Logarithmic functions are frequently used in numerical

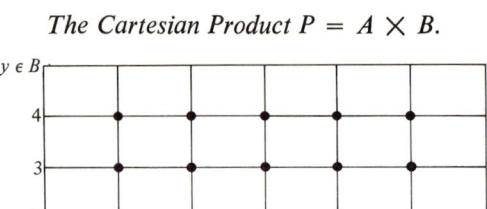

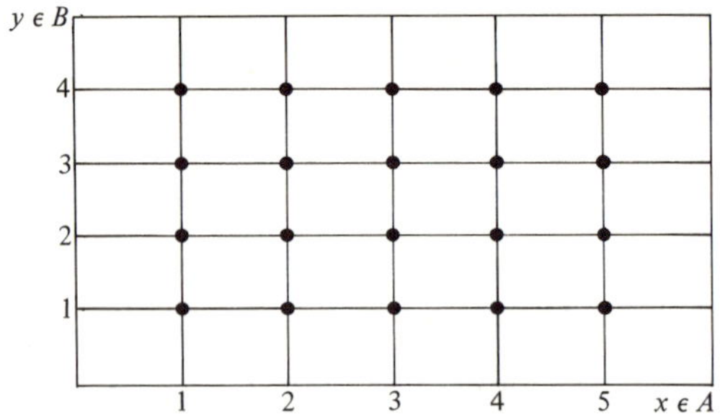

Figure 6.1

The Cartesian Product P = A × B.

calculations (to find cube roots and higher roots, for example) and to transform data into analytical forms that facilitate further study. Some of the many applications of these functions are illustrated later in this chapter and in the chapters to follow.

In this chapter the meaning of inverse functions is discussed in Section 6.1. The concept of inverse functions is explored in detail because it is key to an understanding of logarithmic functions, and to the development of the rules for derivatives of exponential functions from those for logarithmic functions, and, perhaps most important of all, because it is crucial to an understanding of the fundamental theorem of the calculus presented in Chapter 7. A basic derivation of a rule for the derivative of a simple logarithmic function is presented in Section 6.6 and the derivation of all other rules is dependent upon it. Theorem 6.1 on the relationship between rules for the derivatives of functions that are inverses of each other provides the link between the rules for the derivatives of exponential functions to those for logarithmic functions, and the chain rule permits basic rules to be extended to more complex cases. The use of exponential and logarithmic functions is illustrated in a number of examples in Sections 6.6 and 6.7.

In Appendix 6A, the characteristics of exponential and logarithmic functions are briefly reviewed, and the rules for manipulation of logarithms (the basis for the slide rule) are discussed. In Appendix 6B, the calculation of the numerical value of the irrational number e is illustrated.

6.1 Inverse Functions

Given the sets A and B, where $A = \{1, 2, 3, 4, 5\}$ and $B = \{1, 2, 3, 4\}$; the Cartesian product set $A \times B$ is designated by P, where P is the set of ordered pairs $\{(1, 1), (1, 2), \ldots, (5, 4)\}$ represented graphically by the circled intersections in Figure 6.1.

A particular function, call it Q, can be defined, consisting of the ordered pairs (x, y), which fit the rule $y = 2x$. For Q to meet the requirements for a function, each element of its domain must be mapped into one and only one element in the range, and its domain set must be exhausted by the mapping. Thus the domain

Figure 6.2

Elements of the Function Q Identified by Crosses in P.

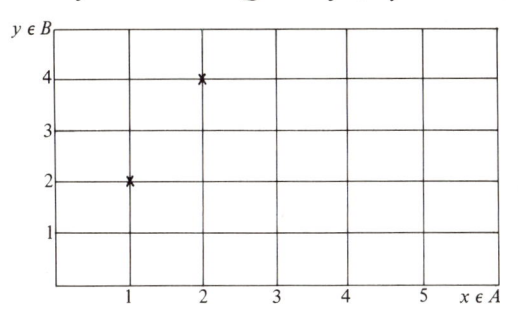

set must be restricted to a subset of A called $A_1 = \{1, 2\}$. This function can be stated in set notation as follows:

$$Q = \{(x, y): x \in A_1 = \{1, 2\}, y \in B, y = 2x\}$$

where Q is a subset of the product set P represented graphically in Figure 6.2. The function Q maps the element 1 in A to the element 2 in B and the element 2 in A to the element 4 in B. The elements 3, 4, and 5 in A have been excluded from the domain of Q, since there are no corresponding points in B under the pairing rule given. For example, the rule for the function requires that a value $2 \cdot 3 = 6$ in B be found to correspond to the element 3 in A, but 6 is not an element of B. The range set for Q is not exhausted; thus Q is an into function.[1]

The subset $B_1 = \{2, 4\}$ of the set B may be used as a range for a new function, call it Q_1, which is an onto function:

$$Q_1 = \{(x, y): y = 2x, x \in A_1, y \in B_1\}.$$

Figure 6.3

The Function Q for A_1 into B, the Function Q_1 for A_1 onto B_1.

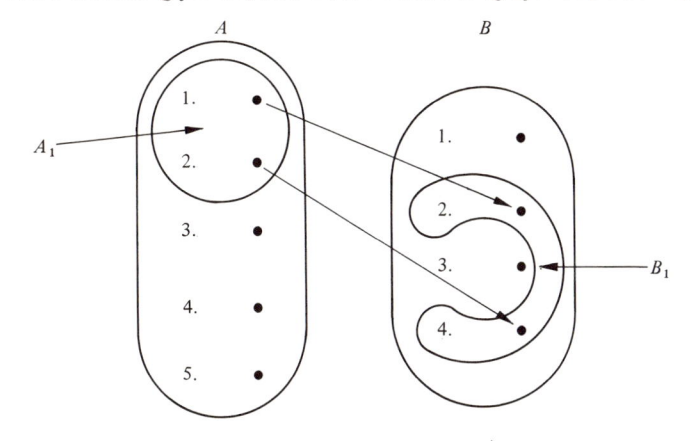

[1] Into and onto functions are defined and discussed in Chapter 1, Section 1.5.

The functions Q and Q_1 are illustrated in another way in Figure 6.3 in which the various sets are shown graphically.

The function Q_1 also meets the criteria for a one-to-one function (a subset of onto functions): each element in the domain is mapped to a distinct element in the range, and the range set is exhausted. Given a one-to-one function such as Q_1, it is possible to define a new function by reversing the direction of the mapping. The domain of the old function becomes the range for the new, and the range of the old becomes the domain for the new. This new function is called the *inverse* of the first function.

> **Definition—An Inverse Function:** *Given the one-to-one function f with domain set A and range set B with rule $y = f(x)$ where $x \in A$ and $y \in B$, then a new function g called the inverse of the function f can be defined with rule $x = g(y)$. The function g, the inverse of f, consists of the same set of ordered pairs as f, but with their order reversed.*

The inverse function of Q_1 (called Q_1 inverse) is a function whose domain is B_1 and whose range is A_1; it is a mapping from B_1 onto A_1, as illustrated in Figure 6.4.

Figure 6.4

The Function Q_1 inverse which maps from B_1 onto A_1.

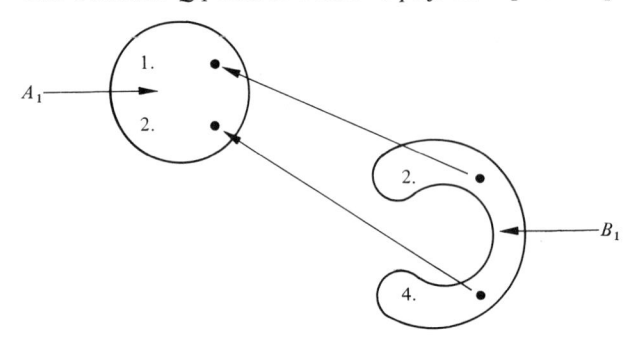

The function Q_1 maps element 1 in A_1 to the element 2 in B_1, while Q_1 inverse maps element 2 in B_1 to element 1 in A_1. Similarly Q_1 maps element 2 in A_1 to element 4 in B_1, while Q_1 inverse maps element 4 in B_1 to element 2 in A_1. The rule for the function Q_1 is $y = 2x$; the rule for the inverse function of Q_1 is $x = \frac{1}{2}y$. The new rule is obtained from the old by solving the equation $y = 2x$ explicitly for x in terms of y.

From the preceding discussion it should have been noted that the necessary and sufficient condition for the function f to have an inverse function g is that f be a one-to-one function. Only then is it possible to reverse the order of the pairs in f and still have the resulting set of ordered pairs constitute a function. Recall that, for a relation to be classed as a function,

1. The domain must be exhausted in the pairing,
2. Each element in the domain must map to one and only one element in the range.

Sec. 6.1]   INVERSE FUNCTIONS

Only in the case of a one-to-one function are these criteria met in both directions for the mapping.

Functions with the following rules have inverse functions, the rules for which are also indicated

Rule for f	Rule for f inverse
$y = a + x$	$x = y - a$
$y = x/b$	$x = by$
$y = cx$	$x = y/c$

The function whose rule is $y = x^2$, for example, does *not* have an inverse, since both $+2$ and -2 map to the range value 4; it is not a one-to-one function. However, if the domain set were limited to, say, the positive reals, then the function would be one-to-one, and it would be possible to define its inverse function. A frequently used symbol for f inverse is f^{-1}. Note that $(f^{-1})^{-1} = f$.

There is an interesting relationship between the derivative of a function f and the derivative of its inverse function f^{-1}, presented here as Rule 6.1 and illustrated earlier in Examples 5.7 and 5.8 of Chapter 5.

Rule 6.1: *If f and g are two inverse functions, whose rules are $y = f(x)$, and $x = g(y)$, then the value of the derivative of f at a point (x_0, y_0) is the reciprocal of the value of the derivative of g at that point, if the derivatives of both f and g exist at the point; that is, $dy/dx = 1/(dx/dy)$ at the point (x_0, y_0).*

Proof:

Given the rules for f and g,

$$y = f(x) \tag{6.1}$$

and

$$x = g(y). \tag{6.2}$$

Substituting equation (6.1) into equation (6.2) gives

$$x = g[f(x)]. \tag{6.3}$$

Differentiating equation (6.3) implicitly with respect to x, using the chain rule, gives

$$1 = \frac{dg}{dy} \cdot \frac{dy}{dx}, \tag{6.4}$$

provided both dg/dy and dy/dx exist. Because dg/dy is simply another way of writing dx/dy, dividing both sides by dx/dy yields

$$\frac{1}{dx/dy} = \frac{dy}{dx}, \quad \frac{dx}{dy} \neq 0 \tag{6.5}$$

which was to be proved. Division by dx/dy is valid only if $dx/dy \neq 0$. However,

Figure 6.5

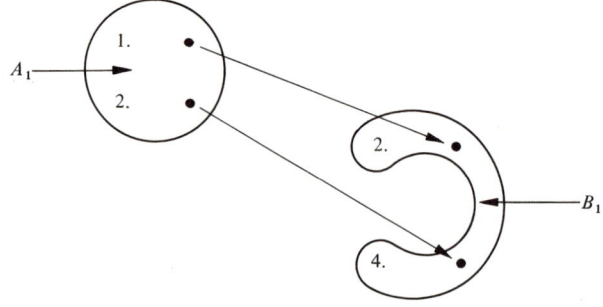

Function: M, A_1 onto B_1 by Rule $y = 2^x$.

if f and g are inverse functions, the derivative of one will equal zero at a particular point in its domain, only if at this point the derivative of the other function does not exist. Thus, if the conditions necessary for equation (6.4) hold, the condition for equation (6.5) will also hold.

6.2 Exponential Functions

If a rule such as $y = a^x$ where a is a specific constant, (say, 2) is now used for defining a function from set A_1 onto B_1, a new function is defined. The rule involves the domain variable x as an exponent, therefore the function, call it M, is known as an *exponential function*. For $a = 2$, a value of y in the range can be found for the given values of x in the domain. When $x = 1$, then $y = 2^1 = 2$; when $x = 2$, then $y = 2^2 = 4$. The *exponential function M* stated in set terms is

$$M = \{(x, y): x \in A_1, y \in B_1, y = 2^x\}.$$

It is graphed in Figure 6.5.

> **Definition—An Exponential Function:**[2] *A function f with domain variable x defined over the real numbers (or a subset thereof) and range variable y specified by the rule $y = a^x$, where a is a positive constant and x is the exponent (power) to which a is raised, is known as an exponential function.*

With x an element of A_1 and y an element of B_1, the function has the same domain and range as the previous function Q_1. As before, A_1 has been mapped onto B_1, but this time in accordance with the rule $y = 2^x$. It is again possible to map from B_1 onto A_1, and this mapping is the inverse of the function M. Recall that the rule for M is $y = 2^x$. But how can the pairing rule be solved for x in terms of y in this case to find the pairing rule for the inverse function? A rule is needed to relate these variables, that is, to specify a function that is the inverse of the exponential function. Fortunately, the values to be derived from this rule

---

[2] This definition holds for all real values of x, but this is stated without proof. It is useful to define an exponential function such that the base a must be greater than one. For a greater than one, y is positive for all values of x. This condition is adopted here.

Figure 6.6
Functions f_1 with Rule $y = 2x$ and f_2 with Rule $y = 2^x$.

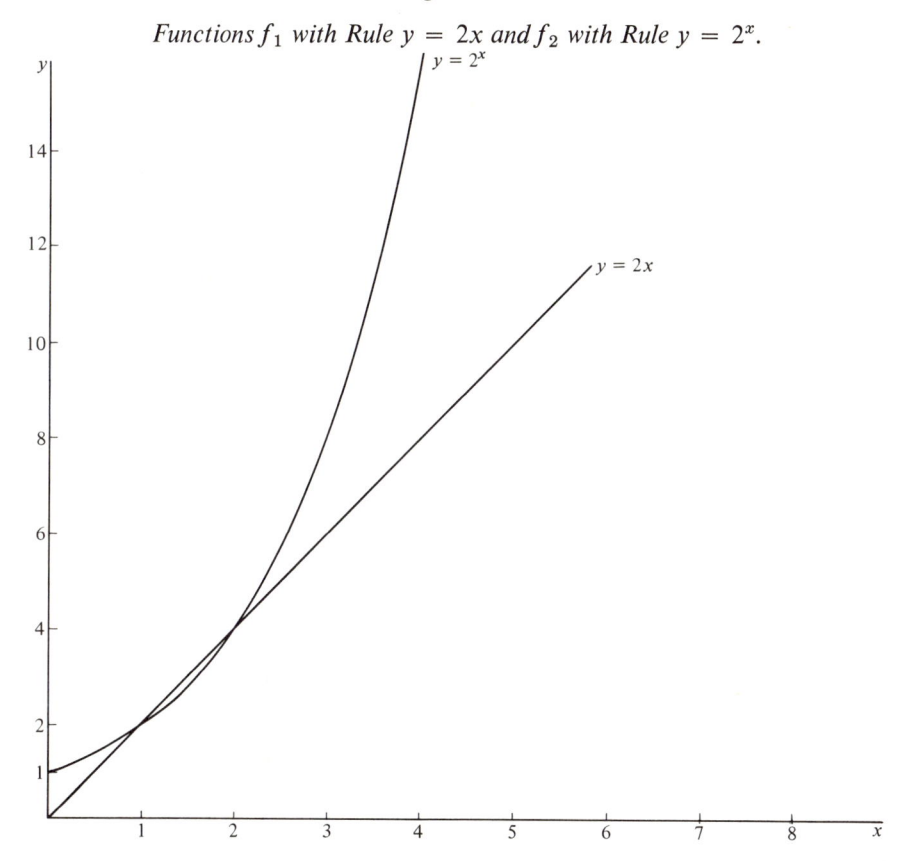

are known in advance. A rule of the form $x = g(y)$ is needed. The relationship that provides the solution to this problem is the *logarithm*, which is further discussed in Section 6.5. For the moment attention is concentrated on the exponential function itself.

Two pairs of inverse functions have been discussed, each mapping to and from the sets A_1 and B_1. These are now redefined over new sets, C and D, and E and H:[3]

$$C = \{x: x \in [0; 8]\} \qquad E = \{x: x \in [0; 4]\}$$
$$D = \{y: y \in [0; 16]\} \qquad H = \{y: y \in [1; 16]\}.$$

Under each of the earlier rules $y = 2x$ and $y = 2^x$, continuous functions mapping C onto D, and E onto H are defined; that is,

$$f_1 = \{(x, y): x \in C, y \in D, y = 2x\}$$
$$f_2 = \{(x, y): x \in E, y \in H, y = 2^x\}.$$

The graphs of the functions f_1 and f_2 are presented in Figure 6.6.

---

[3] These are read, for example: "C is the set of all elements x, such that x is a point (real number) in the closed interval between zero and eight."

172 EXPONENTIAL AND LOGARITHMIC FUNCTIONS [Ch. 6

The derivative function for f_1 is arrived at readily, but that for f_2 is more difficult to arrive at. Given the nature of an exponential function, the values of its derivative are not readily estimated intuitively. This makes the development of formal rules for derivatives of exponential functions even more useful than for polynomials. These statements are illustrated below.

The following relationships can be summarized for the simple linear function f_1 with rule $y = 2x$.

1. The average rate of change of y with respect to x is 2, that is, $\Delta y/\Delta x = 2$.
2. The instantaneous rate of change of y with respect to x is 2; that is, $\lim_{\Delta x \to 0} (\Delta y/\Delta x) = 2$.
3. The derivative of y with respect to x is 2. (By definition, it is equal to the instantaneous rate of change of y.)
4. The slope of the curve is 2 everywhere.

For the function f_2 with rule $y = 2^x$, similar relationships can be derived, but it is necessary to work through from the general definitions to see how they apply in this case. It is possible to carry out a series of calculations for the average rate of change for f_2 for various values of Δx_i, beginning at $x_0 = 1$, as summarized in Table 6.1; but the limit approached by the average rate of change is not obvious

Table 6.1

Data from Which Average Rate of Change of f with Rule $y = a^x$ Can Be Calculated for $x_0 = 1$, $a = 2$ and $f(x_0) = 2^1$ and Various Values of Δx_i.

i	Δx_i	$x_0 + \Delta x_i = 1 + \Delta x_i$	$f(x_0 + \Delta x_i) = 2^{(1+\Delta x_i)}$	$\Delta y = f(x_0 + \Delta x_i) - f(x_0) = (2^{(1+\Delta x_i)} - 2)$	Average Rate of Change $\Delta y_i/\Delta x_i$
1	3.0	4.0	16.000	14.000	4.667
2	2.5	3.5	12.726	10.726	4.288
3	2.0	3.0	9.000	7.000	3.500
4	1.5	2.5	5.656	3.656	2.650
5	1.0	2.0	4.000	2.000	2.000
6	0.5	1.5	2.828	0.828	1.637
7	−0.5	0.5	1.414	−0.586	1.172

from these data. All that can be said is that it is between 1.172 and 1.637. The values of the average rate of change of the function with rule $y = 2^x$ at $x_0 = 1$ for decreasing Δx_i, does not lead to an obvious value for its limit (the instantaneous rate of change); its value can be found formally by use of the definition of a derivative, however.

The derivative of a function at a point x_0 is given by

$$\frac{dy}{dx} = \lim_{\Delta x \to 0} \frac{f(x + \Delta x) - f(x)}{\Delta x}$$

which for the given function with rule $y = a^x$ is equal to

$$\frac{d}{dx}(a^x) = \lim_{\Delta x \to 0} \frac{a^{(x+\Delta x)} - a^x}{\Delta x}$$

since at the point at which the value of x is $x + \Delta x$, the value of the function $f(x + \Delta x)$ is equal to $a^{(x+\Delta x)}$, as shown also in the Table 6.1.

The point at which the derivative is evaluated has been specified to be $x = x_0 = 1$ where the value of a^x is a constant, that is, $a^1 = a$. The limit of a constant is that constant itself. Under the limit sign, there is only one variable, and that is Δx. To simplify the expression for this limit, a^x can be factored out in the numerator, and because it is a constant, it can be moved outside the limit sign;[4] thus,

$$\frac{d}{dx}(a^x) = \lim_{\Delta x \to 0} \frac{a^x(a^{\Delta x} - 1)}{\Delta x}$$
$$= a^x \lim_{\Delta x \to 0} \left(\frac{a^{\Delta x} - 1}{\Delta x}\right) \quad (6.6)$$

If a substitution is made

$$k = \lim_{\Delta x \to 0} \frac{a^{\Delta x} - 1}{\Delta x} \quad (6.7)$$

then

$$\frac{dy}{dx} = \frac{d}{dx}(a^x) = a^x(k). \quad (6.8)$$

This is a fairly simplified expression for the derivative of a^x. If there were an alternative way to express k, the expression might be simplified further. In fact this limit can be stated in several forms, as illustrated in the next three sections and highlighted in the proof of Rule 6.4 and then in Example 6.6 in Section 6.6. First the very special case in which $k = 1$ is discussed.

6.3 The Irrational Number *e*

If some value, say a_*, of a could be found for which k were equal to 1, then the result would be a startling relationship: the derivative d/dx of $(a_*)^x$, where a_* is equal to a particular value of a, would be equal to $(a_*)^x(1)$; that is,

$$\frac{d}{dx}(a_*)^x = (a_*)^x \quad \text{for all } x. \quad (6.9)$$

---

[4] It may be helpful to refresh the following characteristics of exponents:
$$x^a \cdot x^b = x^{a+b}$$
$$x^a \cdot x^a = (x^a)^2 = x^{2a}$$
$$(x^a)^n = x^{na}$$
$$x^a/x^b = x^{a-b}$$
$$x^a/x^a = x^{a-a} = x^0 = 1.$$
Some characteristics of exponential functions are reviewed further in Appendix 6A.

Figure 6.7

Function with Rule $y = e^x$.

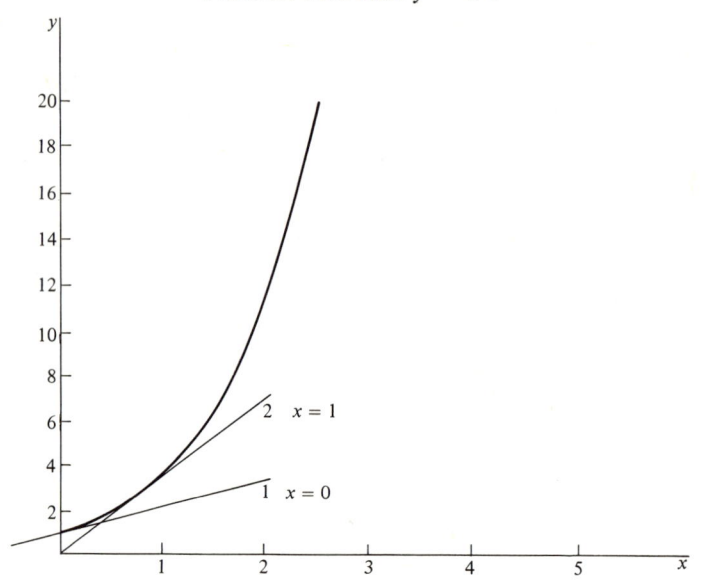

The value for a is sought for which $\lim\limits_{\Delta x \to 0} (a^{\Delta x} - 1)/\Delta x$ is equal to 1; that is,

$$\lim_{\Delta x \to 0} \frac{a^{\Delta x} - 1}{\Delta x} = 1.$$

In fact, this is the case. For a particular value of a, the value 2.71828..., an irrational number called e, the value of which is given by the limit [5]

$$e = \lim_{n \to \infty} \left(1 + \frac{1}{n}\right)^n, \tag{6.10}$$

the expression

$$\lim_{\Delta x \to 0} \frac{e^{\Delta x} - 1}{\Delta x}$$

reduces to 1.

When a is set equal to this irrational number e,

$$\frac{d}{dx}(e^x) = e^x \cdot (1) = e^x. \tag{6.11}$$

---

[5] The numerical estimation of the value for e for several values of n is shown in Appendix 6B. The estimate for $n = 2$ is

$$f(2) = (1 + \tfrac{1}{2})^2 = 2.25,$$

and for $n = 1000$ is

$$f(1000) = (1 + \tfrac{1}{1000})^{1000} = 2.718,$$

for example. The expression $\lim\limits_{n \to \infty} [1 + (1/n)]^n$ exists (although this is not proved here) and is defined to be the value of e.

The function and its derivative function are identical! The function f with rule

$$y = e^x,$$

has a derivative function with rule

$$y' = f'(x) = e^x \quad \text{for all } x.$$

This result is proved in Section 6.6 but in the intervening sections it is merely assumed.

The implications for the graph of the exponential function e^x and its derivative; that is, that

$$\frac{d}{dx}(e^x) = e^x \quad \text{for all } x,$$

is that the *slope* of the function at every point and the *value* of the function at that point are always equal! If this relationship can be shown to hold, then the intuitive reasonableness of this result can be established. The relationship in fact is shown in Figure 6.7 as plotted from a table of values of the number $e = 2.71828$ raised to various powers. In the illustration, the value of e is first raised to the first power.

Table 6.2
Values of e^x for Selected Values of x

x	e^x	x	e^x
0.00	1.00	2.00	7.39
0.70	2.00	2.20	9.03
1.00	2.72	2.30	9.97
1.10	3.00	2.50	12.18
1.38	3.97	2.80	16.44
1.60	4.95	3.00	20.09

If $x = 1$, then $f(x) = f(1) = e^1 = 2.71828$. A line drawn tangent to the curve at the point $f(1) = 2.71828$ must pass through the origin; the point $(1, 2.718^+)$ on the curve is 2.718^+ units from the origin in the y direction and one unit from the origin in the x direction. Because the *slope* of the line tangent to the curve is also equal to 2.718^+, it rises by this amount over a run of 1.0. Thus in this case the line from the origin to the point $(1, 2.718^+)$ on the curve and the tangent to the curve at that point are coincident. (The reader should verify this for line 2 in Figure 6.7, noting also the differences in scales on the x and the y axes.) If $x = 0$, a tangent to the curve at the point $y = e^0 = 1$ has the slope 1 as seen from line 1 in Figure 6.7; again note the scale differences for the axes. Thus for these two test points, the statement that the slope of the curve at a given point is equal to the value of the function at that point has been verified; the value of the first derivative and the value of the function are identical for the function with rule $y = e^x$. The reader should attempt to place tangents at other points for which the value of $f(x)$ may be read. If he does so, he will see that the slope of the tangent at that point is, in fact, equal to the value of $f(x)$ at that point.

The number e, like another irrational number, the familiar $\pi = 3.1416\ldots$ seen frequently in plane geometry, occurs in many different contexts in the literature of mathematics and the physical sciences. It proves to be very useful, in that it simplifies many mathematical operations and expressions, such as that for the derivative of an exponential function just discussed.

6.4 Rules for the Derivative Functions of Exponential Functions

In the discussion in Section 6.3, the rule for the derivative function f' of the exponential function f with rule $y = f(x) = e^x$ was implied. This rule is now formally stated although its proof is delayed to Section 6.6.

Rule 6.2: *Given the function f with rule $y = f(x) = e^x$, the rule for the derivative function of f is given by $y' = e^x$; that is,*

$$\frac{d}{dx}(e^x) = e^x. \tag{6.12}$$

This rule can be extended by use of the chain rule; several more complex exponential functions with rules such as $y = e^{g(x)}$ can also be differentiated. This is shown in the proof of Rule 6.3.

Rule 6.3: *Given the function f with rule $y = f(x) = e^{g(x)}$, the rule for the derivative function of f is given by $y' = e^{g(x)} \cdot g'(x)$; that is,*

$$\frac{d}{dx}(e^{g(x)}) = e^{g(x)} \cdot g'(x). \tag{6.13}$$

Proof:

Given $y = e^{g(x)}$, let $u = g(x)$.
Then, by the chain rule and Rule 6.2 (equation (6.12)):

$$\frac{dy}{dx} = \frac{dy}{du} \cdot \frac{du}{dx}$$

thus

$$\frac{d}{dx}(e^u) = e^u \cdot \frac{du}{dx}.$$

Substituting $g(x)$ for u and $g'(x)$ for du/dx gives

$$\frac{d}{dx} e^{g(x)} = e^{g(x)} \cdot g'(x). \tag{6.13}$$

Rule 6.3 can now be extended to the most general case in which the rule is sought for the derivative function for an exponential function with any constant

Sec. 6.4] DERIVATIVES OF EXPONENTIAL FUNCTIONS

a as base. This is readily accomplished once it is observed that any positive constant a can be equated to the constant e raised to an appropriate power. For example, using the data in Table 6.2, if a were equal to 3.00, e raised to the 1.10 power would equal 3.00. If a were equal to 2.00, e raised to the .70 power would equal 2.00. In general then, it can be said that

$$a = e^b;$$

for any real number $a > 1$, a real valued exponent b can be found such that $a = e^b$ and therefore by the rules for exponents,

$$a^x = (e^b)^x = e^{bx}. \tag{6.14}$$

The rule for the derivative of a function f with rule $y = a^x$ is now derived.

Rule 6.4: *Given the function f with rule $y = f(x) = a^x$, the rule for the derivative function of f is $y' = a^x \cdot b$; that is,*

$$\frac{d}{dx}(a^x) = a^x \cdot b, \tag{6.15}$$

where b is a real number such that $a = e^b$.

Proof:

The number a may be replaced by its equivalent e^b in the rule of the function f:

$$y = f(x) = a^x = e^{bx}$$

which is in the form

$$y = e^{g(x)}.$$

By Rule 6.3 (equation (6.13)),

$$y' = e^{g(x)} \cdot g'(x).$$

Since $g(x) = bx$, $g'(x) = b$; then substituting gives

$$y = e^{bx} \cdot b.$$

But $e^b = a$ and $e^{bx} = a^x$, so

$$y' = a^x \cdot b. \tag{6.15}$$

This rule is the same as that derived directly from the definition of the rule for a derivative function in the discussion in Section 6.3, except that the letter k was used instead of b. Rule 6.4 is the most useful of the results derived thus far. Once it is stated in its final form (as is done at the end of Section 6.5) the other rules for the derivatives of exponential functions follow from it immediately. For example, Rule 6.4 equation (6.15) is now extended by the chain rule for the case in which the rule for f is expressed as

$$y = a^{g(x)}.$$

Letting $u = g(x)$ and using the chain rule gives

$$\frac{dy}{dx} = \frac{dy}{du} \cdot \frac{du}{dx}.$$

The rule for the derivative function is found to be

$$\frac{d}{dx}(a^{g(x)}) = \frac{d}{dx}(a^u)\frac{du}{dx} = a^u \cdot b \cdot \frac{du}{dx}$$

but $u = g(x)$ and $du/dx = g'(x)$, so

$$\frac{d}{dx}(a^{g(x)}) = a^{g(x)} \cdot b \cdot g'(x) = a^{g(x)} \cdot g'(x) \cdot b. \qquad (6.16)$$

This is the most general case of the rule for the derivative function of an exponential function. Examples of the use of these differentiation rules are postponed until Section 6.6 following the discussion of the function that is the inverse of the exponential function, the logarithmic function.

6.5 Logarithmic Functions

A logarithm *is* an exponent. The equation $x = \log_2 4 = 2$ says, "The exponent to which the base 2 must be raised to get the number 4 is the exponent 2." The equation $x = \log_a y$ says, "x is the exponent to which the base a must be raised to get the number y." Such an equation can be the rule for a logarithmic function; a logarithmic function is the inverse of an exponential function and vice versa.

For the function M in the discussion in Section 6.2, two values were specified:

$$y_1 = 2^{(x_1=1)} = 2$$
$$y_2 = 2^{(x_2=2)} = 4.$$

The values for the function M inverse, the logarithmic function, are:

$$x_1 = \log_2 y_1 = \log_2 2 = 1$$
$$x_2 = \log_2 y_2 = \log_2 4 = 2.$$

The function that is the inverse of the exponential function is the logarithmic function.[6] The inverse mapping of the function M shown in Figure 6.8 must be accomplished by a logarithmic function. The function M is

$$M = \{(1, 2), (2, 4)\}$$

---

[6] Logarithms are exponents; thus several characteristics follow from those discussed for exponents in earlier footnotes:
$$\log c + \log d = \log (c \cdot d)$$
$$\log c - \log d = \log (c/d)$$
$$n(\log c) = \log (c^n).$$

The relationship between these facts and those presented in footnote 4 for exponents may be made more clear by the following illustration. Assume $\log_x c = a$ and $\log_x d = b$, then the equations for the inverses are $x^a = c$ and $x^b = d$. Further

$$c \cdot d = x^a \cdot x^b = x^{a+b}$$

or, taking the logarithms of both sides,

$$\log_x (c \cdot d) = \log_x (x^{a+b}) = a + b.$$

Substituting for a and b,

$$\log_x (c \cdot d) = \log_x c + \log_x d.$$

These relationships hold for *any* base for the logarithm, not just the base x given in the illustration.

Sec. 6.5]   *LOGARITHMIC FUNCTIONS*

Figure 6.8

Function M^{-1}, B_1 onto A_1 Whose Rule Is?

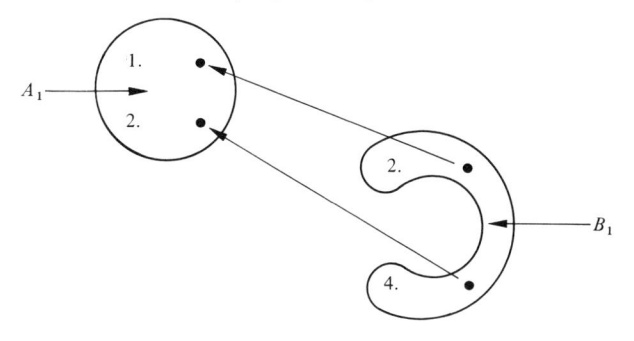

which represents the mapping from A_1 onto B_1 by the rule $y = 2^x$. The function M^{-1} is the set,

$$M^{-1} = \{(2, 1), (4, 2)\}.$$

The rule for the mapping of y onto x is expressed as

$$x = \log_2 y$$

which is read: "x is the logarithm to the base 2 of the number y"; or "x is the exponent to which the base 2 must be raised to give the number y."

When mapping from B_1 to A_1, one chooses a value of y first (it will be either 2 or 4); then the number x is the power to which the base 2 must be raised in order to obtain y. This x is the logarithm of y to the base 2. Thus if the exponential function whose rule is $y = 2^x$ maps A_1 onto B_1, then the logarithmic function, whose rule is $x = \log_2 y$ is the inverse of the exponential function and maps B_1 onto A_1. The formal definition follows.

> **Definition—Logarithmic Function:** *A function g with domain variable y over the reals (or a subset thereof) and range value x specified by the rule,*
>
> $$x = \log_a y, \tag{6.17}$$
>
> *where a is a positive constant greater than one, is known as a* logarithmic function. *The rule is read: "x equals the logarithm to the base a of the value y."*

A given logarithmic function is the inverse of an exponential function. The range value x of the logarithmic function is the exponent to which the given base a must be raised to reach the domain value y.

A few simple examples of the use of logarithms in calculations are now presented.

Example 6.1

To determine the logarithm of 1000 to the base 10, that is,

$$x = \log_{10} 1000,$$

the exponent to which 10 must be raised for the result to be equal to 1000 has to be found. Since $10^3 = 10 \cdot 10 \cdot 10 = 1000$, the answer is $x = 3$; thus

$$x = \log_{10} 1000 = 3.$$

Example 6.2

For $x = \log_3 9$, what is the exponent to which 3 must be raised in order to get 9? The answer is 2, since $3^2 = 3 \cdot 3 = 9$; therefore,

$$x = \log_3 9 = 2.$$

What is the logarithm of 27 to the base 3? The answer is 3. (Why?)

Example 6.3

Now consider a new use of a logarithm (that is, of an exponent). Take some number, such as 5.656, and find $\log_2 (5.656)$. The exponent to which the base 2

Figure 6.9

Functions f_3 with Rule $x = \tfrac{1}{2}y$ and f_4 with Rule $x = \log_2 y$.

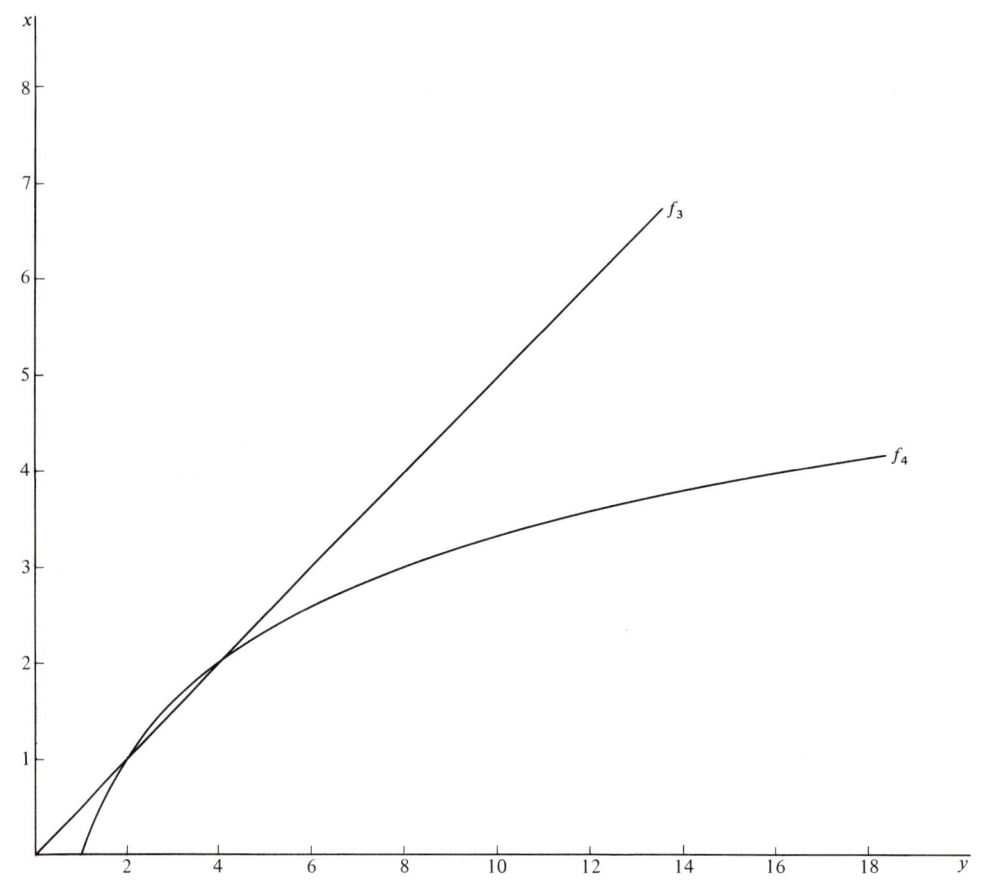

Sec. 6.5] LOGARITHMIC FUNCTIONS

must be raised in order to yield a value of 5.656 must be found. It is asserted that this value is approximately equal to 2.5. To test this, take the base 2 and raise it to the 2.5 power. This is equivalent to multiplying 2 by itself two and one-half times.

$$2^{2.5} = 2^1 \times 2^1 \times 2^{1/2}$$
$$= (2)(2)(1.414)$$
$$= 5.656.$$

That is, 2 raised to the 2.5 power means that 2 is multiplied by itself twice and by its square root once: $2^{2.5} = 2^1 \times 2^1 \times 2^{1/2}$. (Note that $2^{1/2}$ is the square root of 2, which is approximately equal to 1.414.) Thus $2^{2.5} = 5.656$; if the base 2 is taken and raised to the power 2.5, the result is approximately 5.656.

The procedure may be generalized: A convenient base can be chosen and raised to an appropriate power to approximate the value of any positive number. However, it must be noted that if the base is defined as positive, a negative value cannot be attained.[7]

In Section 6.2, the functions f_1 and f_2 were defined as follows:

$$f_1 = \{(x, y): x \in C, y \in D, y = 2x\}$$
$$f_2 = \{(x, y): x \in E, y \in H, y = 2^x\}$$

It is now possible to define functions f_3 and f_4, the inverse functions of f_1 and f_2, graphed in Figure 6.9. The functions f_3 and f_4 are:

$$f_3 = \{(y, x): y \in D, x \in C, x = \tfrac{1}{2}y\}$$
$$f_4 = \{(y, x): y \in H, x \in E, x = \log_2 y\}.$$

In Figure 6.9 the variable x is shown on the verticle axis and the variable y on the horizontal—since for these inverse functions the domain variable is y and the range variable x.[8]

Because f_4 is the inverse of f_2, the rule for the derivative function for either can be derived from that of the other by applying Rule 6.1. Given the rule of the derivative function of f_4, the rule for the derivative function of its inverse, that is, f_2 can be found since

$$\frac{dy}{dx} = \frac{1}{dx/dy}, \quad \frac{dx}{dy} \neq 0. \tag{6.5}$$

The rule for the derivative function of f_2 has been stated. Given the function f_2 with rule $y = 2^x$, the rule for the derivative function f_2' is given by equation (6.15):

$$y' = a^x \cdot b$$

or

$$y' = 2^x \cdot b.$$

---

[7] The reader who requires further refreshing on the meaning of exponential and logarithmic functions is referred to Appendix 6A.

[8] The values of the variables that make up the curve f_4 are identical to those that make up f_2; the only difference is in the order of the numbers in each pair. Thus, if Figure 6.6 were cut out along the boundary of f_2 and up the vertical axis and the piece of paper flipped over and fitted onto Figure 6.9, it would coincide with the area between the boundary of f_4 and the horizontal axis. (This should not be surprising since the vertical (y) axis of Figure 6.6 *is* the horizontal (y) axis of Figure 6.9.)

But this is a rather awkward expression, resulting from equating the general constant a to a power of the specific constant e; that is,
$$a = e^b.$$
The number b can not be evaluated directly in this form. But it is now possible to rewrite it in a form that can be evaluated. If the *inverse* of the above equation is written, as is now possible, then
$$b = \log_e a$$
and the value for b can now be read from a table of logarithms for the base e.[9] Thus the rule for the derivative function f'_2 can be rewritten
$$y' = 2^x \cdot \log_e 2$$
by substituting the number $\log_e 2$ for the number b. The reader should make similar substitutions in each of the equations (6.12)–(6.16). Equation (6.15) for Rule 6.4 is rewritten

$$\frac{d}{dx}(a^x) = a^x \log_e a. \tag{6.15}$$

The extension of this key rule by the chain rule in equation (6.16) is rewritten

$$\frac{d}{dx}(a^{g(x)}) = a^{g(x)} \cdot g'(x) \cdot \log_e a. \tag{6.16}$$

Equation (6.12) can also be developed from equation (6.15) by substituting the number e for the number a:

$$\frac{d}{dx}(e^x) = e^x \log_e e = e^x(1) = e^x, \tag{6.12}$$

since $\log_e e$ is 1 ($e^1 = e$). Similarly equation (6.13) can be developed from the extension of (6.15), the equation (6.16):

$$\frac{d}{dx}(e^{g(x)}) = e^{g(x)} \cdot g'(x) \cdot \log_e e = e^{g(x)} \cdot g'(x). \tag{6.13}$$

The substitution is made in these rules for the remainder of this chapter.

By Rule 6.1, the rule for the derivative of the logarithmic function $x = \log_2 y$ (where $y = 2^x$) is

$$\frac{dx}{dy} = \frac{1}{dy/dx} = \frac{1}{2^x \cdot \log_e 2} = \frac{1}{y}\left(\frac{1}{\log_e 2}\right).$$

The number $1/\log_e 2$ can be rewritten $\log_2 e$;[10] thus the rule for the derivative of

---

[9] The logarithm of a number to the base e is called the "natural" or "naperian" logarithm and is frequently written "ln" rather than $\log_e$.

[10] This can be shown as follows: Let $y = \log_a b$, and $x = \log_b a$; then the inverses are also valid; that is,
$$a^y = b \qquad b^x = a.$$
Taking the yth root of both sides of $a^y = b$ gives
$$a = a^{y(1/y)} = b^{1/y}$$
or since a also equals b^x
$$b^x = b^{1/y}$$
that is $1/y = x$. Substituting for $1/y$ and x gives
$$\log_b a = \frac{1}{\log_a b}$$

Sec. 6.6] RULES FOR DERIVATIVES OF LOGARITHMIC FUNCTIONS 183

f_4 can be rewritten

$$x' = \frac{1}{y} \log_2 e \quad \text{or} \quad \frac{d}{dy}(\log_2 y) = \frac{1}{y} \log_2 e.$$

6.6 Rules for Derivatives of Logarithmic Functions

The preceding discussion forms the basis for deriving the rules for the derivative functions of logarithmic functions. The general form of the rule is now derived. For the general logarithmic function f with rule $y = \log_a x$ the rule for the derivative function f' (with domain variable now x and range variable y) is

$$y' = \frac{1}{x} \log_a e.$$

After this rule is formally derived, then it is used along with Rule 6.1 to verify Rule 6.2 on which the derivatives of the exponential functions are based.

Rule 6.5: *Given the logarithmic function f with rule $y = \log_a x$, where a is a positive constant the rule for its derivative function f' is given by*

$$\frac{dy}{dx} = y' = \frac{1}{x} \cdot \log_a e. \tag{6.18}$$

Proof:

Given the function f with rule $y = \log_a x$ determine the quotient $\Delta y / \Delta x$.

$$y + \Delta y = \log_a (x + \Delta x)$$
$$\Delta y = \log_a (x + \Delta x) - y.$$

Substituting for y gives,

$$\Delta y = \log_a (x + \Delta x) - \log_a x.$$

Combining terms under the rules for manipulating logarithms gives

$$\Delta y = \log_a \left(\frac{x + \Delta x}{x} \right);$$

dividing the term in brackets through by its denominator x gives

$$\Delta y = \log_a \left(1 + \frac{\Delta x}{x} \right).$$

If both sides of the equation are divided by Δx, this gives

$$\frac{\Delta y}{\Delta x} = \frac{1}{\Delta x} \log_a \left(1 + \frac{\Delta x}{x} \right).$$

The right-hand side is then restated as follows:

$$\frac{\Delta y}{\Delta x} = \frac{1}{\Delta x} \cdot \frac{x}{x} \log_a \left(1 + \frac{\Delta x / \Delta x}{x / \Delta x} \right)$$
$$= \frac{1}{x} \left(\frac{x}{\Delta x} \right) \log_a \left(1 + \frac{1}{x / \Delta x} \right).$$

By the rules for logarithms,

$$\frac{\Delta y}{\Delta x} = \frac{1}{x} \log_a \left(1 + \frac{1}{x/\Delta x}\right)^{x/\Delta x}.$$

The term $x/\Delta x$ can be replaced by n

$$\frac{\Delta y}{\Delta x} = \frac{1}{x} \log_a \left(1 + \frac{1}{n}\right)^n$$

and as $\Delta x \to 0$, it can be seen that $n \to \infty$. The rule for the derivative function is by definition the limit of this expression

$$\lim_{\Delta x \to 0} \frac{\Delta y}{\Delta x} = \frac{1}{x} \lim_{n \to \infty} \log_a \left(1 + \frac{1}{n}\right)^n$$

or restated slightly

$$\lim_{\Delta x \to 0} \frac{\Delta y}{\Delta x} = \frac{1}{x} \log_a \lim_{n \to \infty} \left(1 + \frac{1}{n}\right)^n.$$

But $\lim_{n \to \infty} [1 + (1/n)]^n \equiv e$, so this reduces to

$$\frac{dy}{dx} = \frac{1}{x} \log_a e. \tag{6.18}$$

Rule 6.5 is used to develop another rule as shown in Example 6.4 and may be extended by the chain rule as shown in Examples 6.5 and 6.6. Rule 6.5 is thus a key rule for logarithmic functions just as Rule 6.4 was for exponential functions.

Example 6.4

Given the function f with rule

$$y = \log_e (x)$$

find the rule for its derivative function f'. By Rule 6.5

$$y' = \frac{1}{x} \cdot \log_e e = \frac{1}{x}. \tag{6.19}$$

Example 6.5

Given a function f with rule

$$y = \log_a (g(x))$$

find the rule for its derivative function. Let $u = g(x)$; thus $y = \log_a (u)$. Then by the chain rule and Rule 6.5,

$$\frac{dy}{dx} = \frac{dy}{du} \cdot \frac{du}{dx}$$

$$\frac{dy}{dx} = \frac{1}{u} \cdot \log_a e \cdot \frac{du}{dx}$$

Sec. 6.6]   RULES FOR DERIVATIVES OF LOGARITHMIC FUNCTIONS

but $du/dx = g'(x)$, and $u = g(x)$; therefore

$$\frac{dy}{dx} = \frac{1}{g(x)} \cdot \log_a e \cdot g'(x)$$

$$\frac{dy}{dx} = \frac{g'(x)}{g(x)} \cdot \log_a e. \qquad (6.20)$$

Example 6.6

Given the function f with rule

$$y = \log_e \big(g(x)\big)$$

find the rule for its derivative function f'. Let $u = g(x)$; then by the chain rule and Rule 6.5,

$$y' = \frac{dy}{dx} = \frac{dy}{du} \cdot \frac{du}{dx}$$

$$y' = \frac{d}{du}(\log_e u) \cdot \frac{du}{dx}$$

$$y' = \frac{1}{u} \log_e e \cdot \frac{du}{dx}.$$

But $du/dx = g'(x)$, $\log_e e = 1$ and $u = g(x)$; therefore

$$y' = \frac{g'(x)}{g(x)} \cdot (1) = \frac{g'(x)}{g(x)}. \qquad (6.21)$$

Now to verify Rule 6.4 for the derivative of the function f with rule $y = a^x$, Rule 6.1 is used. Since the logarithmic function f^{-1} with rule $x = \log_a y$ is the inverse of the function f, the verification begins with the derivative of f^{-1}:

$$\frac{d}{dy}(x) = \frac{d}{dy}(\log_a y) = \frac{1}{y} \log_a e.$$

By Rule 6.1

$$\frac{dy}{dx} = \frac{1}{dx/dy}, \qquad \frac{dx}{dy} \neq 0;$$

or

$$\frac{dy}{dx} = \frac{d}{dx}(y) = \frac{d}{dx}(a^x) = \frac{1}{dx/dy} = \frac{1}{(1/y) \log_a e},$$

which simplifies to:

$$\frac{d}{dx}(a^x) = y \cdot \log_e a = a^x \cdot \log_e a \qquad (6.15)$$

confirming Rule 6.4. Also when $a = e$,

$$\frac{d}{dx}(e^x) = e^x \log_e e = e^x \qquad (6.12)$$

confirming Rule 6.2.

At the end of Section 6.2 and in Rule 6.4, the following results were given:

$$\frac{d}{dx}(a^x) = a^x \cdot \lim_{\Delta x \to 0} \left(\frac{a^{\Delta x} - 1}{\Delta x} \right) = a^x \cdot k,$$

and

$$\frac{d}{dx}(a^x) = a^x \cdot b.$$

It has just been shown that

$$\frac{d}{dx}(a^x) = a^x \log_e a.$$

These are different expressions for the same thing, thus:

$$\lim_{\Delta x \to 0} \frac{a^{\Delta x} - 1}{\Delta x} = \log_e a = b = k.$$

The mathematical procedures now available make possible the differentiation of expressions previously too difficult to differentiate as illustrated in Example 6.7.

Example 6.7

Given the function f with rule

$$y = x^x,$$

find the rule for the derivative function f'. Taking the $\log_e$ of both sides of the equation

$$\log_e y = \log_e (x^x) = x \log_e x$$

differentiating implicitly gives:

$$\frac{1}{y} \cdot \frac{dy}{dx} = x \left(\frac{1}{x} \right) + \log_e x = 1 + \log_e x.$$

Substituting for y and cross multiplying gives

$$\frac{dy}{dx} = x^x(1 + \log_e x).$$

Several further examples are now presented to illustrate the use of logarithmic and exponential functions. Practical applications requiring these functions are presented in Section 6.7.

In each of the following examples, Examples 6.8–6.15, the rule for a function f is given and it is necessary to find the rule for the derivative function f'.

Example 6.8

Given $y = 7^{(2x+2)}$, find y'. The rule

$$y = 7^{(2x+2)}$$

is in the form

$$y = a^{[g(x)]}$$

and by equation (6.16),
$$y' = a^{g(x)} \cdot g'(x) \cdot \ln a$$
so
$$= 7^{(2x+2)}(2) \ln 7 = 2(7^{(2x+2)}) \ln 7.$$

Example 6.9

Given $y = \log_5 (x^2 - 2x + 7)$, find y'. The rule
$$y = \log_5 (x^2 - 2x + 7)$$
is in the form
$$y = \log_a g(x)$$
and by equation (6.20)
$$y' = \frac{1}{g(x)} \cdot g'(x) \log_a e,$$
so
$$y' = \frac{1}{x^2 - 2x + 7} (2x - 2) \log_5 e.$$

Example 6.10

Given $y = \ln (x^4 + 3x^3 - 736)$, find y'. By equation (6.21)
$$y' = \frac{g'(x)}{g(x)} = \frac{4x^3 + 9x^2}{x^4 + 3x^3 - 736}.$$

Example 6.11

Given $z = 3e^x + 5 \log_2 y$, find $\partial z/\partial x$, $\partial x/\partial y$. Since
$$z = 3e^x + 5 \log_2 y,$$
$$\frac{\partial z}{\partial x} = 3e^x$$
$$\frac{\partial z}{\partial y} = 5\left(\frac{1}{y}\right) \log_2 e = \frac{5 \log_2 e}{y}.$$

Equations (6.12) and (6.20) are applied.

Example 6.12

Given $y = \frac{1}{2}(3)^{3x^2}$, find y'. Since
$$y = \tfrac{1}{2}(3)^{3x^2},$$
then by equation (6.16)
$$y' = \tfrac{1}{2}(3^{3x^2}) \cdot (6x) \cdot \ln 3$$
$$y' = 3x(3^{3x^2}) \ln 3.$$

Example 6.13

Given $y = \log_7 x^{13}$, find y'. Since

$$y = \log_7 x^{13},$$

then by equation (6.20)

$$y' = \frac{1}{x^{13}} \cdot 13x^{12} \cdot \log_7 e = \frac{13}{x} \cdot \log_7 e.$$

Example 6.14

Given $y = e^{-x} \ln x$, find y' and y''. Since

$$y = e^{-x} \ln x,$$

then

$$y' = e^{-x}\left(\frac{1}{x}\right) + \ln x(-e^{-x})$$
$$y' = \frac{e^{-x}}{x} - \ln x \cdot (e^{-x}) = e^{-x} \cdot x^{-1} - \ln x(e^{-x}).$$

The rules in equations (6.14) and (6.19) were applied. Then,

$$y'' = e^{-x}(-x^{-2}) + x^{-1}(-e^{-x}) - \left[\ln x(-e^{-x}) + e^{-x}\left(\frac{1}{x}\right)\right]$$
$$y'' = -e^{-x}(x^{-2} + x^{-1} - \ln x + x^{-1})$$
$$y'' = -e^{-x}(x^{-2} + 2x^{-1} - \ln x)$$

applying the same rules.

Example 6.15

Given $y = \log_6 \sqrt{x + 2}$, find y'. Since

$$y = \log_6 \sqrt{x + 2},$$

then by equation (6.20)

$$y' = \frac{\frac{1}{2}(x + 2)^{-1/2}}{(x + 2)^{1/2}} \log_6 e$$
$$y' = \frac{\log_6 e}{2(x + 2)}.$$

Tabulations of the rules for derivative functions of exponential and logarithmic functions are presented in Tables 6.3 to 6.6.

Sec. 6.6]   RULES FOR DERIVATIVES OF LOGARITHMIC FUNCTIONS   189

Table 6.3

Three-Part Rules for the Derivatives of Exponential Functions.

Rule for Function	Rule for Derivative		
	Part 1	Part 2	Part 3
$y = a^{g(x)}$	$a^{g(x)}$ ·	$g'(x)$ ·	$\log_e a$
$y = a^x$	a^x ·	$d/dx(x) = 1$ ·	$\log_e a$
$y = e^{g(x)}$	$e^{g(x)}$ ·	$g'(x)$ ·	$\log_e e = 1$
$y = e^x$	e^x ·	$d/dx(x) = 1$ ·	$\log_e e = 1$

Table 6.3 can be written more simply as Table 6.4.

Table 6.4

Rules Simplified.

Rule for Function	Rule for Derivative
$y = a^{g(x)}$	$\dfrac{dy}{dx} = a^{g(x)} \cdot g'(x) \cdot \log_e a$
$y = a^x$	$\dfrac{dy}{dx} = a^x \log_e a$
$y = e^{g(x)}$	$\dfrac{dy}{dx} = e^{g(x)} \cdot g'(x)$
$y = e^x$	$\dfrac{dy}{dx} = e^x$

Table 6.5

Three-Part Rules for the Derivatives of Logarithmic Functions.

Rule for Function	Rules for Derivative		
	Part 1	Part 2	Part 3
$y = \log_a [g(x)]$	$\dfrac{1}{g(x)}$ ·	$g'(x)$ ·	$\log_a e$
$y = \log_a x$	$\dfrac{1}{x}$ ·	$d/dx(x) = 1$ ·	$\log_a e$
$y = \log_e [g(x)]$	$\dfrac{1}{g(x)}$ ·	$g'(x)$ ·	$\log_e e = 1$
$y = \log_e x$	$\dfrac{1}{x}$ ·	$\dfrac{d}{dx}(x) = 1$ ·	$\log_e e = 1$

Table 6.5 can be written as Table 6.6.

Table 6.6

Rules Simplified.

Rule for Function	Rule for Derivative
$y = \log_a [g(x)]$	$\dfrac{g'(x)}{g(x)} \cdot \log_a e$
$y = \log_a x$	$\dfrac{1}{x} \log_a e$
$y = \log_e [g(x)]$	$\dfrac{g'(x)}{g(x)}$
$y = \log_e (x)$	$\dfrac{1}{x}$

6.7 Applications to Business and Economics

Two useful applications of logarithms are presented in this section; the first is an extension of the regression technique introduced in Appendix 5A (which the reader new to the topic should review before proceeding), and the second is the relationship of logarithmic functions to the economic concept of elasticity.[11] Additional applications are included in Chapter 7 after the procedures for integration are introduced.

Example 6.16

In order to determine the general form of its total-cost function, a company made cost estimates for a production rate q of 100, 200, 300, units, and so on, under its present scale of operation. The data appear in Table 6.7. From the graph of

Table 6.7

Total Cost at Various Levels of Output

q	C
100	300
200	400
300	500
400	400
500	450
600	600
700	600
800	750

---

[11] The symbol log in these applications is used consistently to imply $\log_e$.

Figure 6.10

Freehand Fit of Function f with Rule $C = \alpha e^{\beta q}$ to Data on Cost of Production.

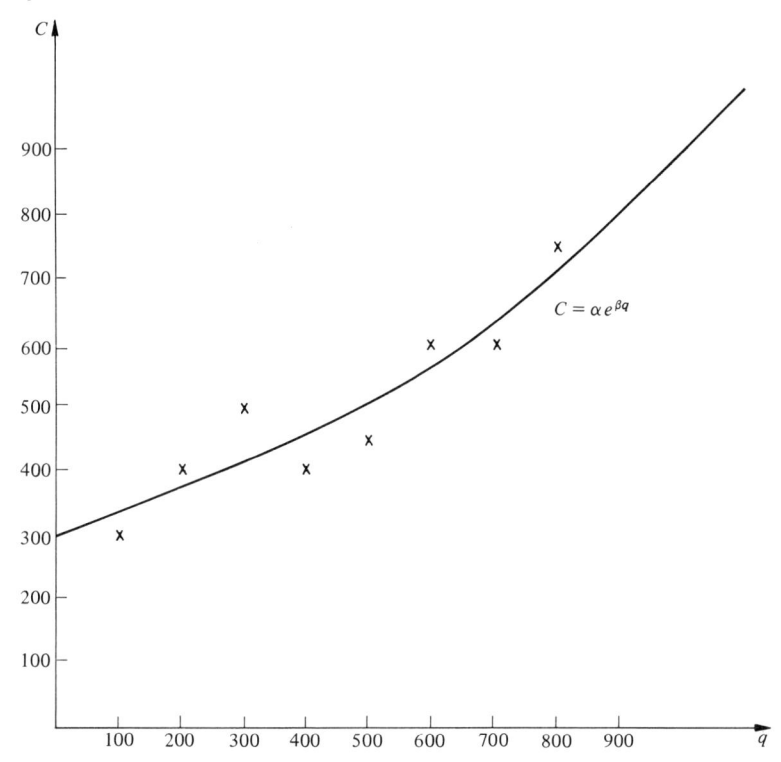

the data shown in Figure 6.10, it appeared that the rule of the cost function might be of the form $C = \alpha e^{\beta q}$ where C = total cost and α and β are constants. An attempt was made to fit this function to the data by means of *linear* regression.

To permit the use of the *linear* regression technique for a curvelinear function such as f, the function must be linearized in some way. This can be done by taking the logarithms of both sides of equation (6.22):

$$C = f(q) = \alpha e^{\beta q} \tag{6.22}$$
$$\log C = \log \alpha + \beta q \log e$$

and, since $\log e = 1$,

$$\log C = \log \alpha + \beta q.$$

The rule $f(q)$ is now in the form of an equation for a straight line.

In order to simplify the computations, the form $C/100 = (\alpha/100)e^{\beta q}$ and $\log (C/100) = \log (\alpha/100) + \beta q$ are used. Table 6.8 presents $C/100$ and $\log (C/100)$; the logarithms of the data are shown in Figure 6.11.

The linear regression technique can now be used to fit a straight line of the form $\log (C/100) = \log (\alpha/100) + \beta q$ to the transformed data. The calculations appear in Table 6.9.

Table 6.8
Logarithms of Costs Stated in Hundreds of Dollars

$\left(\dfrac{C}{100}\right)$	$\log\left(\dfrac{C}{100}\right)$
3	1.0986
4	1.3863
5	1.6094
4	1.3863
4.5	1.5041
6	1.7918
6	1.7918
7.5	2.0149

Table 6.9
Calculations of Data to be Used in Fitting Regression Line

q	q^2	$\log\left(\dfrac{C}{100}\right)$	$q \log\left(\dfrac{C}{100}\right)$
100	10,000	1.10	110
200	40,000	1.39	278
300	90,000	1.61	483
400	160,000	1.39	556
500	250,000	1.50	750
600	360,000	1.79	1074
700	490,000	1.79	1253
800	640,000	2.01	1608
3600	2,040,000	12.58	6112

Let b be the estimate of β:

$$\begin{aligned}
b &= \frac{n \sum q \log (C/100) - (\sum q)[\sum \log (C/100)]}{n \sum q^2 - (\sum q)^2} \\
&= \frac{8 \cdot (6112) - 3600(12.58)}{8 \cdot 2040000 - (3600)^2} \\
&= \frac{48896 - 45288}{16320000 - 12960000} \\
&= \frac{3608}{3360000} \\
&= .00107.
\end{aligned}$$

Sec. 6.7] APPLICATIONS TO BUSINESS AND ECONOMICS 193

Figure 6.11

Freehand Fit of Linear Function to Logarithms of Cost and Production Data.

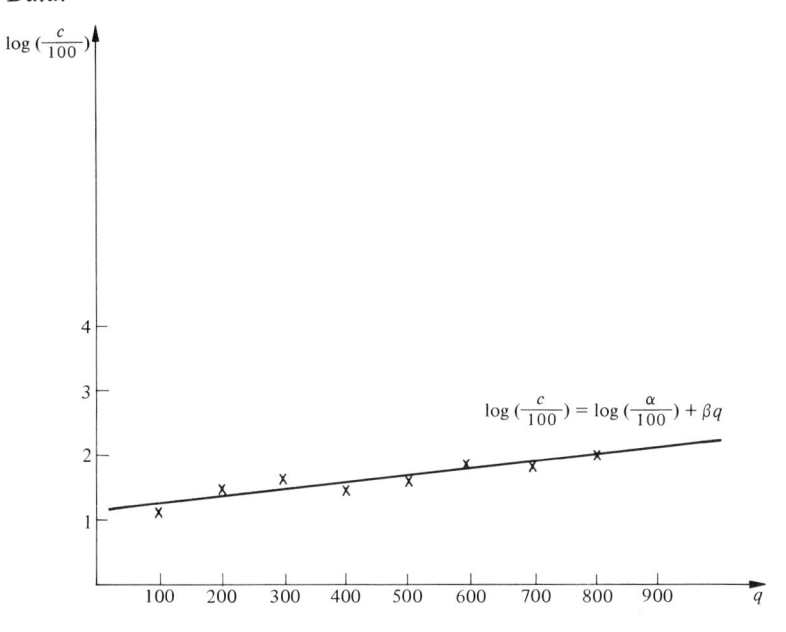

Let $\hat{\alpha}$ be the estimate of $\log(a/100)$

$$\begin{aligned}
\hat{\alpha} &= \overline{\log(C/100)} - b\bar{q} \\
&= \frac{\sum \log(C/100)}{n} - b\frac{\sum q}{n} \\
&= \frac{12.58}{8} - b\left(\frac{3600}{8}\right) \\
&= 1.5725 - .00107(450) \\
&= 1.5725 - .4815 \\
&= 1.0910.
\end{aligned}$$

Thus $\hat{\alpha} = \log(a/100)$ is estimated to be 1.091, and b to be .00107. Hence the estimate of the linear rule for the function is:

$$\log\left(\frac{C}{100}\right) = 1.091 + .00107q.$$

If $\hat{\alpha} = \log(a/100) = 1.091$, then $(a/100) = 2.98$ and $a = 298$. Hence the exponential function that fits the data has the form $C/100 = a/100(e^q)$

$$\frac{C}{100} = 2.98e^{.00107q},$$

or restating the result in units rather than in terms of hundreds of units:

$$C = 298e^{.00107q}.$$

The curvelinear function has been estimated by applying logarithms and solving for a and b by simple linear regression as if the function were linear.

Elasticity is a measure of the degree of responsiveness of the range variable for a function to changes in the domain variable; it measures the percentage change in the range variable for a given percentage change in the domain variable. A managerial use of logarithmic and exponential functions is illustrated by a discussion of the elasticity of a curve, specifically the price elasticity of a demand curve. Elasticity is an important concept for pricing decisions under conditions of imperfect competition; that is, when the firm faces a downward sloping demand curve. Given the differentiable function f with rule $q = f(p)$, where q denotes quantity and p price, then the price elasticity of demand is defined to be

$$\eta = \frac{dq/q}{dp/p} = \frac{p}{q} \cdot \frac{dq}{dp}. \qquad (6.23)$$

This equation defines the point elasticity η of the function f at the point (p_0, q_0). A function generally has different elasticity at different points. One can also measure arc elasticity, the elasticity over a certain interval of f. If dq/dp is approximated by $\Delta q/\Delta p$, where $\Delta q/\Delta p = (q_1 - q_0)/(p_1 - p_0)$, then equation (6.23) can be expressed as

$$\eta = \frac{\Delta q/q_0}{\Delta p/p_0}, \qquad \Delta q \text{ and } \Delta p \text{ small.} \qquad (6.24)$$

Expressed in words, this says that the price elasticity is the relative change in quantity demanded resulting from a small relative change in price. If $|\eta| > 1$, then demand is said to be elastic and if $|\eta| < 1$, inelastic.[12] When $|\eta| = 1$, then a price change will result in such a change in quantity demanded that gross revenue remains unchanged. When demand is elastic, a decrease in price leads to an increase in total revenue since the quantity demanded increases more than proportionally to the price change; a price increase decreases total revenue. If demand is inelastic, then a price increase results in a relatively smaller drop in demand so that gross revenue is increased; a price decrease would lead to a decrease in gross revenue. The situation is summarized in Table 6.10.

Table 6.10

The Relation of Price Changes, Gross Revenue, and Elasticity

	$|\eta| < 1$	$|\eta| = 1$	$|\eta| > 1$
Price increase	Gross revenue increase	Gross revenue same	Gross revenue decrease
Price decrease	Gross revenue decrease	Gross revenue same	Gross revenue increase

---

[12] Because a price decrease usually leads to an increase in the quantity demanded (and vice versa), the value of η is usually negative for a demand curve. A question of major interest is whether the change in quantity demanded is at least proportional to the change in price and attention is thus focused on the *absolute value* of η.

Sec. 6.7] APPLICATIONS TO BUSINESS AND ECONOMICS   195

Figure 6.12

The Demand Curve f with Its Associated Marginal Revenue Curve.

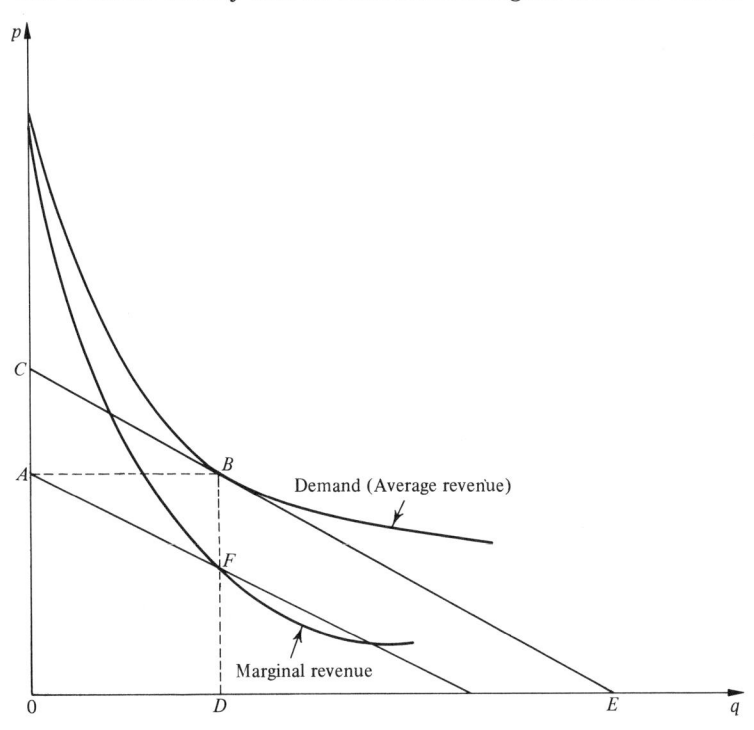

A graphic exposition is illuminating (Figure 6.12). By making use of the geometric relationships, the elasticity at the point B is obtained as

$$\eta = \frac{p}{q} \cdot \frac{dq}{dp} = \frac{OA}{AB} \cdot \frac{AB}{AC} = \frac{OA}{AC}.$$

Drawing AF parallel to CB gives $AC = FB$ (and $OA = BD$). Hence

$$\eta = \frac{OA}{AC} = \frac{BD}{FB} = \frac{p}{p-m},$$

if p denotes average revenue and m marginal revenue. Thus there exists an important relation between price, elasticity, and marginal revenue. The geometric relations are absolute magnitudes and will not reflect sign. In order to have η negative, it is therefore necessary to write

$$\eta = -\frac{p}{p-m}.$$

This expression can be rearranged as follows:

$$m = p\left(1 + \frac{1}{\eta}\right).$$

It is seen from this equation that for $-1 < \eta < 0$, marginal revenue is negative.

Thus the optimal price could not be in this range of the demand curve, but must be where $\eta < -1$[13]; the optimal price is on the elastic portion of the demand curve. (For $\eta = -1$, m equals zero.) For smaller and smaller η, that is, larger negative values or greater elasticity), m approaches p; that is, marginal revenue approaches price, since $1/\eta$ approaches 0. This happens for $q = 0$.

The characteristics of the price elasticity measure can be illustrated by investigating a linear, downward sloping demand function. Such functions are frequently stated with price as the dependent variable, for example, the function f with rule $p = 16 - q/2$. However, the definition of point elasticity in equation (6.23) is given for price as the independent variable:

$$\eta = \frac{p}{q} \cdot \frac{dq}{dp}, \tag{6.23}$$

that is, it is stated for f-inverse. The rule for f-inverse is $q = 32 - 2p$ and dq/dp is

$$\frac{d}{dp}(q) = \frac{d}{dp}(32 - 2p)$$

$$\frac{dq}{dp} = -2.$$

Although the *slope* of this linear inverse function is constant, the *elasticity* is not— the elasticity changes at various points with the ratio of p to q. The values of p and q as well as the ratio of p to q are tabulated in the following informal table.

p	16	14	12	10	8	6	4	2	0
q	0	4	8	12	16	20	24	28	0
p/q	—	$\frac{7}{2}$	$\frac{3}{2}$	$\frac{5}{6}$	$\frac{1}{2}$	$\frac{3}{10}$	$\frac{1}{6}$	$\frac{1}{14}$	0

Thus, the elasticity which is equal to $(p/q) \cdot (dq/dp)$ is given at each point by multiplying the bottom row of the above tabulation by -2 $(= dq/dp)$:

η	—	-7	-3	$-\frac{5}{3}$	-1	$-\frac{3}{5}$	$-\frac{1}{3}$	$-\frac{1}{7}$	0

At a price of $8 a quantity of 16 would be sold and the price elasticity of demand is -1. The function f-inverse is inelastic for prices below $8 and elastic for prices above that value.

The relationship between elasticity, marginal revenue and optimal price can also be illustrated for the function f-inverse. Total revenue r is equal to the quantity sold times the price at which it was sold:

$$r = p \cdot q$$

$$= \left(16 - \frac{q}{2}\right)q = 16q - \frac{q^2}{2}.$$

[13] This is the reason behind the claim that imperfect competition tends to be output limiting.

Sec. 6.7]   APPLICATIONS TO BUSINESS AND ECONOMICS

Marginal revenue is given by dr/dq:

$$\frac{dr}{dq} = \frac{d}{dq}\left(16q - \frac{q^2}{2}\right)$$
$$= 16 - q.$$

At the point $q = 16$ at which f-inverse changes from elastic to inelastic, that is when $\eta = -1$, marginal revenue is zero. For larger quantities sold (lower prices), it is negative, for smaller quantities sold (higher prices), it is positive. Note that this implies that *total revenue* would be reduced if price were to drop below \$8, and that for $p > 8$, total revenue can be increased by lowering price. Another way to reach this same conclusion is to find that price for which revenue is maximized

$$r = p \cdot q = p(32 - 2p).$$

Taking the first derivative with respect to price

$$\frac{d}{dp}r = \frac{d}{dp}(32p - 2p^2)$$
$$\frac{dr}{dp} = 32 - 4p = 0$$

and setting it equal to zero, gives the extreme point

$$32 = 4p$$
$$8 = p$$
$$16 = q.$$

Checking to confirm this as a maximum, the second derivative is taken

$$\frac{d^2r}{dp^2} = \frac{d}{dp}(32 - 4p)$$
$$= -4,$$

a negative value confirming the extreme point as a maximum point.

Thus, at the point of maximum total revenue ($p = 8, q = 16$), marginal revenue equals zero, and $\eta = -1$. For higher prices (lower quantity sold), marginal revenue is positive, and $\eta < -1$. For lower prices (larger quantity sold), marginal revenue is negative, and $\eta > -1$.

Example 6.17

The elasticity of a function can also be expressed in another way, involving what is known as the logarithmic derivative of the function and that of its inverse. The logarithmic derivative of f at a point is the ratio of the derivative f' to the function itself at that point

$$\frac{f'(p)}{f(p)} = \frac{dq/dp}{q} = \frac{1}{q} \cdot \frac{dq}{dp}, \tag{6.25}$$

which is seen to be $(d/dp)(\log q)$. This also explains the name "logarithmic derivative." Given the inverse function of f, that is, f^{-1}, with rule $p = f^{-1}(q)$, then the logarithmic derivative of f^{-1} equals

$$\frac{(f^{-1}(p))'}{f^{-1}(p)} = \frac{dp/dq}{p} = \frac{1}{p} \cdot \frac{dp}{dq}. \tag{6.26}$$

The elasticity of f can be defined as the ratio of the logarithmic derivative of f to that of f^{-1}, but with the derivative of f^{-1} also taken with respect to p (the domain variable of f).

$$\eta = \frac{(d/dp)(\log q)}{(d/dp)(\log p)} = \frac{(1/q) \cdot (dq/dp)}{(1/p) \cdot (dp/dp)}. \tag{6.27}$$

Clearing of fractions gives

$$\frac{p}{q} \cdot \frac{dq}{dp}$$

which is η by definition.

Example 6.18

Given the demand function f with rule

$$p = f(q) = ae^{-bq}, \quad a > 1, b > 0, \tag{6.28}$$

find the rule for determining the price elasticity of the curve at a point.

Since the equation for the price elasticity is given by $\eta = (p/q) \cdot (dq/dp)$, equation (6.23) is first solved for q in order to find dq/dp.[14] Taking the logarithm of both sides of equation (6.28) gives

$$\log p = \log a - bq \log e. \tag{6.29}$$

Hence

$$q = \frac{\log a - \log p}{b},$$

and

$$\frac{dq}{dp} = -\frac{1}{bp}.$$

Thus

$$\eta = \frac{p}{q} \cdot \left(-\frac{1}{bp}\right) = -\frac{1}{bq}. \tag{6.30}$$

Two things appear from equation (6.30). First, the elasticity is actually negative reflecting the basic relationship that a decrease in price leads to an increase in quantity; that is, the demand curve is downward sloping. Second, the elasticity is different at different points along the demand curve, η is smaller, the larger q.[15]

---

[14] This could be found by Rule 6.1, that is, from $dq/dp = 1/(dp/dq)$. However, to illustrate the use of logarithms, the alternate procedure is used in the text.

[15] It is worth noting that this is not always the case. For instance, if the demand function is $p^a q = c$, for $a, c > 0$ and $dq/dp = -acp^{-(a+1)}$, then

$$\eta = (p/q) \cdot (dq/dp) = (p/cp^{-a}) \cdot (-ac/p^{a+1}) = -a;$$

the elasticity is constant throughout the demand curve. For $a > 1$ it would be elastic, and for $a < 1$ inelastic.

Sec. 6.7] APPLICATIONS TO BUSINESS AND ECONOMICS 199

The same result is obtained by using the concept of elasticity as the quotient of logarithmic derivatives:

$$p = ae^{-bq}. \tag{6.28}$$

The log derivative of p by equation (6.26) is:

$$\frac{d}{dp} \log p = \frac{dp/dp}{p} = \frac{1}{p}.$$

To find $(d/dp) \log q$, take logarithms in the original function:

$$\log p = \log a - bq$$
$$q = \frac{\log a - \log p}{b} = \frac{1}{b} \log a - \frac{1}{b} \log p$$
$$\frac{dq}{dp} = -\frac{1}{bp}.$$

By equation (6.24):

$$\frac{d}{dp} \log q = \frac{dq/dp}{q} = \frac{-1/bp}{q} = -\frac{1}{bpq}$$

therefore,

$$\eta = \frac{d/dp \log q}{d/dp \log p} = \frac{-1/bpq}{1/p} = -\frac{1}{bq}.$$

For the case just discussed with $p = ac^{-bq}$, η is equal to $-1/bq$. This leads to the following expression for marginal revenue

$$m = p\left(1 - \frac{1}{1/bq}\right) = p(1 - bq) = ae^{-bq} - bqae^{-bq}.$$

This can be verified by taking the derivative of the gross revenue function:

$$r = pq = qae^{-bq}$$
$$\frac{dr}{dq} = ae^{-bq} + q(-abe^{-bq})$$
$$= ae^{-bq} - bqae^{-bq}.$$

To find the optimal price, and the price elasticity at a point, marginal revenue is set equal to marginal cost.[16] Assume that the rule for the cost function is

$$c = aqe^{-bq} - a^{-1}e^{bq} + aq$$
$$\frac{dc}{dq} = -abqe^{-bq} + ae^{-bq} - a^{-1}be^{bq} + a.$$

Hence

$$-abqe^{-bq} + ae^{-bq} - a^{-1}be^{bq} + a = ae^{-bq} - abqe^{-bq}.$$

[16] The functions are so chosen that second-order conditions are satisfied. (There is only one intersection.)

This simplifies to
$$a = a^{-1}be^{bq}$$
$$\frac{a^2}{b} = e^{bq}$$
$$2\log a - \log b = bq$$
$$q = \frac{2\log a - \log b}{b} = \frac{1}{b}\log\left(\frac{a^2}{b}\right).$$

This gives
$$p = ae^{-\log(a^2/b)}.$$

The elasticity then becomes
$$\eta = \frac{-1}{bq} = \frac{-1}{2\log a - \log b}$$
$$\eta = \frac{1}{\log b - 2\log a}.$$

It is known that this has to be < -1. It therefore follows that $-1 < \log b - 2\log a < 0$, that is, $-1 < \log(b/a^2) < 0$. As long as this is satisfied, and $(1/b)\log(a^2/b)$ gives a reasonable value for q, any a and b would specify possible demand and cost curves.

Another important elasticity concept is that of (price) cross elasticity, which shows the relation between the relative change in quantity demanded of one good (product) and the relative change in the price of *another* good (ceteris paribus), that is

$$\eta_{ij} = \frac{\frac{\Delta q_i}{q_i}}{\frac{\Delta p_j}{p_j}} = \frac{p_j}{q_i} \cdot \frac{\partial q_i}{\partial p_j} \qquad (6.31)$$

where i and j refer to two products, and p and q to price and quantity demanded.

Cross elasticity is thus a measure of the demand relationship between two products. Products are *substitute* products if the quantity of the first that is demanded *increases* with a *price increase* for the second (consumers substitute the first product for the second, which is now more costly). Two products are *complementary* products if the quantity of the first that is demanded *decreases* with a price increase in the other. (The price increase for the second product leads to fewer units of it being sold, and sales of the first product move in the same direction.) A traditional example of substitute products is butter and margarine, of complementary products, automobiles and tires.

To illustrate, suppose two products X and Y have demand functions with rules

$$p_x = 8 - x + 2y \qquad (6.32)$$
$$p_y = 6 - y \qquad (6.33)$$

where x, y denote the quantities of products X and Y, and p_x and p_y their respective prices. Note that demand for Y is independent of the demand for X, but that demand for X is influenced by the demand for Y. The cross-elasticity formula is

$$\eta_{xy} = \frac{p_y}{x} \cdot \frac{\partial x}{\partial p_y} \qquad (6.31)$$

To find $\partial x/\partial p_y$, equation (6.32) is rewritten as $x = 8 - p_x + 2y$. Equation (6.33) is rewritten as $y = 6 - p_y$ and substituted into (6.32) to obtain

$$\begin{aligned} x &= 8 - p_x + 2(6 - p_y) \\ &= 20 - p_x - 2p_y \end{aligned} \quad (6.34)$$

Then $\partial x/\partial p_y = -2$. Using the following illustrative values:

$$\begin{array}{cc} p_y = 2 & p_x = 10 \\ y = 4 & x = 6 \end{array}$$

then

$$\eta_{xy} = \frac{p_y}{x} \frac{\partial x}{\partial p_y} = \frac{2}{6}(-2) = -\frac{2}{3}.$$

Note that the (price) cross elasticity depends on p_y and x and therefore will be different for different values of these variables.

Because the cross elasticity is negative, Y is a complementary product to X, since an increase (decrease) in p_y will cause a decrease (increase) in x, the quantity of product X demanded. (For a substitute good the cross-price elasticity would have been positive.)

In the present example there are two ways to see that relative to X, Y is in fact complementary. Firstly, if $p_x = 8 - x + 2y$, then $x = 8 - p_x + 2y$, and consequently x, the quantity of X demanded, goes up with an increase in the quantity of Y demanded, i.e., the demand for both products moves together. An increase in the quantity of Y demanded results from a lowering of its price; the cross-price elasticity will be negative. Secondly, if $p_x = 8 - x + 2y$, we have

$$\begin{aligned} p_x &= 8 - x, & \text{if } y = 0 \\ &= 10 - x, & \text{if } y = 1 \\ &= 12 - x, & \text{if } y = 2 \end{aligned}$$

that is, the demand curve shifts upward for an increase in y. The slope remains the same but the intercept is higher. (Draw a sketch of this relationship.) Thus the greater the demand for Y the more is demanded of X.

Relative to Y, X is neither complementary nor a substitute as $\eta_{yx} = 0$, resulting from $\partial y/\partial p_x$ being zero. The demand for Y is unaffected by product X, as x does not appear in the demand equation for Y.

6.8 Summary

Given a function f, its inverse function f^{-1} maps the range set of f into the domain set of f. In order for this inverse mapping to meet the requirements of the definition of a function, the original function f must be one-to-one: each element in its domain set A must map to a distinct element in its range set B and both A and B must be exhausted by the mapping.

Given functions with the following equations as rules:

$$\begin{aligned} y &= x + 4 \\ y &= x - 4 \\ y &= 4x \\ y &= x/4 \end{aligned}$$

the rules for their inverse functions can be found by solving these equations for the current domain variable in terms of the range variable; that is,

$$x = y - 4$$
$$x = y + 4$$
$$x = y/4$$
$$x = 4y.$$

Also the inverse of f^{-1} is f; solving the rule for f^{-1} for y gives the rule for f.

The function f with rule $y = a^x$ is an exponential function in which the variable x over the domain set is used to raise the base a (a positive constant) to a power. To write the rule for f^{-1}, the inverse function of f, a new type of function, the logarithmic function must be defined.

The rule for a logarithmic function is expressed in terms of logarithms. A logarithm is an exponent. The inverse function of f requires that for each (given) value y, the value x be determined. Thus given the value $y = 8$ and assuming that $a = 2$, the appropriate value of x is sought. The rule for f, $y = a^x$, with the known values substituted becomes

$$8 = (2)^x.$$

The question that must be answered for the rule for f^{-1} is "what exponent is required to raise the base 2 to the value 8?" Stated in terms of logarithms the question becomes: "What is the logarithm to the base two of the number 8?" The answer to both questions is 3; the number 3 is both the logarithm and the exponent; the logarithm is an exponent. The general rule for f^{-1} is written

$$x = \log_a y$$

where y is now the variable of the domain and x the variable over the range of f^{-1}.

Two rules are most important to the development of derivative functions in this chapter. The first is the rule for the relationship between derivatives of inverse functions as expressed in Rule 6.1. Given the inverse functions f and f^{-1} with rules $y = f(x)$ and $x = f^{-1}(y)$, then the rules for their (first) derivative functions are related in the following way:

$$y' = dy/dx = \frac{1}{dx/dy} = \frac{1}{x'}, \qquad dx/dy \neq 0.$$

Thus once the rule for the derivative of a logarithmic function has been found, the rule for the derivative function of its inverse exponential function has also been found. This fact is demonstrated in the chapter by first asserting that given the function with rule $y = e^x$, the derivative function itself has as its rule $y' = e^x$ (equation (6.11)). For such a relationship to hold, y' the slope of the tangent to f at any point, must be equal to y, the value of f at that point. This presumption is illustrated to be true in Figure 6.7, Section 6.3.

Since any constant $a > 1$ can be expressed as the constant e raised to an appropriate power, b, the rule $y = a^x$ can be expressed as $y = e^{bx}$. By the chain rule and Rule 6.4, this implies that the rule for the derivative function is $y' = e^{bx} \cdot b = a^x \cdot b$ (equation (6.15)). In the more general case of a function with rule $y = a^{g(x)}$, which can be rewritten $y = e^{b(g(x))}$, by the chain rule the rule for the derivative function becomes $y' = e^{b(g(x))} \cdot g'(x) \cdot b = a^{g(x)} \cdot g'(x) \cdot b$ (equation (6.16)). The rules for derivative functions of exponential functions begin with the proof of Rule 6.5 for the function with rule $y = \log_a x$. The rule for the derivative function is $y' = \dfrac{1}{x} \log_a e$ (equation (6.18)). For the function with rule $y = \log_e x$ the derivative function simplifies to $y' = \dfrac{1}{x} \log_e e = \dfrac{1}{x}$ (equation (6.19)). For the rule $y = \log_a (g(x))$, that for the derivative function is

$$y' = \frac{1}{g(x)} \cdot g'(x) \cdot \log_a e = \frac{g'(x)}{g(x)} \log_a e$$

(equation (6.20)). The rule for the derivative of exponential function $y = a^x$ is then verified by the use of Rule 6.1. The rules for the derivatives of other exponential functions are derived from this Rule 6.4.

The use of exponential and logarithmic functions is illustrated in several examples and applications in Sections 6.6 and 6.7. One of the most useful applications of logarithmic functions is in the transformation of curvelinear functions into linear functions so that they may be subjected to analysis by relatively simple techniques. Elasticity and cross-elasticity concepts are also employed.

Key Concept

Concept	Reference Section
Inverse Functions	6.1
Derivatives of Inverse Functions	6.1
Exponential Functions	6.2
Irrational Number e	6.3
Rules for Derivatives of Exponential Functions	6.4
Derivative of e^x	
Derivative of $e^{g(x)}$	
Derivative of a^x	
Derivative of $a^{g(x)}$	
Logarithmic Functions	6.5
Rules for Derivatives of Logarithmic Functions	6.6
Derivative of $\log_a x$	
Derivative of $\log_e x$	
Derivative of $\log_a g(x)$	
Derivative of $\log_e g(x)$	
Application to Regression	6.7
Normal Equations	
Application to Elasticity	6.7
Point Elasticity	
Arc Elasticity	
Logarithmic Derivative	
Elasticity from Logarithms	
Cross Elasticity	

Appendix 6A: A Brief Review of Some Characteristics of Exponential and Logarithmic Functions

6A.1 Review of Exponential Functions

The rule for a general exponential function is $y = a^x$. Although the function can be defined for any real value of a, it is convenient to restrict the discussion to the case for which a is greater than 1. This is the case of greatest practical significance. For a greater than 1, y is positive for all values of x. When x is negative, y is between 0 and 1; when x is 0, y is 1; when x is positive, y is greater than 1. The exact shape of the function depends on the value of a, as shown in Figure 6A.1.

Figure 6A.1
Exponential Functions with Rules $y = a^x$, $y = b^x$ for $a < b$, and $y = 1^x$.

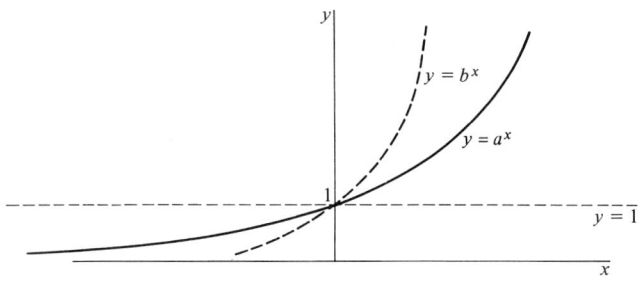

As x increases from 0 to larger positive values, the function rises rapidly. As x changes from 0 through increasingly negative values, the value of y approaches, but never quite reaches 0. The rate of increase as x increases or the rate of decrease as x decreases will depend upon the particular value of a. If the value of a is very close to 1, then the function will lie relatively close to the line $y = 1$ for a large part of its length. As the value of a increases, the function rises or falls more rapidly. Figure 6A.1 shows two exponential functions for a case in which the base of the exponential function b is larger than the base a.

The mapping from one set X to another set Y under an exponential rule defines an exponential function f from values of x on the entire x axis onto the values of y on the positive half of the y axis. Values of x from the positive half of the x axis are mapped onto values of y that are greater than $y = 1$. Similarly, values of x from the entire negative half of the x axis are mapped onto values of y which are in the interval between 0 and $+1$. The exponential function is, then, a one-to-one mapping from $x \in R$ onto $y \in R^+$.

Now consider any two points x_1 and x_2 and the corresponding values of y which make up the ordered pair in f as illustrated in Figure 6A.2. Under the mapping rule $y = a^x$, corresponding to the point x_1 is the point $y_1 = a^{x_1}$ on the y axis; corresponding to the point x_2 is the point $y_2 = a^{x_2}$. Now consider the point which is matched with $(x_1 + x_2)$, that is, $y_2 = a^{(x_1 + x_2)} = a^{x_3}$, the point labeled x_3 in Figure 6A.2. Since x_3 is equal to the sum of $x_1 + x_2$ it follows that a^{x_3} is equal to $a^{(x_1 + x_2)} = (a^{x_1})(a^{x_2})$.

From this example, one of the most important characteristics of the exponential function can be generalized. Given a value of x_0 equal to the sum of any other two values of x, x_1 and x_2, then the value y_0 of the exponential function corresponding to x_0 will be the *product* of the values of the exponential function for the other two points. That is, for the exponential function, addition in the domain of the function implies multiplication in the range of the function, because the domain variable *is* an *exponent* which

APPENDIX 6A: EXPONENTIAL AND LOGARITHMIC FUNCTIONS 205

Figure 6A.2

The Exponential Function f with Rule $y = a^x$ (for $a = 2$).

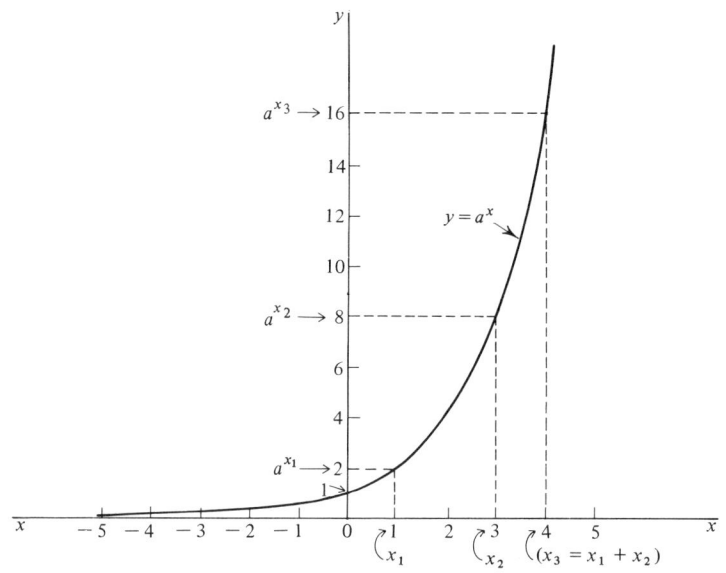

operates on a fixed base, the constant a. Multiplication of two numbers each represented by a common base raised to a different exponent is equivalent to raising that base to the power equal to the sum of their exponents.

Consider also the relationship between the points x_1 and $\frac{1}{2}x_1$ in Figure 6A.2 and the corresponding values of y. The value of $\frac{1}{2}x_1$ on the x axis corresponds to $a^{(1/2 x_1)}$ on the y axis, and $a^{(1/2 x_1)}$ is the square root of a^{x_1}. Multiplying (or dividing) values in the domain of an exponential function is the same as taking powers (roots) of numbers on the x axis. As another example, $2(x_1)$ would be mapped into $y = a^{(2 x_1)} = (a^{x_1})^2$. The value x_1 corresponds to a^{x_1}; then if x_1 is multiplied by 2, this is equivalent to raising a^{x_1} to the second power.

The use of the exponential function to find the square root of the number can be illustrated in another example. In numerical calculations it is usually most convenient to use an exponential function whose base a is equal to 10, but Figure 6A.2 is graphed for the base 2, so the base 2 is used. The steps in the process can be followed graphically on this diagram. To find the square root of the number 8, for example, first find 8 on the y axis. Next locate the value of x in the domain that corresponds to a value of 8. If the base a is equal to 2, the value of x corresponding to 8 will be 3, since $2^3 = 8$. To find the square root of 8, divide 3 by 2 (or multiply by $\frac{1}{2}$) to get 1.5 and then find the value of $y = a^{(x = 1.5)}$ corresponding to $x = 1.5$. If the scale is sufficiently detailed, this can be read to be 2.83; that is, $2^{1.5} = 2.83$. Thus 2.83 is the square root of 8.

In the preceding example, when 2.83 is identified as the value of y corresponding to the x value 1.5, this was making use of the exponential function $y = 2^x$. However, if the value $x = 3$ was to be obtained from the value $y = 8$, this would be relying on the inverse function of $y = 2^x$, or the logarithmic function. In this example, the problem was solved graphically; similar problems are solved by use of tables of logarithms in Section 6A.2 to solve equations of the form $x = \log_2 y$.

Since the exponential function defines a one-to-one mapping between the values of x and the values of y, it follows that there will also be a function (mapping) from the values of y as domain onto values of x as range, the logarithmic function (in this illustration with base 2). The graphs of the two functions—the exponential and the logarithmic—are identical. They differ only in the direction of the mapping.

6A.2 Review of Rules for Use of Logarithms in Computation

Many computational tasks, such as multiplication, division, taking powers of numbers, and finding roots of numbers, can be simplified by using logarithms. No attempt is made in this short appendix, however, to show the reader how to use logarithms in computation. The objective is merely to illustrate how the results of the discussion of exponents can be employed in computation, and to review the elementary theory of logarithms.

The so-called "common logarithms" are all powers of 10.[17] They can be combined according to the rules for the manipulation of exponents. For example, to multiply two numbers, a and b, the numbers should first be written in terms of the base 10 raised to the proper exponent:

$$a = 10^x \quad b = 10^{x'}$$

Then

$$ab = (10^x)(10)^{x'} = 10^{x+x'}.$$

Once this result is obtained, the problem is one of dealing with an exponential relationship, the inverse of the logarithm. The number for y, which corresponds to the exponential expression $10^{(x+x')}$ must be found; that is,

$$y = 10^{(x+x')}.$$

The value of y that fits this relationship is called the *antilog* for the base 10 of $(x + x')$. A useful way to think about these factors is to rewrite the equation for an exponential function as follows:

$$\text{antilog} = (\text{base})^{(\text{log})}.$$

The antilog is the number that results from raising a given base to the power specified by the logarithm; for example,

$$y = 10^{(3+2)} = 10^5.$$

Hence the antilog (for the base 10) of the logarithm 5 is 100,000. When the base 10 is used, 5 is the log of the number 100,000 ($10^5 = 100,000$); therefore 100,000 is the antilog of 5. In Table 6A.1, it is seen that the exponent 5 corresponds to the number 100,000.

Table 6A.1

Selected Logarithms to the Base 10.

Number (antilog)	Logarithm
10	1
100	2
1,000	3
10,000	4
100,000	5
1,000,000	6
10,000,000	7

Rule 6A.1—Logarithmic Multiplication: To multiply two numbers a and b, use a table of common logarithms to find the logarithm x of a, and x' of b; add the two logs $x + x'$ and find the antilog of the sum in an appropriate table.

---

[17] When the base used is 10, the subscript is generally omitted from the expression. Thus $x = \log_{10} y$ may be written simply $x = \log y$.

APPENDIX 6A: EXPONENTIAL AND LOGARITHMIC FUNCTIONS 207

Suppose that $\log a = x = \log b = x' = 1$. Therefore $a \cdot b = 10^{1+1} = 10^2$. That is, 100 is the antilog to the base 10 of the exponent 2; or in other words, 100 is the base 10 raised to the exponent 2. This antilog can be read directly from Table 6A.1.

Similarly, to raise any number k to the power a, note first that k can be expressed as 10 raised to an appropriate power x''; that is, $k = 10^{x''}$; then $k^a = (10^{x''})^a$ and, using rules for exponents, this equals $10^{ax''}$. That is, $\log_{10} k = x''$; $\log_{10}(k^a) = a(x'') = a(\log_{10} k)$. Therefore, $\log_{10}(k^a) = a \log_{10} k$.

Rule 6A.2—Logarithmic Calculation of Powers: *To raise any number k to any power a, find the logarithm of k, multiply that logarithm by a and then find the antilog of the product from an appropriate table.*

For example, if $k = 100 = 10^2$ (that is, $\log 100 = 2$) and $a = 3$, then $k^3 = (10^2)^3$. But since $\log k = 2$, then $\log(k^3) = 3(\log k) = 3(2) = 6$. This is the logarithm to the base 10 of k^3; that is, $6 = \log_{10}(k^3)$. A table of logarithms to the base 10 is necessary to find the antilog of 6. From Table 6A.1, it can be seen that 1,000,000 is the number for which the log to the base 10 is 6.

To illustrate the economy that these rules permit, consider what sort of effort would be involved if logarithms were not used in calculating $100(1.05)^{25}$. This could represent the value, after 25 years, of $100 invested at 5 per cent, with interest compounded annually, for a case in which the interest is allowed to accumulate. (Carry out this calculation.)

The slide rule consists of rulers ruled off to correspond with the logarithms to the base 10 of the numbers 1 to 10 and carefully assembled together. To multiply two numbers, the logarithm of one number is added to the logarithm of the second number by adding the respective lengths physically. The sum of the lengths of the two logarithms is the length of the logarithm of their product; that is, $\log 3 + \log 2 = \log(3 \cdot 2) = \log 6$ and the antilog of the sum is their product:

$$(3)(2) = 6.$$

The various other characteristics of exponential and logarithmic functions discussed earlier in this appendix apply also to the slide rule and are directly transferable to its use.

Finally, in numerical calculations, a table of logarithms to the base e is often used, but the most readily available tables are to base 10. If $a = 10$, then since $\log_e 10 = 2.3026$, if $a \neq 0$, the following relationships can be used:

$$\log_e a = \log_{10} a \cdot \log_e 10$$
$$= (\log_{10} a)(2.3026).$$

The proof of this follows. From the definition of a logarithm,

$$a = e^{\log_e a}$$

and

$$a = 10^{\log_{10} a}.$$

Also

$$10 = e^{\log_e 10}.$$

Substituting $e^{\log_e 10}$ for 10 in the equation $a = 10^{(\log_{10} a)}$ gives

$$a = (e^{\log_e 10})^{\log_{10} a} = e^{(\log_e 10)(\log_{10} a)}.$$

Thus, since

$$a = e^{\log_e a}$$

and

$$a = e^{(\log_e 10)(\log_{10} a)}$$

it must be true that

$$\log_e a = \log_e 10 \cdot \log_{10} a.$$

Appendix 6B: Calculation of the Numerical Value of the Constant *e*

The value of *e* can be approximated as closely as desired by using the formula derived in Section 6.3, $e = \lim_{n \to \infty} [1 + (1/n)]^n$. The first approximation for $n = 1$ is rather poor, but later ones become increasingly close. The function $f(n) = [1 + (1/n)]^n$ is used for various values of *n* to demonstrate the improved approximations:

$$f(1) = (1 + \tfrac{1}{1})^1 = 2,$$
$$f(2) = (1 + \tfrac{1}{2})^2 = 2.25,$$
$$f(3) = (1 + \tfrac{1}{3})^3 = 2.370,$$
$$f(4) = (1 + \tfrac{1}{4})^4 = 2.441,$$
$$f(10) = (1 + \tfrac{1}{10})^{10} = 2.594,$$
$$f(100) = (1 + \tfrac{1}{100})^{100} = 2.692,$$
$$f(1000) = (1 + \tfrac{1}{1000})^{1000} = 2.718.$$

Logarithms are very helpful in the evaluation of numbers such as these, once a table of logarithms has been developed. The value 2.718 above is computed using a table of logarithms to the base 10. The logarithm to the base 10 of $(1 + \tfrac{1}{1000})^{1000}$ is that exponent to which 10 must be raised in order to reach $(1 + \tfrac{1}{1000})^{1000}$; that is,

$$y = 10^x = (1 + \tfrac{1}{1000})^{1000}$$

or

$$\log y = \log_{10}[(1 + \tfrac{1}{1000})^{1000}] = x.$$

Collecting terms gives

$$\log y = \log_{10}[(1.001)^{1000}] = 1000 \log_{10}(1.001) = x.$$

From the table of logarithms,

$$\log_{10}(1.001) = 0.0004341,$$

then

$$1000 \log_{10}(1.001) = 1000(0.0004341) = 0.4341000 = \log y.$$

Taking the antilog of both sides gives

$$y = 2.717\ldots.$$

Problems

Simplify.

1. $y = 2^{\log_2 7}$
2. $y = 4^{\log_2 6}$
3. $y = e^{\ln x}$
4. $y = \ln(e^{x^2})$
5. $y = e^{-3 \ln x}$
6. $z = e^{3 \ln y - \ln x}$

Find dy/dx for functions with the following rules:

7. $y = e^{ax^2}$
8. $y = e^{[(ax+g)/x]}$
9. $y = (e^{x^2} - e^{-x^2})(e^{x^2} + e^{-x^2})$

PROBLEMS

10. $y = \dfrac{e^x + e^{-x}}{e - e^{-x}}$

11. $y = (x^3 - 4e^{2x})^{1/2}$

12. $y = \dfrac{\log_e ax}{e^{ax}}$

13. $y = \ln \sqrt{ax^2 + b}$

14. $y = \ln\left[\dfrac{2x}{\sqrt{1+x}}\right]$

15. $y = (\ln x)^4 \cdot x$

16. $x = \ln y$

17. $y = \ln (\ln x^2)$

18. $y = \ln (4x\sqrt{x^2 + 2})$

Find $\log_e y$ and then take its derivative with respect to x.

19. $y = e^{ax^2}$

20. $y = e$ raised to the $\left[\dfrac{\sqrt{ax}}{\sqrt{x^2+1}}\right]$ power.

Find dy/dx.

21. $y = 5^{x^2} + 2^{x-3}$

22. $y = \ln\left(\dfrac{x^3}{1+x^5}\right) + 7x^{1/4}$

23. $y = a^{b\sqrt{x^2-1}}$

24. $y = a^{(x+x^2)\cdot a(1-x)}$

25. $y = \log_e x^x$

26. $y = x^{x^x}$

27. $y = e^{x^x}$

28. $y = e^{e^{x^e}}$

29. $y = e^{\ln x e^x}$

30. Given the following data:

Price	Demand	Gross Revenue
11	0	0
10	2	20
9	4	36
8	6	48
7	8	56
6	10	60
5	12	60
4	14	56
3	16	48

(a) Identify the elastic and inelastic portions of this demand curve.
(b) Consider a price reduction from $8 to $7, and compute the elasticity using the formula $\eta = p_0/q_0 \cdot (q_1 - q_0)/(p_1 - p_0)$.
(c) Could $p = 7$ possibly be the optimal price?

31. Let the demand curve be specified by $p = 20 - 2q$.
   (a) Find η at $p = 10$.
   (b) Compute MR at $p = 10$.
   (c) In what range should the optimal price be found?
   (d) What is η at the point of maximum gross revenue?
   (e) Discuss the size of η at different points along a straight line demand curve.
   (f) Find the optimal price if MC $= 4$.

32. Let $c = pq$ be a demand function.
   (a) Find η for $p = e^c$.
   (b) Find η using the logarithmic derivative.

33. Why is the elasticity concept important to a businessman? In what way would it help him make decisions? How could he use it?

34. A corporation is interested in the form of its total-cost function in the range 3000 to 10,000 units. The data gathered appear in the table below.

Quantity in 1000 Units	Total Cost in 1000 Units
3	5.5
3.5	4.5
4	4.2
4.5	4.0
5	4.5
5.5	4.2
6	3.7
7	4.1
8	3.6
9	4.0

Find a function that appears to fit the data, and apply the least squares regression technique to obtain the parameter values of the function.

35. A firm is interested in developing a function that allows it to predict sales as function of the number of salesmen. From past experience, the management has the feeling that sales are exponentially determined by a relation of the type $s = a \cdot e^{bx}$, where s is sales, x the number of salesmen, and a and b are constants. The management's beliefs are based on the following data:

x	s
0	100
1	122
2	149
3	182
4	223
5	272
6	332

Use logarithms to transform the rule for s ($s = a \cdot e^{bx}$) into a linear relation, and then estimate a and b using the regression technique presented in Chapter 5.

Find all *first* and *second* partial derivatives:

36. $f(x, y, z) = ye^x + xe^{2y+z^2}$

37. $f(x, y, z) = e^x \log_e (zx) + \dfrac{x + y^2 + z^3}{xyz}$

Find f''_{xy} and f''_{yx}, that is, first differentiating with respect to x, then with respect to y and first with respect to y, then with respect to x. Compare the results.

38. $f(x, y) = e^{-ax^2 + by^3} + e^{xy}$

39. $w^3 = e^{x+2y}$

40. $wx + x^2 + y^2 = 4e^x$

41. Find dy/dx, $\partial z/\partial x^2$, and $\partial z/\partial y^2$ by explicit differentiation. Find dy/dx by implicit differentiation

$$x^4 e^x + \ln(xy) + 4^{2x} + 5^{xy} + \ln(x+y) + e^{\ln x - \ln y} = 0$$

42. Find dy/dx by implicit and explicit differentiation.

$$(x^2 + y)(x + y^2) + \ln\left(\frac{1}{x-y}\right) + \ln(e^{xy}) + 4^{\log 4^{16}} + a^{\log_a e} = 0$$

43. The following function represents the demand curve for product A:

$$Q = 268 - .1Y + .2P_b - 2P_a^2$$

where Q is the quantity demanded, Y is the consumer personal disposable income, P_b is the price of a related product and P_a is the price of A.

(a) Compute the price elasticity of demand $[\partial Q/\partial P_a \cdot P_a/Q]$ when $P_a = 5$, $P_b = 60$, and $Y = 300$.
(b) Compute the cross elasticity of demand for A with respect to P_b, $[\partial Q/\partial P_b] \cdot [P_b/Q]$ when $P_a = 5$, $P_b = 60$, and $Y = 300$.
(c) Compute the income elasticity of demand for A, $[\partial Q/\partial Y] \cdot [Y/Q]$ when $P_a = 5$, $P_b = 60$, and $Y = 300$.
(d) What is our intuitive feeling about the characteristics of product A when we considered its income elasticity? (That is, what happens to the demand for A when income changes? What type of product might be affected this way?)
(e) How is the demand for A affected by the prices of A and B? Would you say A and B are complementary or substitute products? Give an example of what product A might be.

44. A corporation is faced with a choice between two products X and Y to market next year. It can choose either or both. The demand curves are the following:

$$p_x = 8 - x + 2y$$
$$p_y = 6 - y$$

where x and y denote the quantity of X and of Y demanded, and p their prices. (Note that Y is independent of X, but that Y influences the demand for X.) The total cost functions are the following:

$$TC(x) = x^2 - 8x + 4, \quad \text{for } x \geq 2$$
$$TC(y) = y^2 + 2y + 6$$

(a) Recommend a choice to the corporation, *including prices to set.* (Support your recommendation by reasoning.)
(b) Relative to X, is Y a complementary or substitute product? (Why?)

CHAPTER 7

Integration

In Chapter 6 inverse functions were defined and discussed in the context of exponential and logarithmic functions. Throughout mathematics there are many processes that are inverses to one another. For example, subtraction is the inverse process to addition, multiplication is the inverse of division, and so on. This chapter deals with integration, the inverse process to differentiation.

Figure 7.1

Function $y' = 2x$ with Area Under Curve Measured by Rectangles with Height Equal to Mean Value in Each Interval.

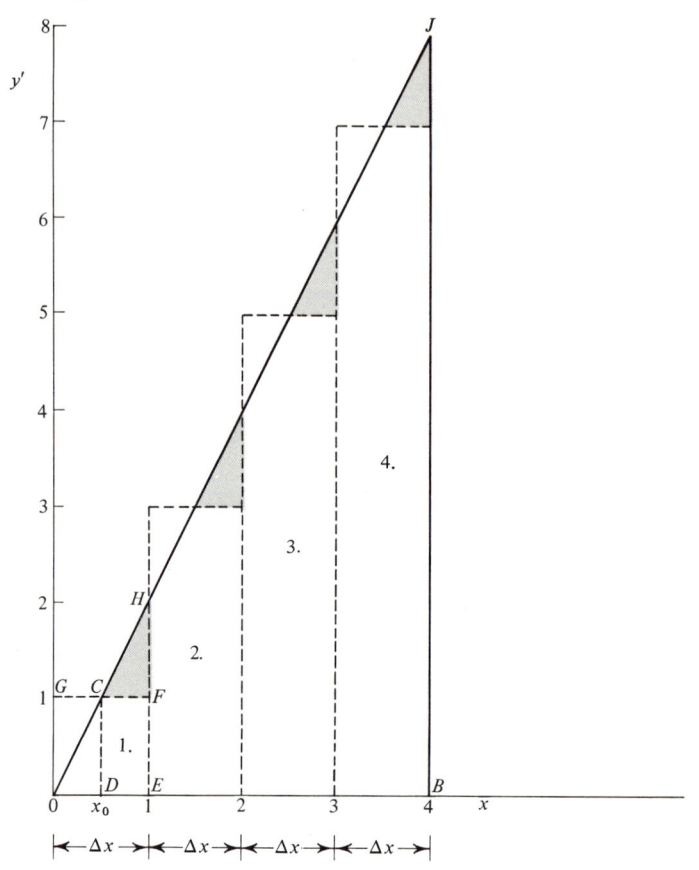

The process of integration is extremely useful in a wide variety of applications. For example, just as the marginal cost and marginal revenue functions can be derived from the functions expressing total cost and total revenue, so also is it possible to determine the value of total costs from functions that express marginal cost, or total revenues from functions that express marginal revenue. One of the most useful applications of integration is found in dealing with probability functions associated with continuous probability distributions, an important topic in the fields of probability, statistics and decision theory.

The process of integration is somewhat more complex than that of differentiation. Although it is generally possible to find the rule for a derivative function given the rule for the original function, this is not the case for integration; that is, there are many functions for which rules can be clearly expressed as equations, but for which it is not possible to find an *equation* for the integral function. Despite this, however, it is generally possible to find a *value* of an integral function, given the rule for the function from which the integral is developed, by the process of numerical

Sec. 1.1] INTEGRATION, THE INVERSE PROCESS TO DIFFERENTIATION 215

computation. The calculation involves finding the limit of a sum of terms that converge to the desired value of the function. With the advent of modern computers, such calculations have been greatly facilitated. The essence of the procedure is illustrated in Figure 7.1.

In Figure 7.1 it is possible to approximate the area A of the triangle OJB by dividing its base into several ($n = 4$) subintervals Δx, then locating a mean height of the function in each subinterval (for example, $DC = f(x_1)$ in the first subinterval), raising a rectangle on the base ($OE = \Delta x$) with height equal to the mean height selected (DC), and then summing the areas of the rectangles (rectangles 1 + 2 + 3 + 4 in Figure 7.1, $A = \sum_{i=1}^{4} f(x_i)\,\Delta x_i$, for example). The area that is not a part of the triangle but is included in each of the rectangles (for example, OCG for rectangle 1) exactly balances the portion of the triangle left out of the rectangle (the shaded area CFH for rectangle 1. This is a procedure similar to using a histogram (a bar chart) to represent the value of a variable in each subinterval of the interval [0; 4] in Figure 7.1, for example. In this chapter, the approaches to evaluating integrals through rules for integral functions and through direct computation are presented, beginning with the subset of functions for which the rule of the integral function can be developed.

7.1 Integration, the Inverse Process to Differentiation

In this section, the first and second derivative functions of a function f with rule $y = f(x) = x^2$ are derived and graphed. Then the areas under the curves of the second and first derivatives, respectively, are calculated, and their significance for the value of their respective integral functions is discussed. A common interpretation of the value of the integral function is that it is equal to the area under the curve of the function from which the integral was derived. Other interpretations are successively introduced in relation to the graphs and the significance of the various calculations is discussed on pages 220 and 221.

Given the function f with rule

$$y = x^2 \tag{7.1}$$

the rule for the first derivative function f' (whose graph is shown in Figure 7.1) is

$$y' = 2x. \tag{7.2}$$

The rule for the second derivative function f'' is

$$y'' = 2. \tag{7.3}$$

Several direct calculations of the areas under portions of the functions f'' and f' are presented in Figures 7.2 and 7.3 along with the approximation of areas under portions of f in Figure 7.4. The function f'' is graphed in Figure 7.2 in which various areas are indicated by hatching.

Starting at the point $x = 1$, the area under the curve $y'' = 2$ from $x = 1$ to $x = 2$ is the rectangle with base one unit long and height of two units. The area of

Figure 7.2

Function f'' with Rule $y'' = f''(x) = 2$.

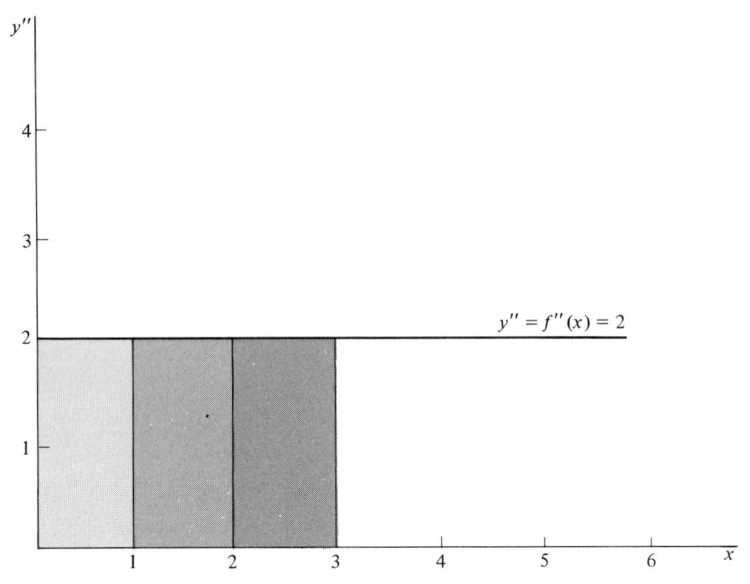

the rectangle is $2 \times 1 = 2$ square units. It is also possible to move back in the domain from $x = 1$ to $x = 0$ and define a new rectangle whose base goes from $x = 0$ to $x = 2$. The base of the rectangle is then two units long, the height unchanged at two, and the area is four square units. If next the upper limit of the base is moved from $x = 2$ to $x = 3$, then the area would be $3 \times 2 = 6$ square units. In each case, the function f'' and its rule remain unchanged, but with different limits to the domain variable, the area under the function does change; from two, to four, to six square units. The area A under the curve $y'' = 2$ for the base of the rectangle representing that area beginning at $x = 0$ and extending to a given value of x, that is, in the interval $[0; x]$, is

$$A = 2x.$$

If the analogy is shifted from physical areas to one of speeds and distances (used also to introduce the concept of the derivative function in Chapter 3), then y'' can be interpreted as the rate of acceleration of a vehicle, the rate at which its speed is changing, over time. In this interpretation, it is useful (and usual) to assume that values begin at $x = 0$ (at time zero, before the vehicle began to move). If the vehicle were at rest and began to accelerate at the constant rate of two miles per hour (mph) for each minute that it is in motion, then at the end of the first minute, $x = 1$, its speed would be two mph, at the end of the second minute $x = 2$, four mph, when $x = 3$, 6 mph, and so on. The speed of the vehicle at each instant of time can be graphed as shown in Figure 7.3, with the speeds at x equals 1, 2, and 3 minutes shown as points on the graph of the new function. The rule for the new function is speed s equals two times the number of minutes in motion

$$s = 2x.$$

Sec. 7.1]   INTEGRATION, THE INVERSE PROCESS TO DIFFERENTIATION   217

Figure 7.3

Function that Maps Time in Minutes to Speed, Given that Acceleration Is at Rate of Two Miles per Hour Each Minute.

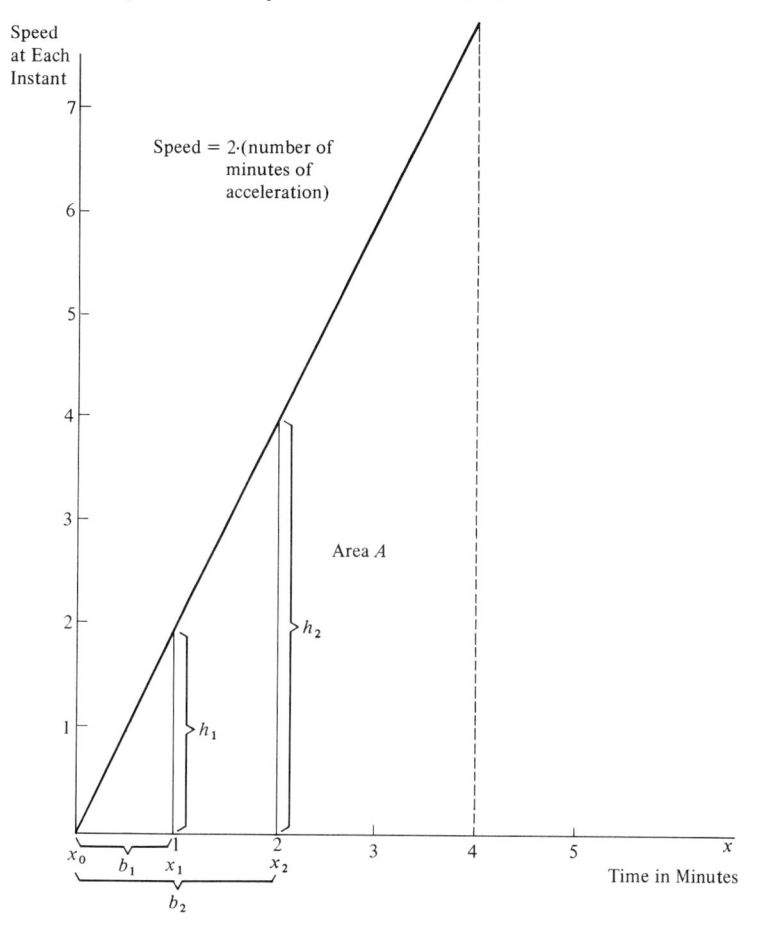

But the last two rules lead to the same ordered pairs as equation (7.2) does for f'

$$y' = 2x. \qquad (7.2)$$

Identical ordered pairs result for a rule showing the $2000 monthly rate of increase (acceleration) in annual revenues r stated in thousands of dollars for an organization:

$$r = 2x.$$

At the end of the second month, the revenues would be proceeding at a $4000 annual rate.

It is possible also to find the area between the curve shown in Figure 7.3 and the x-axis interpretating this curve now as the graph of the function f' with rule

$$y'(=s) = 2x.$$

The area under this curve between $x_0 = 0$, and $x_1 = 1$ in Figure 7.3 is a triangle with base b_1 equal to one unit, $(x_1 - x_0) = x_1 - 0 = 1$ and height h_1 equal to $2(x_1) = 2(1) = 2$.[1] The area then is $\frac{1}{2}b_1h_1$:

$$A = \tfrac{1}{2}b_1h_1 = \tfrac{1}{2}(x_1)(2x_1) = \tfrac{1}{2}(1)(2) = 1.$$

The area for the triangle formed under the curve (the line $y' = 2x$) for the interval $0 \leq x \leq 2$ with base $b_2 = (x_2 - x_0) = x_2 = 2$ and height $h_2 = 2(x_2) = 2(2) = 4$ in Figure 7.3 is:

$$A = \tfrac{1}{2}b_2h_2 = \tfrac{1}{2}(x_2)(2x_2) = \tfrac{1}{2}(2)(4) = 4.$$

The general rule for the area under the curve $y' = 2x$ between 0 and a point x_i on the x axis is

$$A = \tfrac{1}{2}(x_i)(2x_i)$$
$$A = x_i^2$$

but this rule leads to ordered pairs identical to those from the rule for the original function f.

$$y = x^2.$$

Interpreted in terms of speed and distance, the vehicle that has accelerated at a constant rate of two miles per hour for each minute of its movement is proceeding at four miles per hour at the end of the second minute and has covered

$$\tfrac{1}{2} \cdot 2(0 + 4) = 4 \text{ miles}$$

in its first two minutes. In the first three minutes it has covered

$$\tfrac{1}{2} \cdot 3(0 + 6) = 9 \text{ miles}.$$

These calculations can be used to provide insight into the process being described. For example, in the last calculation the instantaneous rate of change at the beginning of the time interval is zero. At the end of three minutes it is six mph for an average rate of change of 3 mph during the interval. The total distance covered, 9 miles, is equal to the time that has passed during the interval, 3 minutes, times the average (of the instantaneous) rate(s) of change from the beginning to the end of the interval, 3 miles per hour.

Interpreted in terms of revenues, over a time interval of 2 months the instantaneous rate of change of revenue (marginal revenue) at the beginning is zero, and at the end of the second month is four. The total revenue over the period is the average revenue during the period times the length of the period. It is equal to the area under the curve (area of the triangle) between $x = 0$ and $x = 2$:

$$\tfrac{1}{2}bh = \tfrac{1}{2}(2)(4).$$

Since the rules for the functions implied by each interpretation all result in the same ordered pairs of numbers as $y = x^2$, the graph of this equation in Figure 7.4

---

[1] The area under the curve could be calculated by using rectangles as was done in Figure 7.1. The result would be the same as using the triangles as is done here. For example, the mean height in the first interval shown is $\tfrac{1}{2}(f'(x_0) + f'(x_1)) = \tfrac{1}{2}(0 + h_1) = (\tfrac{1}{2})h_1$ and the base is b_1; thus the area of the first rectangle would be $b_1(h_1/2) = (\tfrac{1}{2})b_1h_1 = \tfrac{1}{2}(1)(2) = 1$, the same result achieved for triangle one.

Sec. 7.1] INTEGRATION, THE INVERSE PROCESS TO DIFFERENTIATION   219

Figure 7.4

Function f with Rule $y = f(x) = x^2$ with the Area under $y = x^2$ Approximated from $0 \leq x \leq 4$ by Three Rectangles.

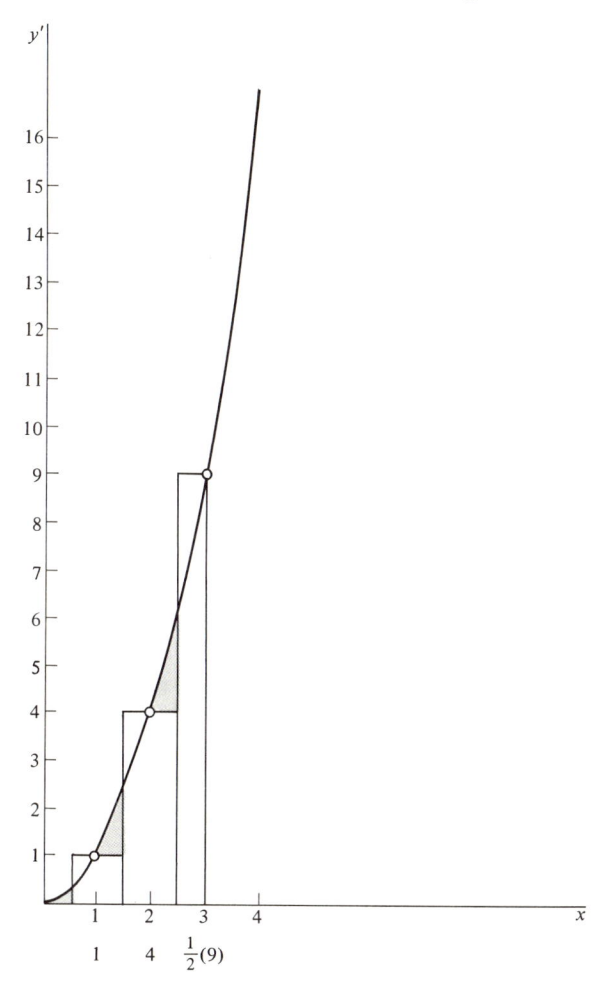

can be used to represent all of them. The numerical values of the areas under f' from Figure 7.3 just calculated can be read at points on the curve in Figure 7.4, at the values of x_i corresponding to the upper limit of each base (upper limit of each interval) in Figure 7.3.

Now it is possible to calculate areas under the graph of f but, since the graph of the equation $y = x^2$ is not a straight line, there is no simple formula for determining the area under f. The area may be approximated by appropriately constructed rectangles, however. In Figure 7.4 three rectangles are drawn to approximate the area under f: the first two each have one unit for the base and the value of f at the integer values of x, $x = 1$, and $x = 2$, respectively, as height, and the third has one-half unit as base and the value of f at $x = 3$ as height. The total area represented by these three rectangles is the sum of their individual areas:

$$\sum (A_1 + A_2 + A_3)$$

where

$$A_1 = b_1 h_1 = 1(1) = 1.0$$
$$A_2 = b_2 h_2 = 1(4) = 4.0$$
$$A_3 = b_3 h_3 = \tfrac{1}{2}(9) = 4.5$$

$$\sum_{i=1}^{3} A_i = 9.5.$$

The function g with rule $y = g(x) = x^3/3$ has a derivative function g' with rule $g'(x) = x^2$. If it is inferred that the area under f from $0 \leq x \leq 3$ is given by the value of this other function g by its rule,

$$y = g(x) = \frac{x^3}{3},$$

then the area under f over the interval [0; 3] should be

$$y_3 = g(3) = \frac{(3)^3}{3} = 9.$$

The approximation is quite close to this value. By choosing smaller subintervals as bases, it should be possible to make the approximation closer still. As is shown later in the chapter, this, in fact, can be done.

In this section the discussion has proceeded on an intuitive basis, and the relationships among the derivatives of various orders has only been implied. In Section 7.2 the relationships are formally defined, and the fundamental theorem of the calculus proved, essentially by use of the basic ideas presented in the discussion in this section. The limit of the sum of rectangles raised on sufficiently small subintervals of the domain is shown to be the value of the integral function over that interval. This forms the basis for the calculation of the value of an integral function either through use of a rule for such a function or through direct calculation of the value of the limit itself. In later sections both processes are illustrated and applied to specific situations.

When the rule for the function results in a graph that is linear, it is possible to calculate exactly the area under the curve for any limit values of the domain variable.[2] When the graph of the function is nonlinear, however, it is possible only to approximate the area, but through the limit process it is possible to approximate the area as closely as desired; the limit converges to the exact area.

Notice that at each stage in the preceding discussion, the new function has the same domain, the values of x over a specified interval, and each successive range is made up of values determined by the *limits* on x, generally from a fixed beginning (lower limit) value. The range values of each new function are the numbers repre-

---

[2] The term "limit" is used here in its two most frequent meanings; first, the endpoints of an interval (its upper and lower limits) and, second, the range value toward which a summation converges.

Sec. 7.2] THE FUNDAMENTAL THEOREM

senting the area under the curves of the preceding function, which, when plotted, lead to new curves of higher order and greater complexity. With a fixed lower limit for the domain variable, changes in the values of the range variable for the new function are related solely to the changing upper limit of the domain variable. The implications of this are demonstrated in the following sections.

Before proceeding to the discussion of integral functions and the fundamental theorem, a few observations may provide insight into the relationships involved.[3] The process of calculating an area under a curve involves the essence of integration. It can be compared to its inverse process, the process of differentiation. Differentiation and integration are alike in that both deal with limits and both deal with small increments Δx in the domain variable x with $\Delta x \to 0$ in the limit. Their significance as inverse processes can be seen in the different ways the values of their range variables are determined. Differentiation of a function requires the use of values of the range variable to determine *average rates of change* over an interval $[x; x + \Delta x]$ and then determination of the *instantaneous rate of change* through the limit for a *quotient* of *differences* in range values divided by Δx, (the difference in the domain variable);

$$\lim_{\Delta x \to 0} \frac{f(x + \Delta x) - f(x)}{(x + \Delta x) - x} = \lim_{\Delta x \to 0} \frac{f(x + \Delta x) - f(x)}{\Delta x}.$$

Integration (as presented thus far) requires the averaging of *instantaneous rates of change* at the end points of an interval to determine an *average rate of change* and then the determination, through the limit of a *sum* of *products* (the average rate times Δx), of the value of a given range value of a function

$$\lim_{\Delta x \to 0} \sum_{i=1}^{n} f(x_i) \Delta x_i$$

where $n \to \infty$ as $\Delta x \to 0$ and

$$f(x_i) = \tfrac{1}{2}(f(x_e) - f(x_b))$$

for each interval i and where e represents the endpoint of the interval and b the beginning point. Not only are the required arithmetic operations inverse to one another, but they also are applied in reverse order: a quotient of differences in one case, a sum of products in the other.

7.2 The Fundamental Theorem

The theorem that establishes that integration is the inverse process of differentiation is known as the fundamental theorem of the calculus. The procedure required for a proof of the fundamental theorem is essentially the same as that required to determine the value of a given integral directly through the limit process, generally with the aid of computers. Therefore, the proof has immediate practical application.

Once the fundamental theorem has been proved, it is then possible to state the rules for integral functions for large groups of functions. The proof that the rules are in fact valid is arrived at by differentiating these functions to see if the rules for

---
[3] It may prove helpful to reread this paragraph after completing the chapter.

the derivative functions are those for the original functions (as was the case in Section 7.1). If it were possible to determine an appropriate value in each interval for the function in Figure 7.4 similar to those determined for the function f' in Figure 7.1, then the procedure suggested for that figure could be used to determine exactly the area under the curve $y = x^2$. At the end of this section it is shown that this can be done, at least in the limit, but first a few definitions are necessary.

The value of the type of function that can be interpreted geometrically as an area is known as the definite integral. It is the limit of the sum of rectangles raised on subintervals of the domain Δx_i as base with height equal to a value $f(x_i)$ in each respective subinterval of the function under which the area is being determined:

$$\lim_{n \to \infty} \sum_{i=1}^{n} f(x_i) \cdot \Delta x_i \qquad (a \leq x \leq b).$$

In the limit, this sum, if it exists, is equal to the area between $f(x)$ and the x axis and between the vertical lines $x = a$ and $x = b$. In symbols, the definite integral is expressed as $\int_a^b f(x)\, dx$:

$$\int_a^b f(x)\, dx = \lim_{n \to \infty} \sum_{i=1}^{n} f(x_i) \cdot \Delta x_i. \qquad (a \leq x \leq b) \qquad (7.4)$$

The lower domain point a is known as the lower limit of integration, the upper domain point b is the upper limit of integration.

Definition—Definite Integral: *Given the function f with rule $y = f(x)$, defined over a closed interval $[a; b]$, with the interval divided into n subintervals Δx_i with mean height in each subinterval designated $f(x_i)$; then*

$$\lim_{n \to \infty} \sum_{i=1}^{n} f(x_i) \cdot \Delta x_i$$

if it exists, is the definite integral of $f(x)$ written in symbols as $\int_a^b f(x)\, dx$; that is,

$$\int_a^b f(x)\, dx = \lim_{n \to \infty} \sum_{i=1}^{n} f(x_i) \cdot \Delta x_i$$

where a and b, the endpoints of the interval, are known as the lower and upper limits of integration, respectively.

Through the examples in Section 7.1 the value of the definite integral has been shown to be a particular number. For example, each specified area is of a given size expressed as a number. But it was also shown that, beginning from a fixed lower limit to an interval, the area under a particular function varied with (was a function of) the upper limit of integration; that is,

$$\int_a^x f(x)\, dx$$

changes with each value of x, the *variable* upper limit of integration. A function whose value y depends on the upper limit of integration is known as an *integral function*.

Sec. 7.2] THE FUNDAMENTAL THEOREM   223

Figure 7.5

The Area under $f(x)$ from a to x_i in the Interval $[a; b]$.

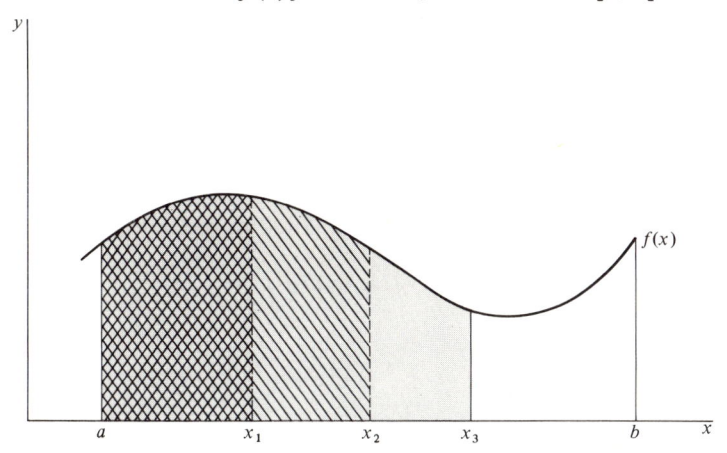

Definition—Integral Function: *Given the function f with rule $u = f(x)$ continuous in a closed interval $[a; b]$, then the function whose rule is in the form of the integral*

$$y = \int_a^x f(x)\, dx$$

with the upper limit a variable x over the domain is an integral function *and $f(x)$ is known as the* integrand.

Integral functions have some interesting characteristics that are now explored. Given the function f graphed in Figure 7.5, the area under f between a and some upper limit x_i is shown for $i = 1, 2$, and 3. The integral over the full interval between a and b is the area under the curve from $f(a)$ to $f(b)$. It is equal to the sum of several areas (integrals).

Because the value of an integral function is determined from a limit, the simple rules for manipulation of limits apply to the value of an integral function. This implies, by Theorem 2.2, that the value of a function that is equal to the (algebraic) sum of two integral functions is equal to the (algebraic) sum of the values of the two integral functions.

Thus the area under f between a and b is equal to the sum of several areas represented below by their integrals:

$$\int_a^b f(x)\, dx = \int_a^{x_1} f(x)\, dx + \int_{x_1}^{x_2} f(x)\, dx + \int_{x_2}^{x_3} f(x)\, dx + \int_{x_3}^b f(x)\, dx$$

that is, the integral from a to x_1 plus the integral from x_1 to x_2 plus the integral from x_2 to x_3 plus the integral from x_3 to b, when summed, give the area under f from a to b. The area between x_1 and x_2 can be expressed in various ways. First, it could be the difference between the integrals from a to x_2 and a to x_1.

$$\int_{x_1}^{x_2} f(x)\, dx = \int_a^{x_2} f(x)\, dx - \int_a^{x_1} f(x)\, dx.$$

By subtracting the second area from the first, the area in the interval $[x_1; x_2]$ is specified. The area between x_1 and x_2 can also be expressed as the sum of the two areas above, the area between a and x_2 plus the *negative* of the area between a and x_1, but this requires the definition of the negative of an integral.

Definition—The Negative of an Integral: Given an integral $\int_a^x f(x)\,dx$, the negative of that integral $-\int_a^x f(x)\,dx$ is the integral with the limits of integration reversed; that is $\int_x^a f(x)\,dx$.

The implication of this definition can be explored in relation to Figure 7.5. Thus

$$-\int_a^{x_1} \text{ is equal to } \int_{x_1}^a$$

and

$$\int_{x_1}^{x_2} f(x)\,dx = \int_a^{x_2} f(x)\,dx + \int_{x_1}^a f(x)\,dx.$$

Minus the integral from a to x_1 equals plus the integral from x_1 to a.

Throughout the preceding chapters, the general function considered as the starting point of a discussion and on which various operations (differentiation, taking of inverses, and so on) have been carried out has been designated f. For integration, the notational convention used in various operations fails in one way or another. Either discussion must begin with f', which is then integrated to give f, or the initial function continues to be referred to as f and its integral is designated by a new symbol, say F. The latter convention is generally used and is adopted here. The function F is called the *primitive function* of f.

Definition—Primitive Function: Given the function f, which is the derivative function of the function F, then F is known as the primitive function of f. In symbols, if

$$\frac{d}{dx}(F) = f,$$

then F is the primitive function of f.

What must now be shown is that the primitive function and the integral function are identical. Once this is done, then one needs only to differentiate the integral function to test whether a given integral is the correct result of integration. If the original integrand results, the integration has been appropriate. Theorems 7.1 and 7.2 deal with the characteristics of primitive functions.

Theorem 7.1: *If F is a primitive function of f, then G, whose rule is $G(x) = F(x) + k$, is also a primitive function of f, where k is a constant.*

Proof:

Given $F(x)$, a primitive function of f, then by definition,

$$\frac{d}{dx}F(x) = f(x).$$

Then taking the derivative of G gives

$$\frac{d}{dx} G(x) = \frac{d}{dx} [F(x) + k] = f(x) + 0 = f(x).$$

Thus, by the definition of a primitive function, $G(x)$ is a primitive function of $f(x)$.

Both $F(x)$ and $G(x)$ have the same derivative, $f(x)$. Theorem 7.2 is a slight variant of this.

Theorem 7.2: *All primitive functions of a given function differ from each other by at most a constant. That is, if $F_1(x)$ and $F_2(x)$ are both primitive functions of $f(x)$, then for k, a constant, $F_1(x) = F_2(x) + k$.*

This theorem is not proved here, but its plausibility is suggested by the following argument. Let $H(x) = F_1(x) - F_2(x)$.
Then

$$\frac{d}{dx} H(x) = \frac{d}{dx} F_1(x) - \frac{d}{dx} F_2(x), \tag{7.5}$$

but by definition, each factor on the right-hand side of equation (7.5) equals $f(x)$; therefore

$$\frac{d}{dx} H(x) = f(x) - f(x) = 0.$$

But if $(d/dx)H(x) = 0$, then $H(x)$ must be a constant, since only constants have derivatives equal to zero. If this is so, $H(x)$, the difference between any two primitive functions of $f(x)$ is a constant.

For the proof of Theorem 7.4, an intermediate step is required, a rule that is called "the first law of the mean for integrals," stated below without proof.

Theorem 7.3—The First Law of the Mean for Integrals: *Given the function f with rule $y = f(x)$, continuous in an interval $[a; b]$, let the minimum value of f in the interval be h and the maximum value H; then,*

$$\int_a^b f(x)\, dx = k(b - a)$$

for a particular value of k, where

$$h \le k \le H.$$

One way to interpret this theorem is as follows. The area under a curve in an interval is at most equal to the area of a rectangle with the maximum value of the function as its height and the width of the interval as the base, and at least equal to the area of a similar rectangle with the minimum value of the function as its height. And there is some mean value of the function such that a rectangle with that value as height and the domain interval as base exactly equals the area in the interval. Only if the function had constant value would the equality in either direction hold, and then both would hold. Another way to state this rule would be $\int_a^b f(x)\, dx = f(x_0)(b - a)$, for some particular value of x, called x_0 where $a \le x_0 \le b$. A simple case of such a situation is shown graphically in Figure 7.6.

Figure 7.6

Rectangle Raised on [a; b] as Base with $f(x_0)$ as Height.

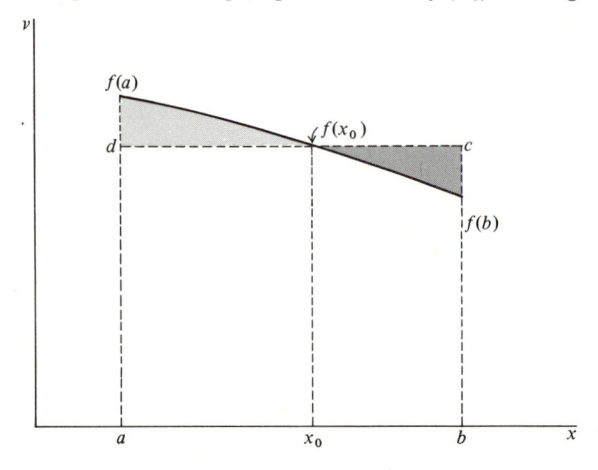

The theorem implies that some value of x, equal to x_0, can be found for which the shaded area left out of the rectangle *abcd* and the shaded area (included in the rectangle *abcd*, but not under the function $f(x)$ in the interval $[a; b]$) are equal, and thus the exclusion of the one balances the inclusion of the other. The true area under $f(x)$ in $[a; b]$ is equal to $f(x_0)(b - a)$; the area *abcd* is given by $f(x_0)(b - a)$.

The fundamental theorem, Theorem 7.4, is now presented and proved.

Theorem 7.4—The Fundamental Theorem: Given an integral function with rule $p(x) = \int_a^x f(x)\,dx$, continuous in an interval $[a; b]$, the integral function is a primitive function. In symbols,

$$p(x) = \int_a^x f(x)\,dx = F(x). \tag{7.6}$$

Proof:

Given $p(x) = \int_a^x f(x)\,dx$, a primitive function, differentiate $p(x)$ by use of the definition of a derivative,

$$\frac{d}{dx}[p(x)] = \lim_{\Delta x \to 0} \frac{p(x + \Delta x) - p(x)}{\Delta x}.$$

Substituting the equivalent integrals for $p(x + \Delta x)$ and $p(x)$ gives

$$\frac{d}{dx}[p(x)] = \lim_{\Delta x \to 0} \frac{\int_a^{(x+\Delta x)} f(x)\,dx - \int_a^x f(x)\,dx}{\Delta x}.$$

If the first integral is divided into two equivalent parts,

$$\frac{d}{dx}[p(x)] = \lim_{\Delta x \to 0} \frac{\int_a^x f(x)\,dx + \int_x^{(x+\Delta x)} f(x)\,dx - \int_a^x f(x)\,dx}{\Delta x}$$

Sec. 7.2] THE FUNDAMENTAL THEOREM

which gives

$$\frac{d}{dx}[p(x)] = \lim_{\Delta x \to 0} \frac{\int_x^{(x+\Delta x)} f(x)\, dx}{\Delta x}.$$

By Theorem 7.3, the first law of the mean for integrals, the numerator (which can be interpreted as an area) can be replaced by $f(x_0) \cdot \Delta x$;[4] therefore

$$\frac{d}{dx}[p(x)] = \lim_{\Delta x \to 0} \frac{f(x_0) \cdot \Delta x}{\Delta x},$$

where $x \leq x_0 \leq x + \Delta x$. The Δx in the numerator cancels that in the denominator and

$$\frac{d}{dx}[p(x)] = \lim_{\Delta x \to 0} f(x_0)$$

where $x \leq x_0 \leq x + \Delta x$. In the limit as $\Delta x \to 0$, $x_0 \to x$ and $f(x_0) \to f(x)$, therefore, the integral function has $f(x)$ as its derivative, and by definition must be a primitive function of $f(x)$

$$p(x) = \int_a^x f(x)\, dx = F(x). \tag{7.6}$$

Given that $F(x)$ is a primitive function of the integral function f, then by Theorem 7.1 there are an infinite number of primitive functions for f whose rules differ from each other only by a constant. Thus the value of the primitive function F at any point is *indefinite;* this is generally recognized by the fact that the rule for F is usually written

$$\int_a^x f(x)\, dx = F(x) + C$$

and is known as the indefinite integral. A constant of integration C has been added to $F(x)$ to indicate the family of functions associated with a given integral. Figure 7.7 shows a family of functions, which differ from one another at each domain point only by a constant.

Stated in set terms, the integral function F defined in the interval $[a; b]$ is

$$F = \{(x, y); a \leq x \leq b, y = \int_a^x f(x)\, dx\}.$$

The function F takes on values only where the limit implied by the integral exists. Just as the rule for the derivative function f' involved the rule of the function f itself

$$y' = \lim_{\Delta x \to 0} \frac{f(x + \Delta x) - f(x)}{\Delta x}$$

with value equal to the value of the limit (a limit of the quotient of differences), so the rule for the integral function F involves the rule for the original function f

---

[4] The area is equal to the base of a rectangle times its height. The base goes from x to $x + \Delta x$ (as the previous one went from a to b) and is thus equal to Δx. The derivative is being evaluated at x.

Figure 7.7

Family of Primitive Functions F.

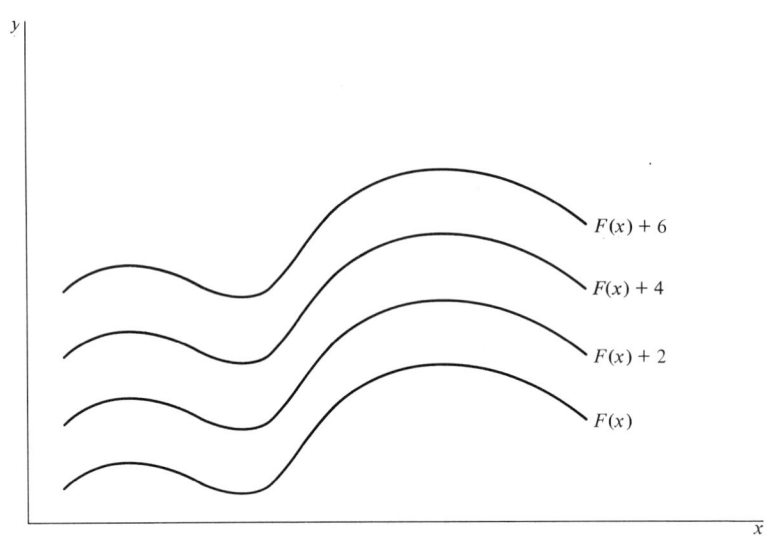

itself in the integral limit

$$y = \int_a^x f(x)\,dx = \lim_{n\to\infty} \sum_{i=1}^n f(x_i)\,\Delta x_i, \qquad (a \le x_i \le x)$$

with value also equal to the value of a limit, but this time the limit of a sum of products.

Given the first law for the mean for integrals, the proof of Theorem 7.4 follows readily. But it is important to note that the logic of the collapsing of the interval (for the average rate of change) in the limit implies that it is not essential to choose the mean value $f(x)$ in each interval. No matter what value of $f(x)$ was selected in an interval, as $\Delta x \to 0$, the value of $f(x_i)$ approaches the value of $f(x)$ at the left-hand endpoint of the interval, and the result given in Theorem 7.4 would obtain. This implies that, in calculating the value of an integral function over a given interval $[a; b]$ by dividing the interval into n equal subintervals, and allowing n to become extremely large, the area under the function can be approximated (to any desired degree of accuracy) by summing the areas of the rectangles raised on the subintervals $\Delta x = (b - a)/n$ as base and with $f(x_i)$ as height for x_i an *arbitrary* point in each interval:

$$A = \lim_{n\to\infty} \sum_{i=1}^n f(x_i)\,\Delta x. \tag{7.7}$$

This is essentially the procedure used in the numerical calculation of the value of an integral.

The discussion of the fundamental theorem thus far has dealt with the indefinite integral, $\int_a^x f(x)\,dx$, whose value is a function of the upper limit of integration. Now the corollary to the fundamental theorem is stated in terms of the definite integral.

Sec. 7.3]   RULES FOR INTEGRATION

Corollary to the Fundamental Theorem: Given the function f continuous in $[a; b]$ with primitive function F, then

$$\int_a^b f(x)\,dx = F(b) - F(a). \tag{7.8}$$

Proof:

By the fundamental theorem,

$$\int_a^x f(x)\,dx = F(x) + A$$

where A is a constant of integration associated with the endpoint a of the interval. But for $x = a$, the area under the point a is zero so

$$0 = F(a) + A$$

or

$$-F(a) = A.$$

For $x = b$,

$$\int_a^b f(x)\,dx = F(b) + A.$$

Substituting $-F(a)$ for A gives

$$\int_a^b f(x)\,dx = F(b) - F(a).$$

Thus, for the evaluation of the definite integral, the additive constant, which is determined by the lower limit of integration a, cancels and only the value of the function F and the endpoints a and b for the interval are relevant. In evaluating definite integrals the rule for F is determined, then evaluated at the points a and b with the following notation:

$$\int_a^b f(x)\,dx = F(x)\Big]_a^b = F(b) - F(a).$$

7.3 Rules for Integration

Although a function f can be continuous in an interval and still have no derivative at one or more points in the interval, this result implies that the integral of a function f, if f is continuous in an interval, must exist and be continuous. That is, given f continuous in $[a; b]$, then for

$$F(x) = \int_a^x f(x)\,dx$$

the function F is continuous for each x of $[a; b]$. The continuity of f in $[a; b]$ is a sufficient condition for the continuity of F in $[a; b]$.

It is now possible to reverse several of the rules for derivative functions in order to establish the rules for integral functions. (Several rules are listed at the end of the chapter.)

Given the function f whose rule is

$$y = f(x) = k \cdot x,$$

where k is a constant, then the rule for the derivative function f' is

$$y' = k.$$

Reversing this rule gives the rule for an integral function: Given the function f with rule

$$y = \int k\, dx \left(= \int_a^x k\, dx \right),$$

where k is a constant, then the rule for the integral function is

$$y = F(x) = k \int dx = kx + C. \qquad (7.9)$$

Similarly, given the function f whose rule is

$$y = x^n$$

then the rule for the derivative function f' is

$$y' = nx^{n-1}.$$

Reversing the process gives the rule for F: Given the integral function with rule,

$$F = \int f(x)\, dx = \int x^n\, dx$$

then the rule for F is

$$F = \frac{x^{n+1}}{(n+1)} + C. \qquad (7.10)$$

Similarly, given u and v as continuous functions with rules $u = f(x)$ and $v = g(x)$, then

$$\int (u \pm v)\, dx = \int u\, dx \pm \int v\, dx + c \qquad (7.11)$$

$$\int (u \pm k)\, dx = \int u\, dx \pm kx + c. \qquad (7.12)$$

Then for the functions f continuous in an interval with rules given as the integrands in equations (7.13) to (7.15), the rules for the primitive functions F are

$$F(x) = \int e^{kx}\, dx = \frac{e^{kx}}{k} + C \qquad (7.13)$$

$$F(x) = \int a^x\, dx = \frac{a^x}{\log_e a} + C \qquad (7.14)$$

$$F(x) = \int \frac{1}{x}\, dx = \log_e x + C. \qquad (7.15)$$

Each of these rules may be checked by differentiating the expression on the right-hand side to insure that the rule for the original function does in fact result. These basic rules are now illustrated in a series of examples.

Example 7.1

Given the function f with rule

$$y = (x^3 + 2x + 4)$$

find the rule for F, its primitive function. The integral of the sum can be found as the sum of the integrals of the individual terms; the rule for the integral of each term is of the form

$$\int x^n \, dx = \frac{x^{n+1}}{n+1} + C$$

then

$$F(x) = \int (x^3 + 2x + 4) \, dx$$

$$F(x) = \int x^3 \, dx + 2 \int x \, dx + 4 \int dx$$

$$= \frac{x^4}{4} + 2\frac{x^2}{2} + 4\frac{x}{1} + C$$

$$= \frac{x^4}{4} + x^2 + 4x + C.$$

That this is the rule for the primitive function F is confirmed by taking the derivative of F

$$\frac{d}{dx} F(x) = \frac{d}{dx} \left(\frac{x^4}{4} + x^2 + 4x + C \right) = x^3 + 2x + 4$$

which is the rule for f.

Example 7.2

Given the function f with rule

$$y = e^{4x}$$

find the rule for its primitive function F. The rule for F is in the form

$$F(x) = \int e^{kx} \, dx = \frac{e^{kx}}{k} + C$$

or specifically

$$F(x) = \frac{e^{4x}}{4} + C.$$

Differentiating gives

$$\frac{d}{dx} F(x) = \frac{d}{dx} \left(\frac{e^{4x}}{4} \right) = \frac{1}{4} (e^{4x} \cdot 4) = e^{4x}.$$

This completes the check.

Example 7.3

Given the function f with rule $y = f(x) = 27/x$, find the rule for its primitive function F:

$$F(x) = \int \frac{27}{x} \, dx.$$

The rule for F is in the form

$$F(x) = k \int \frac{1}{x} \, dx = k \ln x + C$$

specifically

$$F(x) = 27 \int \frac{1}{x} \, dx = 27 \ln x + C.$$

Checking by differentiation gives

$$\frac{d}{dx} F(x) = \frac{d}{dx} (27 \ln x + C)$$

$$F'(x) = \frac{27}{x} = f(x).$$

The proper rule has been found.

Example 7.4

Given the function f with rule $y = f(x) = (x^2 + 4)$, find the rule for its primitive function F and evaluate it from 0 to 3.

$$F(x) = \int_0^3 (x^2 + 4) \, dx$$

$$F(x) = \left[\frac{x^3}{3} + 4x \right]_0^3.$$

By the corollary to the fundamental theorem, the value F is equal to $F(3) - F(0)$.

$$F(x) = \frac{(3)^3}{3} + 4(3) - \frac{0^3}{3} + 4(0)$$

$$= 9 + 12 - 0 = 21.$$

Example 7.5

Given the function f with rule $y = (x^3 - 3x + 4)/\sqrt{x}$, find the rule for its primitive function F:

$$F(x) = \int \left(\frac{x^3 - 3x + 4}{\sqrt{x}} \right) dx.$$

The rule for F can be found by dividing each term in the numerator by the denominator and then integrating each term individually.

$$F(x) = \int (x^{5/2} - 3x^{1/2} + 4x^{-1/2}) \, dx$$

$$= \int (x^{5/2}) \, dx - 3 \int (x^{1/2}) \, dx + 4 \int (x^{-1/2}) \, dx.$$

Each integrand is now in the form x^n, and rule $F(x) = [1/(n+1)]x^{(n+1)}$ applies:

$$F(x) = \frac{1}{7/2} x^{7/2} - 3 \left(\frac{1}{3/2} \right) x^{3/2} + 4 \left(\frac{1}{1/2} \right) x^{1/2} + C$$

$$F(x) = \tfrac{2}{7} x^{7/2} - 2x^{3/2} + 8x^{1/2} + C.$$

Sec. 7.3]   RULES FOR INTEGRATION   233

Testing by differentiation gives
$$\frac{d}{dx}F(x) = x^{5/2} - 3x^{1/2} + 4x^{-1/2}.$$

The proper rule has been found.

Two additional rules for integral functions of logarithmic functions are given below. Given the functions f with rules given as the integrands below, the rules for F are

$$F(x) = \int \log_e x \, dx = (x \log_e x - x) + C \tag{7.16}$$

$$F(x) = \int \log_a x \, dx = (x \log_a x - x \log_a e) + C. \tag{7.17}$$

The test of these rules is left to the reader, but their use is illustrated in Examples 7.6 and 7.7.

Example 7.6

Given the function f with rule $y = 3x^2 + \log_3 x$, find the rule for its primitive function F.

$$F(x) = \int (3x^2 + \log_3 x) \, dx = 3 \int x^2 \, dx + \int \log_3 x \, dx$$

The rule
$$F(x) = \int \log_a x \, dx = (x \log_a x - x \log_a e) + C$$

applies to the term $\int \log_3 x \, dx$.

$$F(x) = x^3 + x \log_3 x - x \log_3 e + C.$$

Testing this by differentiation:

$$\frac{d}{dx}F(x) = \frac{d}{dx}(x^3 + x \log_3 x - x \log_3 e + C)$$

$$F'(x) = 3x^2 + x\left(\frac{1}{x}\right)\log_3 e + \log_3 x - \log_3 e + 0$$

$$F'(x) = 3x^2 + \log_3 x = f(x).$$

Example 7.7

Given the function f with rule $y = \log_e x + 3^x$, find the rule for the primitive function F;

$$F(x) = \int (\log_e x + 3^x) \, dx.$$

The rule for F is of the form

$$F(x) = \int \log_e x \, dx + \int a^x \, dx$$

$$= x \log_e x - x + \frac{a^x}{\log_e a} + C.$$

Substituting 3^x for a^x

$$F(x) = (x \log_e x - x) + \frac{3^x}{\log_e 3} + C.$$

Checking by differentiation gives

$$\frac{d}{dx} F(x) = \frac{d}{dx}\left[(x \log_e x - x) + \frac{3^x}{\log_e 3} + C\right]$$

$$F'(x) = \left[x \cdot \frac{1}{x} + \log_e x(1) - 1\right] + \frac{3^x(1)(\log_e 3)}{\log_e 3} + 0$$

and simplifying gives

$$F'(x) = 1 + \log_e x - 1 + 3^x$$
$$F'(x) = \log_e x + 3^x = f(x).$$

7.4 Integration by Substitution and Integration by Parts

Two very useful techniques for determining the rules for integral functions from functions whose rules may be quite complicated are presented next: integration by substitution and integration by parts. The rule for integration by substitution is a counterpart to the chain rule for differentiation. A complex function $g(x)$ can be replaced by a substitute function u and integrated as $f(u) \, du$.

$$F(u) = \int f(u) \, du = \int f(u) g'(x) \, dx \tag{7.18}$$

Since $u = g(x)$, $du = g'(x) \, dx$ by the chain rule,

$$\int f(u) \, du = \int f[g(x)] g'(x) \, dx. \tag{7.18a}$$

Skilled use of substitution can lead to the integration of functions that appear quite difficult to integrate. The first example is straightforward; the next two are increasingly complex.

Example 7.8

Given the function f with rule $y = [4/(x + 1)] \, dx$, find the rule for its primitive function F.

$$F(x) = \int \frac{4}{(x + 1)} \, dx.$$

The rule for F will be of the form

$$F(x) = \int \frac{k}{u} \, du$$

where $u = (x + 1)$ and $du = dx$, or

$$F(x) = k \log_e (u) + C = k \log_e (x + 1) + C.$$

Specifically,

$$F(x) = 4 \int \frac{1}{(x + 1)} \, dx = 4 \log_e (x + 1) + C$$

and checking by differentiation gives
$$F(x) = 4\log_e(x+1) + C$$
$$\frac{d}{dx}F(x) = 4\left[\frac{1}{x+1} \cdot 1 \cdot 1\right] = \frac{4}{x+1} = f(x).$$

Example 7.9

Given the function f with rule
$$y = \frac{x}{(x^2+4)^{1/2}},$$
find the rule for its primitive function F.
$$F(x) = \int \frac{x\,dx}{(x^2+4)^{1/2}}.$$

In this case the integral is close to the form $\int u^n\,du$ where $u = (x^2 + 4)$, where $n = -\frac{1}{2}$ and $du = 2x\,dx$. But since $du = 2x\,dx$, and $x\,dx$ is all that appears in the original integrand, a multiplication factor 2 must be included in du, with its influence reversed by multiplying the integral by $\frac{1}{2}$:

$$F(x) = \frac{1}{2}\int u^n\,du$$
$$F(x) = \frac{1}{2}\left[\frac{1}{(n+1)}u^{n+1}\right] + C.$$

Substituting gives:
$$F(x) = \frac{1}{2}\left[\frac{(x^2+4)^{1/2}}{\frac{1}{2}}\right] + C$$
$$= (x^2+4)^{1/2} + C.$$

Checking by differentiation gives:
$$\frac{d}{dx}F(x) = \frac{d}{dx}[(x^2+4)^{1/2} + C]$$
$$= \frac{1}{2}(x^2+4)^{-1/2} \cdot 2x$$
$$= \frac{x}{(x^2+4)^{1/2}} = f(x).$$

Example 7.10

Given the function f with rule
$$y = f(x) = \frac{1+3x}{2\sqrt{x}},$$
find the rule for its primitive function F:
$$F(x) = \int \frac{1+3x}{2\sqrt{x}}\,dx.$$

Let $x = u^2$, then $dx = 2u\,du$. Substituting gives

$$F(x) = \int \frac{1 + 3u^2}{2u} \cdot 2u\,du = \int (1 + 3u^2)\,du = u + u^3 + C$$

and substituting back the value of $u = \sqrt{x}$ gives

$$F(x) = \sqrt{x} + x\sqrt{x} + C = \sqrt{x}(1 + x) + C.$$

This result can be checked by differentiation.

Although the rule for the derivative of a product of two functions is very useful, it cannot be reversed directly. However, an analogous rule that is very useful in integration is the rule for integration by parts (the proof of which depends on the product rule).

Rule 7.1: *Given the functions f and g differentiable in an interval $[a; b]$ with rules $u = f(x)$ and $v = g(x)$, then*

$$\int f(x) \cdot g'(x) = f(x) \cdot g(x) - \int g(x) \cdot f'(x). \tag{7.19}$$

In terms of u and v, where

$$\begin{aligned} u &= f(x) & v &= g(x) \\ du &= f'(x) & dv &= g'(x) \end{aligned}$$

this rule is more commonly stated as

$$\int u\,dv = u \cdot v - \int v\,du. \tag{7.19a}$$

Proof:

Given the functions f and g with rules

$$\begin{aligned} u &= f(x) \\ v &= g(x) \end{aligned}$$

and their product

$$f(x) \cdot g(x)$$

the derivative of this product is given by:

$$\frac{d}{dx}[f(x) \cdot g(x)] = f(x)\frac{d}{dx}g(x) + g(x)\frac{d}{dx}f(x)$$

$$\frac{d}{dx}[f(x) \cdot g(x)] = f(x) \cdot g'(x) + g(x) \cdot f'(x).$$

Solving this for $f(x) \cdot g'(x)$ gives

$$-f(x) \cdot g'(x) = -\frac{d}{dx}[f(x) \cdot g(x)] + g(x) \cdot f'(x).$$

Changing signs and taking the integral of both sides gives

$$\int f(x) \cdot g'(x) = \int \frac{d}{dx}[f(x) \cdot g(x)] - \int g(x) \cdot f'(x). \tag{7.20}$$

Sec. 7.4]   INTEGRATION BY SUBSTITUTION AND INTEGRATION BY PARTS   237

By the fundamental theorem, the first term on the right-hand side of equation (7.20) reduces to $f(x) \cdot g(x)$:

$$\int f(x) \cdot g'(x) = f(x) \cdot g(x) - \int g(x) \cdot f'(x).$$

This is the same as equation (7.19) and, when equivalents are substituted, becomes:

$$\int u \, dv = u \cdot v + \int v \, du,$$

which is identical with equation (7.19a).

This very useful rule is now illustrated in a series of examples.

Example 7.11

Given the function f with rule

$$y = x^2 e^x$$

find the rule for the primitive function F.

$$F(x) = \int x^2 e^x \, dx.$$

Select

$$u = x^2 \qquad v = e^x$$
$$du = 2x \, dx \qquad dv = e^x \, dx.$$

Then by the rule of equation (7.19a),

$$\int u \, dv = u \cdot v - \int v \, du.$$

Substituting gives

$$\int x^2 e^x \, dx = x^2 e^x - \int e^x \cdot 2x \, dx. \qquad (7.21)$$

To find the integral of $-\int e^x \cdot 2x \, dx$, again the rule of equation (7.19a) must be used. Select

$$u = x \qquad v = e^x$$
$$du = dx \qquad dv = e^x \, dx$$

then

$$-2 \int x e^x \, dx = -2 \left(x e^x - \int e^x \, dx \right) + C = -2(x e^x - e^x) + C$$
$$= 2e^x - 2x e^x + C.$$

Substituting this result into equation (7.21) gives

$$\int x^2 e^x \, dx = x^2 e^x + 2e^x - 2x e^x = e^x(x^2 - 2x + 2) + C.$$

The next example is Example 7.10 repeated to illustrate its integration by parts.

Example 7.12

Given the function f with rule

$$y = \frac{1 + 3x}{2\sqrt{x}}$$

find the rule for its primitive function through integration by parts. Selecting

$$u = (1 + 3x) \qquad v = x^{1/2}$$
$$du = 3\,dx \qquad dv = \tfrac{1}{2}x^{-1/2}\,dx$$

and substituting into equation (7.19a) gives

$$\int u\,dv = u \cdot v - \int v\,du \qquad (7.19a)$$

$$\int \frac{(1+3x)}{2\sqrt{x}}\,dx = \sqrt{x}\,(1+3x) - 3\int \sqrt{x}\,dx$$
$$= \sqrt{x}\,(1+3x) - 3(\tfrac{2}{3}x^{3/2}) + C$$
$$= \sqrt{x}\,(1+3x) - 2x^{3/2} + C$$
$$= \sqrt{x}\,(1 + 3x - 2x) + C$$
$$= \sqrt{x}\,(1 + x) + C.$$

The use of some of the simpler rules for integral functions derived thus far is now illustrated in an extensive example in an economic context.

Example 7.13

Given the functions c with rule $c(q) = 4 + (q - 4)^2$ for the marginal cost and r with rule $r(q) = 20 - 2q$ for the marginal revenue of a firm where q is the quantity of product produced, find:

(a) The profit-maximizing output for the firm.
(b) The total profit for the firm at that output.
(c) The profit that would be achieved if output were increased by two units beyond the profit maximizing output.
(d) The rule for the demand function for the firm.
(e) The rule for the average cost function of the firm.

(a) Profit maximization requires that $r = c$, or $r - c = 0$. Because profit equals total revenue minus total cost, as derived in Chapter 4 from the profit function, then

$$r - c = 0$$
$$(20 - 2q) - [4 + (q - 4)^2] = 0 \qquad (7.22)$$

Solving for q, this gives:

$$20 - 2q - (4 + q^2 - 8q + 16) = 0$$
$$6q - q^2 = 0$$
$$q(6 - q) = 0$$
$$q = 0, 6.$$

The maximum profit output occurs at one (or both) of these points. Equation (7.22) is the first derivative $d\pi/dq$ of the profit function. The second derivative is found by

Sec. 7.4] INTEGRATION BY SUBSTITUTION AND INTEGRATION BY PARTS 239

Figure 7.8

The Marginal Revenue (r) and Marginal Cost (c) Functions and the Areas between Them.

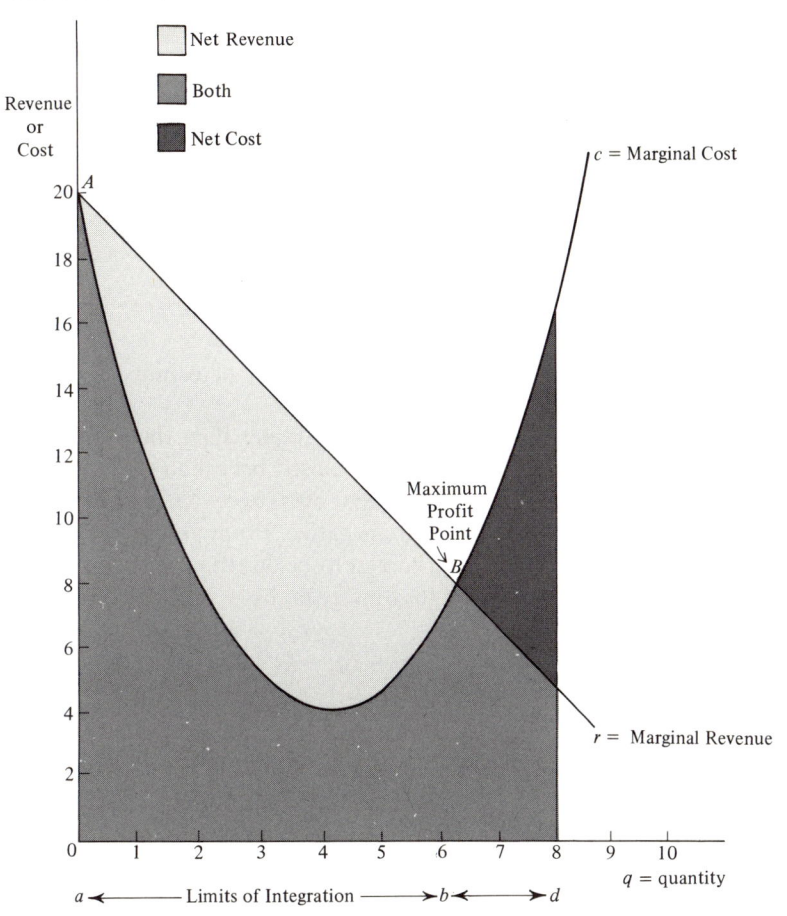

taking its derivative in turn:

$$\frac{d^2\pi}{dq^2} = \frac{d}{dq}(6q - q^2)$$

$$\frac{d^2\pi}{dq^2} = 6 - 2q.$$

At $q = 0$, it equals 6 identifying this as a minimum. At $q = 6$, it equals -6; an output of 6 units will provide maximum profit.

(b) Total revenue is found from the integral of marginal revenue, and total cost is found from the integral of marginal cost. Profit is the difference between total revenue and total cost. Graphically it is the area between the curve for marginal profit and the curve for marginal revenue shown in Figure 7.8. Dark shading indicates net cost; light shading indicates net revenue.

Total profit π_1 is thus given by the integral of (area under) r minus the integral of (area under) c for the appropriate limits of integration, in Figure 7.8 from $a = 0$ to $b = 6$. Thus,

$$\pi_1 = \int_a^b r\, dq - \int_a^b c\, dq = \int_0^6 (20 - 2q)\, dq - \int_0^6 [4 + (q - 4)^2]\, dq.$$

Since the limits are the same for both integrals and both involve the variable q, the integrals can be combined and simplified before integrating as follows:

$$\pi_1 = \int_0^6 (6q - q^2)\, dq = \left(3q^2 - \frac{q^3}{3}\right)\Big|_0^6$$
$$\pi_1 = [3(36) - \tfrac{6}{3}(36)] - [0]$$
$$\pi_1 = 36.$$

The total profit from the profit maximizing output of 6 units is 36 (thousand dollars, perhaps).

(c) The total profit π_2 from an output of 2 greater than the profit maximizing output is the sum of the two areas shown enclosed between r and c in Figure 7.8. From the point B on in Figure 7.8, marginal cost exceeds marginal revenue, and profits for output in excess of 6 units are negative. From part (b) of this example, total profits in the interval [0; 6] are known to be 36 (thousand dollars, perhaps). The (negative) profit in the interval [6; 8] is given by

$$\pi_2 = \int_b^d c\, dq - \int_b^d r\, dq = \int_6^8 [4 + (q - 4)^2]\, dq - \int_6^8 (20 - 2q)\, dq$$

or

$$\pi_2 = \int_6^8 (q^2 - 6q)\, dq = \left(\frac{q^3}{3} - 3q^2\right)\Big|_6^8$$
$$= [\tfrac{8}{3}(64) - 3(64)] - [\tfrac{6}{3}(36) - 3(36)]$$
$$= [-\tfrac{1}{3}(64)] - [-36]$$
$$= (-21\tfrac{1}{3} + 36) = 14\tfrac{2}{3}.$$

Thus total profit would be

$$36 - 14\tfrac{2}{3} = 21\tfrac{1}{3}$$

if output were increased from 6 to 8 units.

(d) The demand function D for a firm is an average revenue function AR. Average revenue is equal to total revenue $\int r\, dq$ divided by quantity q; thus the rule for the demand function can be found from

$$D = AR = \left(\int r\, dq\right) \div q$$
$$= \left[\int (20 - 2q)\, dq\right] \div q = (20q - q^2) \div q$$
$$= 20 - q.$$

Observe that the rule for the demand curve is a straight line with the same intercept as the rule for r, but with slope only $\tfrac{1}{2}$ the slope of the rule for r.

Figure 7.9

Area under the Graph of the Function f with Rule $y = f(x)$ for Values of f that Are Both Positive and Negative.

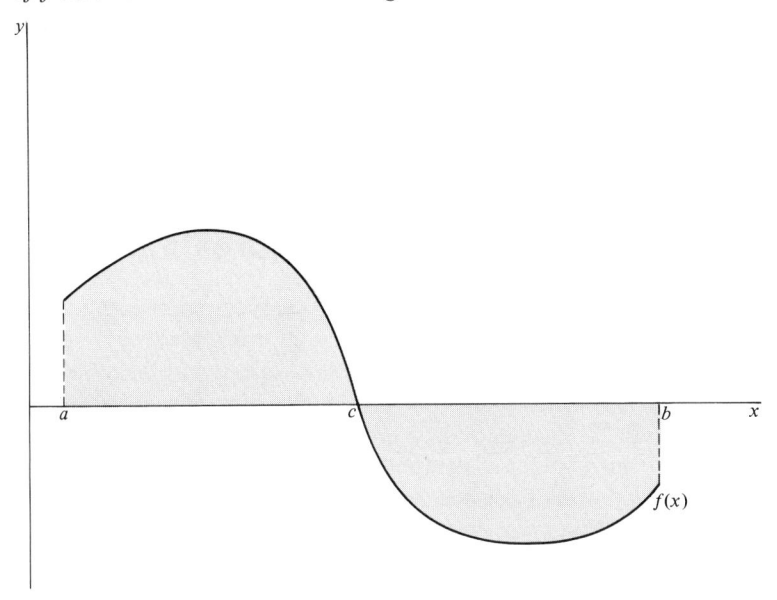

(e) The rule for the average cost function AC is found from the rule for the total cost function divided by the quantity produced; thus $AC = (\int c\, dq) \div q$ or

$$AC = \left(\int [4 + (q-4)^2]\, dq\right) \div q$$
$$= \left[\int (20 - 8q + q^2)\, dq\right] \div q$$
$$= \left(20q - 4q^2 + \frac{q^3}{3}\right) \div q$$
$$= 20 - 4q + \frac{q^2}{3}.$$

Profit could be determined from the average cost and average revenue functions which can also be graphed (but have not been here).

Throughout the preceding discussion the function $f(x)$ has been assumed to be everywhere positive, but this does not imply that it is not possible to find the area under a curve for values of the function that are negative. The question arises when an area that is positive is bounded by the endpoints of an interval, the x axis, and a *negative* value of the function in the interval. From the definition of an integral as the limit of a sum of products, the negative value of $f(x)$ in each product leads to a negative number that would imply a negative area, an illogical result. The usual convention for dealing with this problem is also adopted here; for those domain intervals over which the function is negative, the integral representing the area is preceded by a negative sign, thus reversing the sign of the product in that interval. The situation is illustrated in Figure 7.9.

The area under $f(x)$ in the interval $[a; b]$ is the sum of the two shaded areas. By the convention, this area is represented by

$$\int_a^c f(x)\,dx - \int_c^b f(x)\,dx.$$

The calculation of such areas is illustrated in examples later in the chapter and is equivalent to what was done in (b) of Example 7.13 above.

Other approaches to integration are now considered.

7.5 Integration by Numerical Approximation

In the discussion of the proof of the fundamental theorem, it was noted that the area under a curve in a given interval could be approximated by a series of rectangles raised on subdivisions of the interval as base and with height equal to any value of the function in that subinterval. For computation of such an area by use of a computer program, a systematic means must be found for determining the size of the subinterval and a height of the function in the subinterval. In such a calculation, equal subintervals are easiest to handle, and the height of the function at the left endpoint of the subinterval is the simplest height to calculate.

Given the function f with rule $y = f(x)$ in the interval $[a; b]$, the interval is divided into n equal subintervals, and each is of size

$$\Delta x = \frac{b-a}{n}.$$

For the height of the first rectangle, $f(a)$ is used, for the second $f\{a + [(b - a)/n]\}$, for the third $f\{a + 2[(b - a)/n]\}$, and so on. For very large n, the approximation to the area under f in $[a; b]$ will be quite close. There are other approaches to numerical approximation of the integral of (area under) f in an interval, using the trapezoidal rule, Simpson's rule, and the Taylor series,[5] all of which give a closer approximation than does the rectangle rule. The trapezoidal rule is considered in detail.

The trapezoidal rule involves a procedure that provides a linear approximation to the mean value of f in each subinterval, and thus generally represents a closer approximation to the area under f than does the use of rectangles with height arbitrarily set equal to the left endpoint in the interval. For this approach to approximation, the base of the trapezoid is the height of the function at the left endpoint of the subinterval (the height of f at $x = 1$ in Figure 7.9, for example), and the height of each trapezoid is the width of the subinterval Δx. That is, the area under f is approximated by the areas of n trapezoids raised on bases equal to the height of f at its left endpoint in each subinterval and with equal heights

---

[5] Simpson's rule requires the use of parabolic arcs to approximate the function f over a series of three values of $f(x)$ for an *even* number of equal subintervals, $\Delta x = (b - a)/n$. The approximation by Simpson's rule for any given n is closer than that by the trapezoidal rule. Either of these rules can be used to approximate the area under a graph of a function whose rule is not known or a function defined from a table of values of unknown rule, or for a function whose rule is known. The approximating technique that relies on Taylor's theorem requires the integration of a Taylor expansion for a given function term by term.

Sec. 7.5]   INTEGRATION BY NUMERICAL APPROXIMATION   243

Figure 7.10

Area under f with Rule $y = (x^2/5) - 2x + 7$ Approximated by Trapezoids.

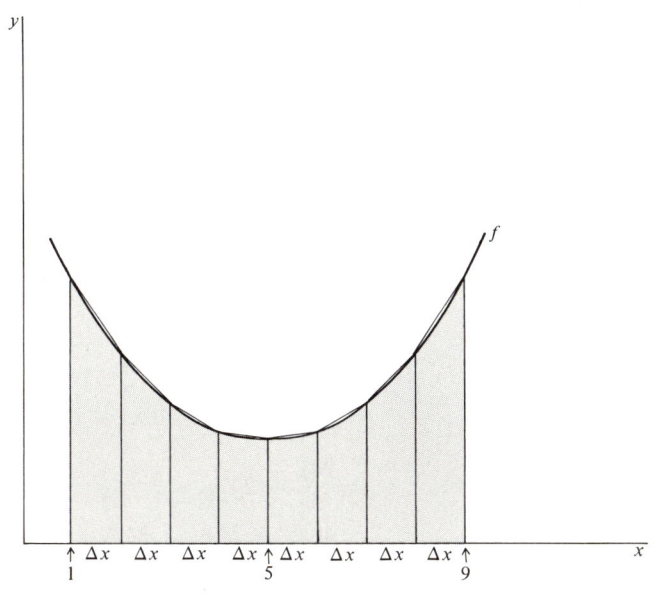

$\Delta x = (b - a)/n$. The area of a given trapezoid in the interval $(x_i - x_{i-1})$ is given by

$$\tfrac{1}{2}[f(x_{i-1}) + f(x_i)] \Delta x.$$

The total area of the n trapezoids is

$$\Delta x \left[\tfrac{1}{2}f(a) + \sum_{i=1}^{n-1} f(x_i) + \tfrac{1}{2}f(b) \right]$$

and by the trapezoid rule,

$$\int_a^b f(x)\, dx \cong \Delta x \left[\tfrac{1}{2}f(a) + \sum_{i=1}^{n-1} f(x_i) + \tfrac{1}{2}f(b) \right]. \qquad (7.23)$$

The calculation procedure using the trapezoid rule is shown in Example 7.14.

Example 7.14

Given the function f with rule $y = (x^2/5) - 2x + 7$, approximate the area under the graph of f by the trapezoid rule over the interval $[1; 9]$ for an n of 8. The function is graphed in Figure 7.10.

The base of each trapezoid is $\Delta x = (b - a)/n = (8/8) = 1$, and the values of $f(x_i)$ are given in Table 7.1; thus

$$\int_1^9 \left(\frac{x^2}{5} - 2x + 7 \right) dx \cong \Delta x[\tfrac{1}{2}f(x_0) + \tfrac{1}{2}f(x_8) + \sum_{i=1}^{7} f(x_i)]$$

$$\cong 1(5.2 + 19.6)$$

$$\cong 24.8.$$

Table 7.1

Values of the Function f at the Endpoints of Each Subinterval in [1; 9].

	\multicolumn{9}{c}{Subinterval}
i	0	1	2	3	4	5	6	7	8
x_i	1	2	3	4	5	6	7	8	9
$f(x_i)$	5.2	3.8	2.8	2.2	2.0	2.2	2.8	3.8	5.2

Given the rule for this function, the area can be checked by integration:

$$\int_1^9 \left(\frac{x^2}{5} - 2x + 7\right) dx = \left(\frac{x^3}{15} - x^2 + 7x\right)\bigg|_1^9$$

$$= \left[\frac{9(81)}{15} - 81 + 63\right] - (6\tfrac{1}{15})$$

$$= \frac{243}{5} - 24\tfrac{1}{15}$$

$$= 48\tfrac{9}{15} - 24\tfrac{1}{15} = 24\tfrac{8}{15}.$$

The approximation by the trapezoid rule is quite close even for an n of 8.

7.6 Improper Integrals

The value of an integral at any given point, say point a, is zero. This is seen from the definite integral,

$$\int_a^a f(x)\, dx = F(a) - F(a) = 0. \qquad (7.24)$$

Thus the value of an integral would not be affected by failure to consider the area under a curve at a single domain point. If a function f is not continuous at either (1) an endpoint of its domain or (2) an interior point in an interval, it may still be possible for the integral of f to exist and be meaningful. For example, if the function f is not continuous at its upper endpoint in the interval $[a; b]$, it is still possible for the integral of f to exist in the interval. If it does exist, such an integral function is known as an *improper integral*.

Definition—Improper Integral: *Given the function f with rule $y = f(x)$ continuous within the interval, $(a; b)$ but discontinuous at either endpoint of the interval, then the definite integral from a to b may exist if the value b at that endpoint is replaced by a domain value β within the continuous portion of the interval and if*

$$\lim_{\beta \to b} \int_a^\beta f(x)\, dx = L.$$

Sec. 7.6] IMPROPER INTEGRALS   245

If this limit does exist, it is defined to be

$$\int_a^b f(x)\,dx$$

and the integral is known as an improper integral.

This can be seen in Example 7.15 in which the function does not exist at its lower domain point.

Example 7.15

Given the function f with rule $y = x^{-2/3}$, find the rule for its primitive function F and evaluate its definite integral over the interval [0; 2]. The function f is continuous everywhere except at $x = 0$, where it is not defined. If the limit

$$\lim_{\alpha \to 0} \int_\alpha^2 x^{-2/3}\,dx$$

exists, then the integral from 0 to 2 can be evaluated. The rule for F is

$$F(x) = 3x^{1/3}$$

$$\lim_{\alpha \to 0} [F(2) - F(\alpha)] = \lim_{\alpha \to 0} [3(2)^{1/3} - 3(0)^{1/3}] = 3\sqrt[3]{2}.$$

Thus the limit exists and, by definition, is an improper integral,

$$\lim_{\alpha \to 0} \int_\alpha^2 x^{-2/3} = \int_0^2 x^{-2/3} = 3\sqrt[3]{2}.$$

A similar approach is used for the problem in which the function is undefined at its upper limit of integration. Consider next a problem involving an infinite lower limit.

Example 7.16

Given the function f with rule $y = f(x) = 1/x^2$, find the rule for its primitive function F and evaluate it as a definite integral from 0 to 2. The function f is not continuous at $x = 0$; thus

$$F(x) = \int_0^2 f(x)\,dx = \int_0^2 \frac{dx}{x^2}$$

may not exist. Replace the lower limit by α and find the rule for the primitive function F.

$$F(x) = \lim_{\alpha \to 0} \int_\alpha^2 \frac{dx}{x^2} = \lim_{\alpha \to 0} \int_\alpha^2 x^{-2}\,dx = \lim_{\alpha \to 0} \left(\frac{x^{-1}}{-1}\bigg|_\alpha^2\right)$$

$$\lim_{\alpha \to 0} [F(2) - F(\alpha)] = \lim_{\alpha \to 0} \left(-\frac{1}{2} + \frac{1}{\alpha}\right).$$

The second term does not exist; therefore, this integral does not exist.

If the function is discontinuous at a point interior to its domain interval, its integral is divided into separate integrals with the point of discontinuity as the upper limit of integration for the first integral and lower limit for the second integral; that is,

$$\int_a^b f(x)\,dx,$$

continuous in $[a; b]$ except at the point c in $[a; b]$, is redefined in terms of the improper integrals,

$$\int_a^b f(x)\,dx = \int_a^c f(x)\,dx + \int_c^b f(x)\,dx \qquad (7.25)$$

and both the integrals on the right-hand side of equation (7.25) must exist, otherwise $\int_a^b f(x)\,dx$ is meaningless.

It also is possible for the interval of integration itself to be infinite, in which case, if

$$\lim_{b \to \infty} \int_a^b f(x)\,dx = L$$

then the integral exists, and the definition

$$\int_a^\infty f(x)\,dx = \lim_{b \to \infty} \int_a^b f(x)\,dx = L \qquad (7.26)$$

results; similarly,

$$\int_{-\infty}^b f(x)\,dx = \lim_{a \to -\infty} \int_a^b f(x)\,dx = L \qquad (7.27)$$

and

$$\int_{-\infty}^\infty f(x)\,dx = \lim_{\substack{a \to -\infty \\ b \to \infty}} \int_a^b f(x)\,dx = L. \qquad (7.28)$$

In equations (7.26) through (7.28), the integral is called an improper integral. If the limit exists, the integral is said to converge and $f(x)$ is integrable; if it does not, the integral is called divergent. Example 7.17 involves a convergent integral.

Example 7.17

Given the function f with rule $y = f(x) = (1/x^2)$, find the rule for the primitive function F and evaluate it as a definite integral in the interval 2 to ∞. In Example 7.16, it was shown that the rule for F is

$$F(x) = \frac{x^{-1}}{-1}.$$

For the definite integral over the interval $[2; \infty]$,

$$F(x) = \lim_{b \to \infty} \frac{x^{-1}}{-1}\bigg]_2^b = \lim_{b \to \infty} \left(-\frac{1}{b} + \frac{1}{2}\right) = \left(0 + \frac{1}{2}\right).$$

The limit L exists and is equal to $\frac{1}{2}$; thus

$$\int_a^\infty \frac{dx}{x^2} = \frac{1}{2}$$

and $\int_a^\infty (dx/x^2)$ is an improper integral. A similar procedure is used for an infinite lower limit $(a = -\infty)$.

7.7 Multiple Integration

Just as there is partial differentiation, there is partial integration or what is called more generally multiple integration. A function with more than one domain variable can be integrated with respect to one variable with all others held constant and then with respect to the next, and so forth. The procedure is illustrated with an example,

$$\int_0^4 \int_0^{x^2} (xy)^{1/2} \, dy \, dx.$$

Integration is from the inside out. The first differential dy indicates the order of integration. The integral is treated first as if it were in the form

$$\int_0^{x^2} (xy)^{1/2} \, dy$$

and integrated with respect to y treating x as a constant

$$\tfrac{2}{3} x^{1/2} y^{3/2} \Big]_0^{x^2}.$$

The limits of integration are expressed in terms of the variable(s) treated as constants at this step. Substituting the values of the limits given for y gives

$$[\tfrac{2}{3} x^{1/2}(x^3) - 0] = \tfrac{2}{3} x^{7/2}. \tag{7.29}$$

The variable y is no longer involved in the integral. Equation (7.29) is now integrated with respect to x in accordance with the outer integral and differential.

$$\int_0^4 \tfrac{2}{3} x^{7/2} \, dx$$

to give

$$\tfrac{2}{3}(\tfrac{2}{9}) x^{9/2} \Big]_0^4 = \tfrac{4}{27} x^{9/2} \Big]_0^4$$

or

$$\tfrac{4}{27}(512 - 0) = \tfrac{2048}{27} = 75\tfrac{23}{27}.$$

Usually double integrals are all that are required for applications. Double integrals can be used to determine physical volumes, for example. Multiple integration to n dimensions is possible, however, should this become necessary. Such an integral is written

$$\underbrace{\int \cdots \int}_{n} f(x_1, \ldots, x_n) \, dx_1, \ldots, dx_n \tag{7.30}$$

The procedure is again illustrated in Example 7.18.

Example 7.18

Given the function f with rule $z = f(x, y) = x + y^2$, find the value of the primitive function of f by multiple integration first with respect to y and then to x from the limits $[0; x]$ and $[0; 2]$, respectively. The integral equal to F is given by

$$F(x, y) = \int_0^2 \int_0^x (x + y^2) \, dy \, dx.$$

Evaluating the first integral,

$$F(x, y) = \int_0^2 \left(xy + \frac{y^3}{3}\right)\Big|_0^x dx$$

or

$$F(x, y) = \int_0^2 \left(x^2 + \frac{x^3}{3}\right) dx.$$

Integrating with respect to x gives

$$F(x, y) = \left(\frac{x^3}{3} + \frac{x^4}{12}\right)\Big|_0^2 = \left(\frac{8}{3} + \frac{16}{12}\right)$$

$$F = \left(\frac{12}{3}\right) = 4.$$

Several rules for integration are summarized in the next section.

7.8 Integration Rules

Here is a list of rules that, for given integrals, provide the corresponding integral functions. The student should use the fundamental theorem to test them (such a test constitutes a proof of each rule). Lengthy tables of primitive functions such as these are available in published form from many sources.

Integration Rules

$$\int_a^b f(x)\,dx = F(b) - F(a) \text{ where } F'(x) = f(x)$$

$$\int x^n\,dx = \frac{x^{n+1}}{n+1} + C, \quad n + 1 \neq 0$$

$$\int kg(x)\,dx = k\int g(x)\,dx, \quad k \text{ a constant}$$

$$\int [f(x) + g(x)]\,dx = \int f(x)\,dx + \int g(x)\,dx$$

$$\int \frac{1}{x}\,dx = \log_e x + C$$

$$\int \log_e x\,dx = [x \log_e x - x] + C$$

$$\int e^{kx}\,dx = \frac{e^{kx}}{k} + C$$

$$\int a^x\,dx = \frac{a^x}{\log_e a} + C$$

$$\int \log_a x\,dx = [x \log_a x - x \log_a e] + C$$

7.9 Summary

The most common intuitive interpretation of an integral is that of an area under the graph of a function in an interval. The area is determined from the limit of the sum of the areas of rectangles raised on subintervals Δx_i as base and with the average value $f(x_i)$ of the function in the subinterval as height. The fundamental theorem of the calculus establishes that integration and differentiation are inverse processes. The significance of their being inverses can be expressed from the definitions of their rules. In symbols the rule for the derivative of the function f is stated:

$$\frac{dy}{dx} = \lim_{\Delta x \to 0} \frac{\Delta y}{\Delta x} = \lim_{\Delta x \to 0} \frac{f(x + \Delta x) - f(x)}{(x + \Delta x) - x} = \lim_{\Delta x \to 0} \frac{f(x + \Delta x) - f(x)}{\Delta x}.$$

It is the limit of the average rate of change in f in the interval Δx, the instantaneous rate of change in f; it is the limit of the quotient of the difference in range values of f divided by the difference in domain values. The rule for the integral function in an interval $[a; b]$ divided into n subintervals Δx_i can be stated as

$$\int_a^b f(x)\, dx = \lim_{\substack{\Delta x \to 0 \\ n \to \infty}} \sum_{i=1}^n f(x_i)\, \Delta x_i$$

where $f(x_i) = \frac{1}{2}(f(x_{ie}) + f(x_{ib}))$ and e stands for the endpoint of subinterval i and b its beginning point. The integral is the limit of the product of the average of instantaneous rates of change at the endpoints of each subinterval times Δx_i, the base of the subinterval. The values a and b, the upper and lower limits of integration, represent the endpoints of the interval over which integration is carried out. When, as here, both endpoints are specified, the integral is known as a definite integral. Its value is a single number equal to $F(b) - F(a)$. The symbol for the rule of an integral function

$$y = \int_a^x f(x)\, dx$$

with the domain variable x as the upper limit of integration is known as the indefinite integral (since the upper limit of integration is indefinite).

Given two functions F and f such that f is the derivative function of F, then F is the primitive function of f. The fundamental theorem of the calculus states that an integral function is a primitive function thus establishing that differentiation and integration are inverse processes, that is:

$$\frac{d}{dx} F(x) = \frac{d}{dx}\left(\int_a^x f(x)\, dx\right) = f(x).$$

If for a given function a rule for a primitive function can be found, then an infinite number of such rules can be found, each differing from the other by a constant (since the derivative of any constant is zero). Thus when the rule for a primitive function is written, a "constant of integration" is added to the rule for the primitive function, for example:

$$\int_a^x e^x\, dx = e^x + C.$$

The rules for several primitive functions are summarized in Section 7.8. The two most important approaches to integration are through integration by substitution (the counterpart to the chain rule for differentiation), and integration by parts (the counterpart to the product rule).

It is not always possible, and it is seldom easy to find the rule for a given primitive function, although once such a rule has been found, the fact that it is the primitive function can be readily verified by differentiation. This does not imply, however, that merely because a rule for a primitive function cannot be found, a given integral cannot be evaluated. It is frequently possible to evaluate a given definite integral through numerical computation and the limit process (if the limit exists). The rectangle rule and the trapezoidal rule are illustrative procedures for doing so.

If a function that is the integrand of an integral function is continuous, then its primitive function is continuous (but not necessarily vice versa). If a function that is being integrated is not continuous, it may be tested as an improper integral and if the limit process can be carried out on the appropriate integral functions, the integral can be evaluated. If not, it cannot.

Just as functions with more than one domain variable can be differentiated, through partial differentiation, so, too, can functions with more than one domain variable be integrated, through multiple integration. An important step in facilitating multiple integration is the appropriate specification of the limits of integration with respect to each variable as suggested in Section 7.7.

Key Concept

Concept	Reference Section
Areas under Curves	7.1
Comparison of Differentiation and Integration	7.1
Approximation of Areas under Curves by Rectangles	7.1
Definite Integral	7.2
Integral Function	7.2
Negative of an Integral	7.2
Primitive Function	7.2
Law of the Mean for Integrals	7.2
Fundamental Theorem of the Calculus	7.2
Constant of Integration	7.2
Rules for Integration	7.3
Integral of a Constant Times a Function	
Integral of a Polynomial	
Integral of a Sum of Functions	
Integral of e^{kx}	
Integral of a^x	
Integral of $1/x$	
Integration by Substitution	7.3
Integration by Parts	7.3
Application of Integration	7.3

Areas between Curves	7.3
Integration by Numerical Approximation	7.4
The Trapezoidal Rule	7.4
Simpson's Rule	7.4
Improper Integrals	7.6
Multiple Integration	7.7
Integration Rules	7.8

Appendix 7A: Applications of Integration

One very important group of applications of the integral and of the exponential function is in computing what are called present values and future values. These concepts will be encountered very frequently in subsequent work, notably in capital budgeting. These pages provide a mathematical basis for such applications, and prove some frequently used results.

7A.1 Future Value of a Sum

Suppose that a certain sum of money A is available today. In each future instant of time, this sum will grow at the instantaneous rate of $(10 = r)$ per cent per year. How much money will have accumulated t years later?

In symbolic terms, the future sum y is a function of t, $[y = f(t)]$ where the function f has the following properties:

$$(1) \quad f(0) = A$$
$$(2) \quad \frac{dy}{dt} = ry.$$

Condition (2) is satisfied by the function $y = f(t) = Ae^{rt}$ since $(d/dt)Ae^{rt} = rAe^{rt} = ry$.
Condition (1) is also satisfied since $f(0) = Ae^{r \cdot 0} = A$.

Example 7A.1

Initial sum is $100. The continuous growth rate is .125 = 12.5 per cent. The future period is eight years.

$$y = \$100 e^{.125t} = 100 e^{(.125)8}$$

if $t = 8$,

$$y = \$100 e^1 = 100(2.718) = \$271.80.$$

7A.2 Present Value of a Future Sum

The future value can be computed if a present sum is given, and the question asked: How much will it be worth at some future date assuming the amount increased at an instantaneous rate of r per period? The inverse of this problem occurs if a future sum, say B, is given which will become available at a specified future date. The question then arises, at an instantaneous growth rate of r: How large a present sum would be needed so that its future value at the specified future date would be equal to B?

In this instance the fact that $(dy/dt) = ry$ and that $y(t_n) = B$ are given. Here time t_n is the specified future date. The problem is to find the value of y at $t = 0$. Let A be the required present sum. The given condition $(dy/dt) = ry$ is satisfied by $y = y(t) = Ae^{rt}$ since t_n, $y(t_n) = Ae^{rt_n} = B$. Solving this equation for A, we have $A = Be^{-rt_n}$. A lender would like to earn 6 per cent on any loan he makes. The borrower agrees to pay a lump sum of \$1500 ten years from now. How much can the lender lend today and still satisfy his objectives?

$$x = (\$1{,}500)e^{-(.06)(10)}$$
$$= \frac{1}{e^{.60}}(\$1{,}500).$$

The antilog of $-.6$ is .549. The amount that could be lent is $(.549)(1{,}500) = \$824$.

7A.3 Present Value of an Annuity

A series of equal future receipts is called an annuity. Suppose that a piece of equipment has a useful life of 10 years. It can be rented at \$10,000 per year (in equal weekly installments), or purchased by an outright cash payment of \$65,000. The salvage value after 10 years is expected to be zero. Is the present value of the rental payments at an 8 per cent interest rate greater or less than the cash outlay?

The present value of the annuity, the series of rental payments, can be approximated by the following integral (since the integral is for continuous compounding and the problem calls for compounding 52 times per year).

$$A = \int_0^n Re^{-rt_n} dt \quad \text{where } R \text{ is the annual payment}$$
$$= \int_0^{10} (10{,}000)e^{-(.08)t} dt.$$

The indefinite integral has the following solution (using Equation (7.13)):

$$R\int e^{-rt} dt = \left(\frac{-e^{-rt}}{r} + C\right)R.$$

Applying the fundamental theorem for $r = .08$, $t = 8$, and $R = 10{,}000$ we have

$$\int_0^{10} (10{,}000)e^{-(.08)t} dt = (10{,}000)\left[\frac{-e^{-.8}}{.08} - \left(\frac{-1}{.08}\right)\right]_0^{10}.$$

Since

$$e^{-.8} = \frac{1}{e^{.8}} = \frac{1}{2.23} = .448; \quad \frac{1}{.08} = \frac{100}{8} = 12.5,$$

the final bracket is equal to:

$$[-.448(12.5) + (12.5)] = (12.5)(1 - .448).$$

Therefore

$$\int_0^{10} (10{,}000)e^{-.08t} dt = (\$10{,}000)(12.5)(.552) = \$69{,}000.$$

The present value of the annuity is \$4000 greater than the present value of the cash purchase price.

In case the annuity has an indefinite life, the greatest value of the annuity has a simple expression:

$$\int_0^\infty e^{-rt} dt = \left.\frac{-e^{-rt}}{r}\right|_0^\infty = 0 - \left(\frac{-1}{r}\right) = \frac{1}{r}.$$

Remember this result. It is especially useful in economics.

Appendix 7B: Table of Values of e^x

x	e^x	e^{-x}
.04	1.0408	.96079
.05	1.0513	.95123
.12	1.1275	.88692
.24	1.2712	.78663
.32	1.3771	.72615
.36	1.4333	.69768
.40	1.4918	.67032
.45	1.5683	.63763
.48	1.6161	.61878
.72	2.0544	.48675
.96	2.6117	.38289
1.44	4.2207	.23693
1.92	6.8210	.14661

Problems

Given the integrals below, find the rules for their primitive functions and evaluate them where appropriate.

1. $\int (ax^2 + b)\, dx$

2. $\int (ax)^{-1/3}\, dx$

3. $\int \dfrac{x^3}{ax}\, dx$

4. $\int (ax + b)^{1/2}\, dx$

5. $\int \dfrac{1}{(ax + b)^2}\, dx$

6. $\int \dfrac{(a + x)(x^2 - b)}{a + b}\, dx$

7. $\int \dfrac{x^2\, dx}{(x^3 + 6)^{1/2}}$

8. $\int \dfrac{a\, dx}{(ax + b)}$

9. $\int \dfrac{x\, dx}{5x^2 + 6}$

10. $\int \dfrac{dx}{x \ln x}$

11. $\int (\ln x + 5^{2x})\, dx$

12. $\int e^{x/a}\, dx$

13. $\int \left(\dfrac{1}{e^{ax}} - \dfrac{1}{x^2} \right) dx$

14. $\int (e^{2x+2t^2})\, dx$

15. $\int \dfrac{dx}{e^{3x}}$

16. $\int 5^{5x}\, dx$

17. $\int x3^{4x^2}\, dx$

18. $\int 5xe^x\, dx$

19. $\int (3x^2 + 4x)\, dx$

20. $\int \tfrac{3}{5} y^{1/2}\, dy$

21. $\int e^{3kx}\, ax$

22. $\int \dfrac{1}{v}\, dv$

23. $\int \left(37u^4 + \dfrac{6}{u} + 233 \right) du$

24. $\int_0^3 \left(4x^2 - \dfrac{x}{6} \right) dx$

25. $\int_3^5 \left(\dfrac{3u^2}{2} + 4 \right) du$

26. $\int \ln(x+4)\, dx$

27. $\int \left(\dfrac{d}{dx} e^x \right) dx$

28. $\int_{-2}^3 (4 + x - x^2)\, dx$

29. $\int_0^1 e^{3x}\, dx$

30. Find the area above the x axis enclosed by the graph of $y = 16 - x^2$.

31. Find the area under the curve $y = (x - 3)^2$, from the origin to the point $x = 6$. Sketch the area.

PROBLEMS

32. Given the functions with rules
$$y = (x - 2)^2$$
$$y = 4$$
find the area between them. Sketch the functions and the area.

33. Find the area bound by the following three linear equations:
$$y = 10$$
$$y = 10 - \tfrac{1}{2}x$$
$$y = -20 + x$$

34. Find the area of $y = -x^2$ between $x = 0$ and $x = 2$.

Find the value of each of the following definite integrals.

35. $\int_0^1 1 \, dx$

36. $\int_{-5}^5 \tfrac{1}{10} \, dx$

37. $\int_{-8}^{17} \tfrac{1}{25} \, dx$

38. $\int_0^1 x \, dx + \int_1^2 (2 - x) \, dx$

Graph the functions with the following rules and find the area under each of them between the limits indicated:

39. $\tfrac{1}{2}x$, between 0 and 2.

40. $(6 - 2x)$, between 2 and 3.

Sketch the functions with the following rules and find an interval within which they integrate to 1 (the area under the curves equals 1):

41. e^x

42. $\ln x$

43. e^{-x}

44. Given the following equations for marginal cost c and marginal revenue r:
$$c = 4 - 2x - x^2; \qquad r = 4 - 6x - x^2$$
(a) If at equilibrium, $c = r$, solve for the equilibrium level of x.
(b) Integrate the equations, and derive the equation for total cost and total revenue. You may assume that fixed cost and fixed revenue (the constant terms) will be zero.

45. The marginal cost function of a shoe manufacturer has the following rule:
$$c = 5 + 10x - 6x^2.$$
The total cost of producing a pair of shoes is $12. Find the total and average cost functions.

46. If the rule for the marginal revenue function of a firm is
$$r = 10 - 6x$$
find the total revenue and demand functions.

47. If a firm knows that a price increase of $1 will cause sales to drop by 3 units, and if the firm also knows that at a price of $10 its sales are 45 units, what is the firm's demand function?

48. The cost of hiring new employees for a firm is approximated by the function $c = \frac{1}{2}n$, where n is the number of new employees hired. The additional personnel would add to sales revenue in accordance with the function $r = 2\sqrt{n}$.
 (a) How many new employees should be hired?
 (b) What would be the net profit produced by the addition of those employees?

49. The after-tax revenue yielded by a machine's output r and its after-tax maintenance cost c at time t can be expressed by functions with the following rules:

$$R = 700 - \tfrac{1}{2}t^2; \qquad C = 100 + t^2$$

 (a) Assuming that the machine has no salvage value, in what year does the break-even point occur?
 (b) What will be the total net earnings after tax yielded by the machine at that time?
 (c) Assuming that the machine has an after-tax salvage value of $1752/2t$ at time t, what will be the break-even time?

50. The repair costs of a machine are given as a function of time by the following equation:

$$c = 4x + 10$$

where x represents number of years of activity.
 (a) Find the total repair costs after three years.
 (b) How many years are necessary for the cumulative costs to be $100?

51. Use the trapezoidal rule to calculate the area under the function $y = 2x + 4$ between 0 and 10, with $n = 5$. Then compare the result with the value of the integral.

52. Using the trapezoidal rule, calculate the area under the curve $y = 4x^3$ within the interval [0, 5], taking $n = 5$.

Do the following integrals exist?

53. $\displaystyle\int_2^\infty \frac{dx}{(x-1)^3}$

54. $\displaystyle\int_0^\infty xe^{-x}\, dx$

55. $\displaystyle\int_1^2 \frac{dx}{\sqrt{x-1}}$

56. $\displaystyle\int_0^1 \frac{dx}{(1-x)^2}$

57. $\displaystyle\int_{-\infty}^\infty \frac{x}{x^2+1}\, dx$

58. $\displaystyle\int_1^2 \int_0^{x^2} \sqrt{x/y}\, dy\, dx$

59. $\displaystyle\int_0^1 \int_0^{\sqrt{y+y^2}} 2x\, dx\, dy$

60. $\displaystyle\int_a^b \int_0^{x^2} (21y^2 + 10y + 3)\, dy\, dx$

61. $\int_0^1 \int_1^2 \int_2^3 x^3 y^2 z \, dz \, dy \, dx$

62. $\int x^3 e^{ax} \, dx$

63. $\int x^2 e^{-ax} \, dx$

64. $\int (\log_b x + x^b + b^{3x}) \, dx$

65. $\int_2^4 (x - bx^2) \, dx$

66. $\int_0^1 e^x \, dx$

67. $\int_2^3 xe^x \, dx$

68. $\int_0^2 \log_e x \, dx$

For a given application each of the functions below must be continuous and finite through the specified interval. If it is discontinuous, at what point is it discontinuous? If it is continuous and finite, evaluate the integral.

69. $\int_{-6}^{10} \dfrac{dx}{(x - 2)^{4/3}}$

70. $\int_{-3}^{3} \sqrt{\dfrac{3 + x}{3 - x}} \, dx$

71. $\int_{-2}^{2} \dfrac{2xax}{(x^2 - 4)^{2/3}}$

72. The rule for the marginal cost function of a product is $5 + Q + 3Q^2 + e^Q$ where Q is the amount of output.
   (a) Find the rule for the total variable cost function.
   (b) If total costs are 301 when output is at zero, what is the rule for the total-cost function?

73. The rule for the marginal revenue function of a product is $5 + 6Q + 6Q^2$ where Q is the amount of output. Find the total revenue function if $TR = 2500$ when $Q = 10$.

74. The sales manager of a department store wants to know how much his sales will increase if he hires 16 new salesmen to add to his 34-man present sales force. If the marginal productivity of an additional salesman is $75 - 3\sqrt{S}$, where S is the number of additional salesmen, find the increase in sales.

75. Baby Manufacturing Company, a producer of baby foods, has advertised in a monthly women's magazine for the last three years and has found that its sales have increased at an estimated rate of 5 per cent per month. The cost of advertising in the magazine has been raised to $8000 for one year (12 issues). The company will undertake the

advertising expenditure if the additional profits (after all costs) from the increase in sales over the year is at least 15 per cent of the advertising investment. The monthly rate of sales for the advertising period can be estimated by the exponential function $25,000e^{.05t}$, where t is the time variable. The sales from the first T months of the advertising campaign can be represented by $\int_0^T 25,000e^{.05t}\,dt$. Assume that the company maintains the present 10 per cent profit margin on all sales. The $8000 spent on magazine advertising should be considered an additional cost.

(a) Determine the first year's sales without the magazine advertising.
(b) Determine the first year's sales with the magazine advertising.
(c) Find the net profits accruing from the increase in sales resulting from the magazine advertising expenditure.
(d) Should the company use the same magazine advertising, with the 15 per cent criterion?
(e) If the magazine publisher increased the cost to $10,000 a year, should the company advertise?

76. Sales from Auto Manufacturing Company have been increasing at a rate of 4 per cent per month. If present sales are at a rate of $15,000 per month, what will the sales be for the first year? Will the first year sales be over $500,000? [*Hint:* $\int_0^t 15,000e^{.04t} =$ total sales for first T months].

77. Sales for Nambler Manufacturing Company have been decreasing for the last four years. Capacity for the plant is $20,000 in sales per year. At present the plant is working at 60 per cent of capacity and the breakeven point for the company is 47.5 per cent of capacity. The Nambler management will continue to produce its product as long as it estimates the company will be profitable. If sales have been declining at a continuous rate of 12 per cent per year, how many more years will the company produce if the present rate of decline continues?

Find the following values:

78. (a) Present value of $100 to be received in six years at 4 per cent continuous discount rate.
(b) Present value of $100 to be received in four years at 6 per cent continuous discount rate.
(c) Future value of $100 now that will earn 4 per cent continuously for six years.
(d) Future value of $100 now that will earn 6 per cent continuously for four years.

Find the following values:

79. (a) Present value of a continuous annuity of $100 per year at 6 per cent for four years.
(b) Present value of a continuous annuity of $100 per year at 4 per cent for six years.
(c) Future value of a continuous annuity of $100 per year at 6 per cent for four years.
(d) Future value of a continuous annuity of $100 per year at 4 per cent for six years.
(e) Find the present value of a continuous annuity of $100 per year starting at the beginning of the ninth year and lasting for 16 years. The appropriate continuous discount rate before the annuity begins is 4 per cent. After the annuity has begun, the continuous rate of interest is 3 per cent.
(f) Find the present value of the annuity in (e) if the appropriate rate had been 3 per cent throughout the period.

80. Find the present value of a stream of $10,000 annual payments which grow at 10 per cent annually for eight years. The appropriate continuous discount rate for these increasing payments is 5 per cent.

CHAPTER 8

Linear Programming

The job of the manager is to allocate scarce resources to their best possible uses. Although this applies to managers in most organizations, it is most clearly recognized in the case of the manager in a business firm. Business managers differ from managers in public, educational, or health organizations in that they often have available to them a quantitative measure of performance, the degree to which they have added to the profit of their company. The availability of a performance measure has contributed to the broad use of linear programming in business today.

Linear programming is one of the most powerful of the modern management tools. It provides the manager with an overall, simultaneous evaluation of the impact of the variables in the program on his objectives. It is used in such varying activities as the simulation of the operation of a petroleum refinery, the blending of feed for cattle, or the selection of media to inform a given public about a new political or commercial development. If a problem can be stated as a continuous, differentiable function, approaches to its solution can be made through the use of the calculus, as suggested in earlier chapters. If it involves nondifferentiable functions and inequality constraints (workers on a given shift can work eight hours or less), then linear programming can often provide a useful approach to the solution.

Although it is a powerful tool, linear programming is subject to a number of limitations, many of them common to most mathematical models. For example, it is necessary to assume that the variables in a linear program are related in a linear fashion. Since, in reality, variables are not generally linearly related, the programming solution can be treated only as an approximation, although in many cases a sufficiently accurate approximation. Also, a linear program may fit conditions only at a particular instant of time and generally is built on the assumption of certainty about the future. Therefore, there may be dynamic changes in the real-world problem that are not directly incorporated into the linear programming model.

There are also particular characteristics of linear programming formulations that imply further limitations upon the use of the solutions resulting from such programs. For example, if the program includes only three scarce resources in the model of a multiproduct firm, then no more than three products would be produced in a linear programming solution to the problem. This is characteristic of this type of model, and the implications of this limitation and similar ones must be understood thoroughly before the manager acts upon the information provided in the program solution. Nevertheless, none of the limitations of this tool are impossible to surmount. As a matter of fact, one of the most useful features of a linear program is that its solution can be tested to determine its sensitivity to changes in any of the major variables. A systematic procedure known as sensitivity analysis, which can be employed with a linear programming solution, makes it possible to test the significance of errors or changes in the coefficients and amounts of the input variables, and the like. Thus, what at first might appear as major limitations in the use of linear programming can, if dealt with properly, constitute significant strengths in many applications.

As a model for management decision making, the linear program is being recognized increasingly by organizations throughout the world. Managers do face limited resources and must attempt to achieve maximum results within the limitations of their resources. The manager who is skillful in interpreting linear programming can identify which of his resources is most limiting in achieving his objectives. Having identified such a resource, he can then bring his talents to bear to expand its availability or minimize the impact of its scarcity. Similarly, there are many variables and many constraints that do not affect significantly the final outcome of a linear programming solution. This, too, is important information for a manager. It identifies the variables that need not consume *his* limited resources (including the resource of his time). For example, if an error of about 100 per cent in the coefficient for an input variable would have no effect on the optimal solution to a linear program, then it is not necessary to expend time or effort in a careful estimation of this coefficient. However, if the program proves to be extremely sensitive to variations of as little as two per cent in the value of the coefficient to another variable, then this is a factor that should indeed command the attention of the manager.

In the preceding chapters, the relationships between logarithmic and exponential functions and between integration and differentiation have been highlighted. A somewhat similar relationship exists in linear programming. Given any linear programming formulation for a problem (termed the "primal program"), there exists automatically and simultaneously a second program known as the dual program, and the solution to the dual program is found simultaneously with the solution of the primal program. Furthermore, through appropriate interpretation of the dual program, it is possible for the manager to make such judgments as, "What would be the effect on the firm's profits if a given constraint is relaxed by one unit?" Then, through the use of sensitivity analysis, it is possible for him to ask further, "For how many units does this effect hold?" Thus, for example, if the relaxation of a particular resource constraint will lead to an additional $5 in profit to the firm over a range of four or more units, then the manager might well consider the expenditure of $20 to achieve such a relaxation. It is through sensitivity analysis of the program that such information is provided. The value of one additional unit of resource required to relax a given constraint is known in the technical literature of linear programming as a shadow price. Similar values have already been encountered in Chapter 5, Section 5.8, in the interpretation of λ for a binding constraint; that discussion itself was tied to concept of the differential presented in Chapter 3, Section 3.3. Given an appropriate set of shadow prices and the sensitivity analysis that indicates that these prices can be employed over a sufficiently wide range, it may be possible to decentralize decision making within an organization and still be assured that the overall result remains optimal. This is an additional powerful factor involved in the use of linear programming, and it is one that has been attractive to managers in Eastern European nations. Frequently, it is through the linear program that such capitalistic ideas as an interest rate for capital—that is, the shadow price appropriate for the use of the scarce resource known as capital—creeps back into Socialist economic thinking.

This chapter and Chapter 9 deal with the formulation and solution of problems as linear programming problems. The solution is first presented graphically, then the graphic result is reproduced algebraically. The outputs of these two approaches to the linear program are then employed to help develop the exposition of the simplex method of solution for a program. Following this, the dual program is introduced, the mechanics of constructing the dual program from a given program are presented, and then the illustrative numerical problem is solved in terms of this dual program. The specific problem is then generalized through the use of general symbols, and the output of the dual program interpreted. Following the more intuitive economic interpretation of the linear program, the formal mathematics that underlies the solution is presented and discussed in Chapter 9.

8.1 The Problem

After consultation with his industrial engineers, the production manager of the Tiny Tot Toy Company concluded that two production resources, the grinding machine and the warehouse, are the main bottlenecks in the production of his major products. There is a total of nine hours of grinding time available on the grinding machine per day, and the warehouse has 10,000 square feet of warehouse space available.

The products produced are xylophones (X) and yo-yo's (Y). A production run for xylophones (X) requires three hours on the grinding machine and 2000 square feet of warehouse space for storage; a production run for yo-yo's (Y) requires only a single hour of grinding machine time, but requires 2000 square feet of warehouse space. The manager desires to utilize the scarce resources to maximize the contribution to the overhead and profit (π) of his company. Each production run of xylophones contributes four dollars to overhead and profit, whereas a run of yo-yo's contributes two dollars. Based on these data, the manager states the problem in linear-programming form consisting of three major components: (1) the objective function, (2) the resource constraints, and (3) the nonnegativities.

1. *Objective function.* The manager's objective is to maximize contribution to overhead and profit by production and sale of the two products X and Y. Each unit of X gives four dollars of contribution and each unit of Y, two dollars; therefore

$$\pi = 4X + 2Y$$

is the rule for the objective function and the manager hopes to maximize its value.

2. *Resource constraints.* The first constraint is the grinding machine constraint. Each run of xylophones requires three machine hours and each run of yo-yo's requires one machine hour from the total of nine hours available. The second constraint is the limited availability of warehouse space. Each run of xylophones requires 2000 square feet of warehouse space and each run of yo-yo's requires 2000 square feet from the 10,000 that are available. In a linear program, this can be expressed as

$$\begin{array}{ll} \text{Grinding machine} & 3X + Y \leq 9 \\ \text{Warehouse} & 2X + 2Y \leq 10 \end{array}$$

(where the second inequality is expressed in units of a thousand).

3. *Nonnegativities.* None of the products produced by the firm can take on negative values. (What might negative yo-yo's be?) These can be expressed as

$$X \geq 0, \quad Y \geq 0.$$

The results are summarized in the program, identified as expression (8.1).

1. Objective function   Max $\pi = 4X + 2Y$
2. Subject to the constraints
    Grinding machine $3X + Y \leq 9$ (8.1)
    Warehouse $2X + 2Y \leq 10$
3. With nonnegativities $X \geq 0, Y \geq 0$

Sec. 8.2]   GRAPHIC SOLUTION   263

The objective is to maximize the contribution to overhead and profit; the value of the objective function depends upon the number of runs of X and Y produced. There is a total of nine hours available on the grinding machine, and production cannot use more than these nine hours. In the warehouse there are 10,000 square feet available and no more can be utilized. It is not possible to produce either negative xylophones or negative yo-yo's, thus the nonnegativity (third) requirement of the linear program. Although the nonnegativity is technically both an inequality and a constraint, it is consistently identified as a "nonnegativity" because if this latter requirement were not explicitly included in the program, then a computer might well introduce negative values for either or both these variables by the logic that if money were saved by not producing them, even more would be saved by producing negative amounts of them![1]

8.2 Graphic Solution

The program (8.1) is presented graphically in Figure 8.1. It should be recognized that when the equality condition of the constraint holds, the constraint can be represented as a straight line. Thus, on a graph in which the ordinate shows the number of runs of Y produced and the abscissa shows the number of runs of X produced, the location of the boundary of the first (machine) inequality is the straight line 1. The line represents combinations of X and Y production that use exactly nine machine hours; combinations of X and Y that require less than nine hours lie to the left of the line. The intercepts on the X and Y axis for line 1 occur at the points $X = 3$ and $Y = 9$; that is, if only xylophones are produced, there is sufficient machine time to produce three production runs; if only yo-yo's are produced, there is enough machine time to produce nine production runs. Figure 8.1 also includes the warehouse constraint, and the nonnegativities which require that the values of X and Y be in the northeast quadrant of the coordinate system. The intersection of these restraining factors leads to the feasible region that is identified by the stippling. The feasible region is determined by the availability of the resources and the number of units of each resource consumed in a production run of each product. Since in the case illustrated the production of either product

---

[1] A two dimensional problem has been chosen to facilitate the discussion throughout the chapter, but it is possible to state the linear program just developed in terms of general notation. It continues to consist of three parts. These parts can be stated in terms of summations in an $m \times n$ system of equations and inequalities as shown in expression (8.2).

$$\text{Max} \quad \pi = \sum_{j=1}^{n} C_j x_j$$

$$\text{Subject to} \quad \sum_{j=1}^{n} a_{ij} x_j \leq b_i \quad \text{for } i = 1, 2, \ldots, m - 1 \quad (8.2)$$

$$\text{for} \quad x_j \geq 0 \quad \quad j = 1, 2, \ldots, n.$$

The objective function is the summation of the product of the individual profits times each variable across the columns of the first row. Then, for the constraints, the summation is across the columns of rows 1 through $m - 1$ (in this case there are $m - 1$ constraints). In Chapter 9 the $m \times n$ problem is discussed by use of linear algebra.

Figure 8.1

Linear Program for Production of X and Y.

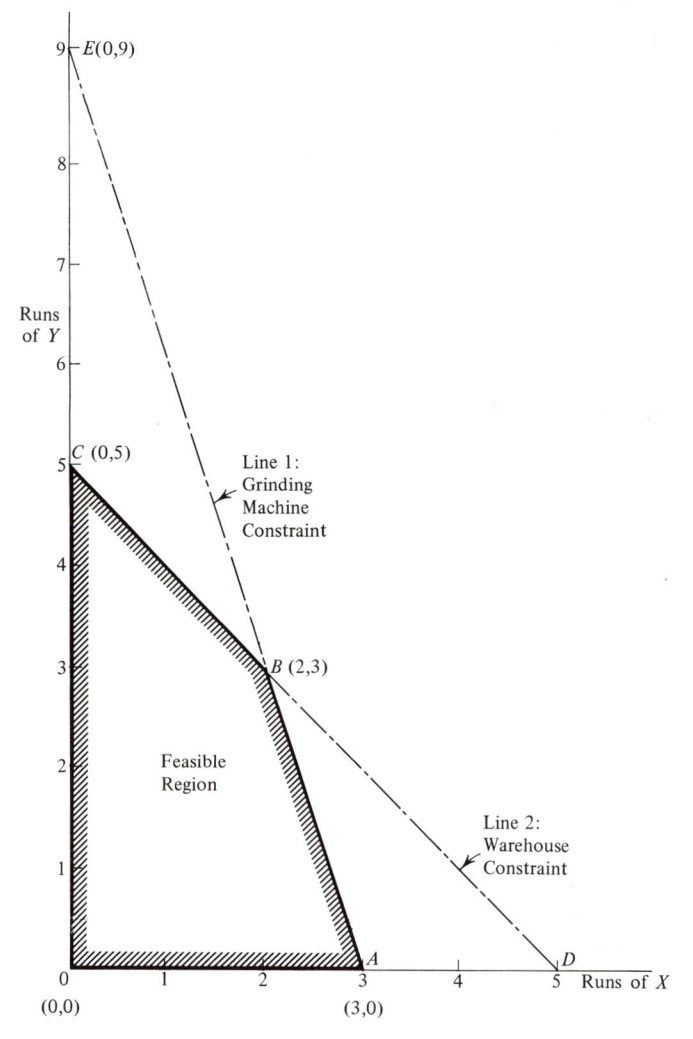

requires both scarce inputs, it is not feasible to produce either product if a constraint is violated.

In terms of point sets, discussed in Section 1.6, a constraint inequality and a nonnegativity each determines a closed half space. The intersection of these sets then constitutes the feasible region. The feasible region is seen to be convex; the set that makes up the intersection set for the constraints is a convex set as it must be for the linear program to be solvable. Each of the corners of the feasible region is an extreme point of the intersection set. The solution to all linear programs occurs at extreme points of the feasible region; it will pay to move the objective function through the region until an extreme point is reached.

Sec. 8.2]   GRAPHIC SOLUTION

Figure 8.1a

Linear Program for Production of X and Y, with Objective Function Included.

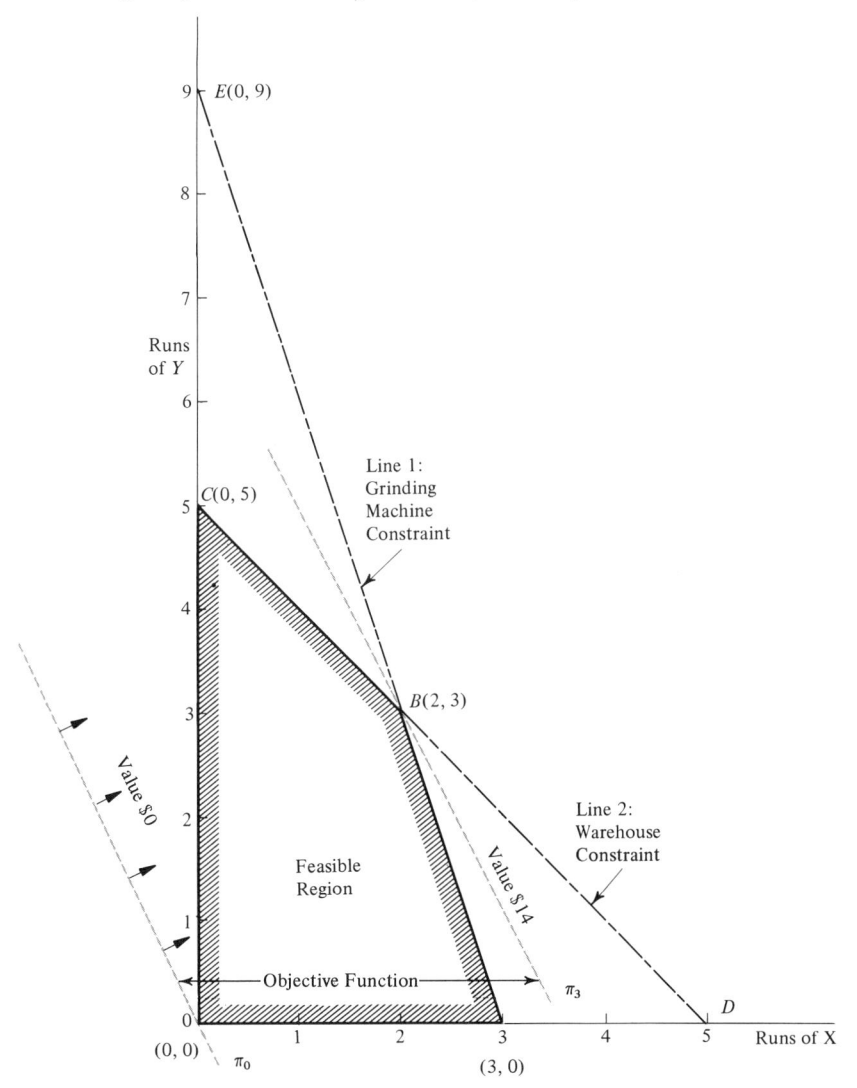

If the production manager were presented with Figure 8.1, after a few minutes' study he could reach some quite useful conclusions. First, if the production of product Y is undertaken, it would be profitable to continue its production until the warehouse constraint is met. Since each unit of Y produced leads to $2 in contribution to overhead and profit, there is no incentive to stop production at two units or three. The production of this product should be continued until there are no longer resources available for its production. At the point $(0, 5)$, point C in Figure 8.1, the warehouse constraint is met; all warehouse space is fully utilized by production of five runs of Y. No further production of product Y is possible and X can be produced only by cutting back on production of Y. Similarly, if the manager were to begin production of product X, it would be desirable to continue producing it until the grinding machine constraint is met—at the point A, with coordinates (3, 0); this occurs after three runs of X are completed. No further output of X is possible, and Y can be produced only by cutting back on current production of X. That is, any changes in output require the trade-off of production of xylophones for the production of yo-yo's. To produce the combination of products represented by point B in Figure 8.1 such a trade-off would be required. If the manager decided that it was profitable to begin the trade-off of xylophones for yo-yo's, he would in effect be moving along the line AB. Once again, if it is profitable to trade a small increment of X for an increment in Y, then it would be profitable to continue the trade off until the point B were reached. At that point, the warehouse constraint also becomes binding.

To determine which of the four (corner) output points is the most desirable for his company in a simple case such as this one, the manager could substitute the values for the outputs represented at each point into the objective function to determine which of the available possibilities provides him with the maximum contribution. At point 0, with coordinates (0, 0) representing output of zero X and zero Y, his profit would be 0 as shown in the following equation:

$$\pi = 4X + 2Y$$
$$\pi_0 = 4(0) + 2(0) = 0. \tag{8.3}$$

At successive points, total contribution to overhead and profit would increase as shown in the following equations: at (0, 5) the point C

$$\pi_1 = 4(0) + 2(5) = 10. \tag{8.4}$$

At (3, 0) the point A, π is even higher:

$$\pi_2 = 4(3) + 2(0) = 12. \tag{8.5}$$

At (2, 3) the point B, profits are highest:

$$\pi_3 = 4(2) + 2(3) = 14. \tag{8.6}$$

It is also possible to solve this simple program graphically. This is done by introducing the objective function in Figure 8.1 (and calling the new figure Figure 8.1a), beginning with π_0 passing through the point (0, 0); the line for π_3 is also shown. A line representing the objective function is an equal-profit line; each point along it represents the same level of contribution to overhead and profit. Thus there is a different line for each level of profit, π_0, π_1 etc., but all such lines have the same slope. The slope is determined by the ratio of the profit from X to that from Y (4/2) as seen from the equation for the objective function solved explicitly for Y:

$$Y = \frac{\pi}{2} - \frac{4X}{2}.$$

Since the ability of the firm to increase profit is constrained only by the limited availability of resources, the optimal profit point must lie on the boundary of the feasible region. It is found by moving the objective function upward and to the

Sec. 8.3] ALGEBRAIC SOLUTION 267

right through the feasible region as far as possible. Increasing levels of profit are represented by lines above and to the right of the line passing through (0, 0), and parallel to that line. The optimal contribution point occurs at the point B. The corners of the feasible region 0, C, A, and B are met successively by the objective function as it sweeps through the feasible region. The implication of the profit line (objective function), is that the products X and Y can be traded off in proportion to the contribution to overhead and profit earned from each; that is, two runs of Y can be replaced by one run of X without change in profit.

8.3 Algebraic Solution

The system of relationships presented in expressions (8.1) are stated (generally) as inequalities. But inequalities are difficult to deal with mathematically; therefore, it proves useful to introduce additional variables into each inequality which allow the inequalities to be restated as equalities. For example, if there were two runs of xylophones and one of yo-yo's, then a total of $6 + 1 = 7$ hours of machine time would be used and there would be two hours remaining unused. Similarly, $4 + 2 = 6$ (thousand) square feet of warehouse space would be used, leaving four (thousand) unused. If a variable representing the amount of unused resource is included in each of the resource constraints, then these inequalities can be restated as equalities. That is, given the illustrative data, a new variable that takes on the value of 2 allows the inequality for the grinding machine to be stated as an equality; a second such variable that takes on a value of 4 for the warehouse constraint allows this inequality to be restated as an equality. These variables are introduced explicitly, and the program restated as follows:

$$\begin{aligned} \text{Max } \pi = \quad & 4X + 2Y + 0S_1 + 0S_2 \\ \text{Subject to:} \quad & 3X + Y + S_1 + 0S_2 = 9 \\ & 2X + 2Y + 0S_1 + S_2 = 10 \\ \text{With nonnegativities:} \quad & X \geq 0, Y \geq 0, S_1 \geq 0, S_2 \geq 0 \end{aligned} \quad (8.1a)$$

Here the new variables S_i, called slack variables, "take up the slack" in each constraint. The slack variables represent the capacity not used by production of the two products. They, too, can take on only positive values; therefore, two additional nonnegatives have been included in expression (8.1a). For symmetry the variable S_2 has been included in the first, and S_1 included in the second constraint equation and both slack variables included in the objective function (in each case with coefficients of zero—they have zero influence on these equations). All variables are included in the objective function and in each constraint. The program thus is expressed as a series of simultaneous equations.

The relationship of the slack variables to the regular or ordinary variables associated with the production of the two products X and Y is useful information. The value of the slack variables is presented at each corner point, as determined by the following equations:

$$\begin{aligned} 3X + Y + S_1 + 0S_2 &= 9 \\ 2X + 2Y + 0S_1 + S_2 &= 10. \end{aligned} \quad (8.7)$$

The relationships are more clear if these equations are solved for S_1 and S_2

$$S_1 = 9 - (3X + Y)$$
$$S_2 = 10 - (2X + 2Y). \qquad (8.8)$$

At the point 0, neither of the basic products is produced and both slack variables have their maximum possible value. At (0, 0), the point 0:

$$S_1 = 9 - (0 + 0) = 9$$
$$S_2 = 10 - (0 + 0) = 10. \qquad (8.9)$$

That is, the grinding machine slack has value 9, and the warehouse slack has value 10. Point C represents five runs of Y and none of X. At (0, 5), the point C:

$$S_1 = 9 - [0 + 1(5)] = 4$$
$$S_2 = 10 - [0 + 2(5)] = 0. \qquad (8.10)$$

In other words, the warehouse is full; there is no slack remaining for that resource. However, there is slack of four hours of grinding machine time still available at the point C.

Point A represents three runs of X and none of Y; at (3, 0), the point A:

$$S_1 = 9 - [3(3) + 0] = 0$$
$$S_2 = 10 - [2(3) + 0] = 4. \qquad (8.11)$$

There is no slack for the grinding machine. At point B both slack variables are 0; the resource in each constraint is used to capacity. At (2, 3), the point B:

$$S_1 = 9 - [3(2) + 1(3)] = 0$$
$$S_2 = 10 - [2(2) + 2(3)] = 0. \qquad (8.12)$$

In the discussion in the last paragraphs, four variables are involved: two variables, X and Y, that are called ordinary variables (those associated with the production of the two products of the firm) and two variables S_1 and S_2, called slack variables. At each of the corner points of the feasible region, two of these variables are nonzero in value, and each of these nonzero variables is positive. This leads to the definition of a *basic*, a *feasible*, and an *optimal* solution.

> **Definition—A Basic Solution:** *Given a linear program with n capacity constraints, a basic solution to the program is one that involves as many nonzero variables as there are constraints.*

> **Definition—Feasible Solution:** *Given a basic solution to a linear program, a feasible solution is one for which all nonzero variables are positive.*

> **Definition—Optimal Solution:** *A basic, feasible solution for which the objective function takes on its optimal value is an optimal solution.*

A computational method known as the simplex method (presented in Section 8.4) allows the program to move systematically from one basic, feasible solution to another—from one corner of the feasible region to another—in such a way that the value of the objective function is consistently improved with each move.

Sec. 8.3]   ALGEBRAIC SOLUTION

Being able to find an initial basic, feasible solution readily is therefore extremely important. By this method, an optimal solution is achieved at the end of a finite number of iterations.

It is useful to follow the algebraic development of the values of the various variables, especially as a reference for the simplex method that is presented in Section 8.4. The algebraic solution should be traced out on the graph of Figure 8.1 and compared with the graphic solution. A move to a new corner is made only if the program solution remains basic and feasible. This requires that there be exactly n nonzero variables and that they all be positive.

In Figure 8.1, an initial basic, feasible solution is found at the origin of the graph, the point (0, 0). Merely by setting the two ordinary variables equal to zero, it is possible to arrive at the initial basic, feasible solution. There are two nonzero variables, both slack variables, that have positive value equal to the full available capacity for each of the bottleneck resources. Each corner of the feasible region represents a basic, feasible solution.

The largest profit per unit is achieved by producing product X. If this variable is made positive, another variable must be reduced to zero for a solution to remain basic. The nonzero variable that reaches zero first if X is produced is S_1. Only three units of X are required to achieve this; five would be required to do so for S_2. This confirms what was shown graphically and can be seen from the equations stated as follows:

$$\pi = 0 + 4X + 2Y \tag{8.13}$$
$$S_1 = 9 - (3X + Y) \tag{8.14}$$
$$S_2 = 10 - (2X + 2Y). \tag{8.15}$$

Resolving these equations in terms of X and S_2 instead of S_1 and S_2 requires resolving equation (8.14) for X:

$$3X = 9 - S_1 - Y$$
$$X = 3 - \tfrac{1}{3}S_1 - \tfrac{1}{3}Y. \tag{8.16}$$

This value of X is now substituted into each of the other equations. For equation (8.15):

$$S_2 = 10 - 2[3 - \tfrac{1}{3}S_1 - \tfrac{1}{3}Y] - 2Y$$
$$S_2 = 10 - 6 + \tfrac{2}{3}S_1 + \tfrac{2}{3}Y - \tfrac{6}{3}Y$$
$$S_2 = 4 + \tfrac{2}{3}S_1 - \tfrac{4}{3}Y. \tag{8.17}$$

For equation (8.13):

$$\pi = 0 + 4[3 - \tfrac{1}{3}S_1 - \tfrac{1}{3}Y] + 2Y$$
$$\pi = 0 + 12 - \tfrac{4}{3}S_1 - \tfrac{4}{3}Y + \tfrac{6}{3}Y$$
$$\pi = 12 - \tfrac{4}{3}S_1 + \tfrac{2}{3}Y. \tag{8.18}$$

The values of S_1 and Y are both zero. The new equations (8.16) to (8.18) thus have the following values:

$$\pi = 12$$
$$X = 3$$
$$S_2 = 4$$

with other variables equal to zero, just as expected from the process of total enumeration of values, as given in Section 8.2 and the graphic solution to the problem.

In equation (8.18) the coefficient of Y is positive, indicating that profits could be improved by producing product Y. If Y is made positive, one of the currently nonzero variables has to be made zero. Three units of Y make S_2 zero, nine units of Y are required to make X zero, and the S_2 limit is reached first. Therefore equation (8.17) must be resolved for Y instead of S_2:

$$S_2 = 4 + \tfrac{2}{3}S_1 - \tfrac{4}{3}Y \qquad (8.17)$$
$$\tfrac{4}{3}Y = 4 + \tfrac{2}{3}S_1 - S_2$$
$$Y = 3 + \tfrac{1}{2}S_1 - \tfrac{3}{4}S_2. \qquad (8.19)$$

Substituting the value for Y into equation (8.16)

$$X = 3 - \tfrac{1}{3}S_1 - \tfrac{1}{3}Y \qquad (8.16)$$
$$X = 3 - \tfrac{1}{3}S_1 - \tfrac{1}{3}[3 + \tfrac{1}{2}S_1 - \tfrac{3}{4}S_2]$$
$$X = 3 - \tfrac{1}{3}S_1 - 1 - \tfrac{1}{6}S_1 + \tfrac{1}{4}S_2$$
$$X = 2 - \tfrac{1}{2}S_1 + \tfrac{1}{4}S_2 \qquad (8.20)$$

and in equation (8.18):

$$\pi = 12 - \tfrac{4}{3}S_1 + \tfrac{2}{3}Y \qquad (8.18)$$
$$\pi = 12 - \tfrac{4}{3}S_1 + \tfrac{2}{3}[3 + \tfrac{1}{2}S_1 - \tfrac{3}{4}S_2] \qquad (8.21)$$
$$\pi = 12 - \tfrac{4}{3}S_1 + 2 + \tfrac{1}{3}S_1 - \tfrac{1}{2}S_2 \qquad (8.22)$$
$$\pi = 14 - 1S_1 - \tfrac{1}{2}S_2 \qquad (8.23)$$

Both S_1 and S_2 are zero valued, so the values of equations (8.19), (8.20), and (8.23) are

$$Y = 3$$
$$X = 2$$
$$\pi = 14$$

which accords with the total enumeration solution and with the graphic solution.

8.4 The Simplex Method

The simplex method is an iterative process that proceeds in a finite number of steps to an optimal solution to a linear program. It is a mechanical calculating technique that has many desirable properties, but its mechanics need be no more mysterious than those in the preceding solution to simultaneous equations at different corner points of the feasible region.

In the earlier discussion it was observed that an optimal solution must occur at one of the extreme points (corners) of the feasible region. In effect, the simplex method is a computational method for moving successively from one corner to another in such a way that the objective function is consistently improved. A simplex is a convex polyhedron, an n-dimensional analog to a convex set, and an n-dimensional intersection of point sets. At each corner of the simplex formed by the constraints, a new set of variables is nonzero and positive. Each simplex iteration

Sec. 8.4] THE SIMPLEX METHOD 271

requires the substitution of a variable currently zero for a variable currently positive in the previous basic, feasible solution, that variable for which the contribution to profit for each *unit* of the variable introduced into the program is the largest. (This may not be the variable that makes the greatest *total* addition to profit.)

The initial basic, feasible solution, achieved by setting each of the ordinary variables equal to zero, thus sets each slack variable at the maximum value of each constraint (for greater than or equal to constraints). In vector terminology, it is possible to select a number of variables that are positive and linearly independent to form a basis for spanning the space suggested in the program. Through the simplex method it is possible to move from one set of linearly independent nonzero variables to another, thus moving from one basis for spanning the space to another, until the optimal solution has been achieved. The mathematics of this is presented in Chapter 9.

The simplex tableau is now presented. The first step is the presentation of the full model with all variables included in each of the equations, as suggested in equation (8.24).

$$\text{Max } \pi = 4X + 2Y + 0S_1 + 0S_2$$

$$\begin{aligned} 3X + Y + S_1 + 0S_2 &= 9 \\ 2X + 2Y + 0S_1 + S_2 &= 10 \end{aligned} \quad (8.24)$$

$$X \geq 0, Y \geq 0, S_1 \geq 0, S_2 \geq 0.$$

From this form of the program, a matrix of the coefficients of each of the variables in the objective function and constraints is prepared as follows:

$$\begin{array}{c|ccccc} & X & Y & S_1 & S_2 & b_i \\ \hline C_j & 4 & 2 & 0 & 0 & 0 \\ \hline S_1 & 3 & 1 & 1 & 0 & 9 \\ S_2 & 2 & 2 & 0 & 1 & 10 \end{array} \quad (8.25)$$

where b_i are the values of the right-hand side (RHS) for each constraint and C_j are the coefficients of the objective function. With only minor rearrangement, this matrix is transformed into the simplex tableau.

In the simplex tableau the coefficients (C_j) of the objective function play an important role. They are tabulated in the first row and the first column of Table 8.1. The nonzero variables are tabulated in the second column as what are known as the basis variables (those nonzero in the basic solution) and their values are tabulated next in the column headed P_0. Then all other variables are arrayed in successive columns.

The coefficients of the nonzero variables S_1 and S_2 form an interesting pattern:

$$\begin{array}{cc} 1 & 0 \\ 0 & 1. \end{array}$$

The Z_j row represents the reduction in the value of the objective function that results from replacement of one of the nonzero variables with a variable currently zero. The total reduction for a given column j is the sum of the reductions from *each*

Table 8.1

First Simplex Tableau

C_j			4	2	0	0
	Basis	P_0	X	Y	S_1	S_2
0	S_1	9	3	1	1	0
0	S_2	10	2	2	0	1
	Z_j	0	0	0	0	0
	$C_j - Z_j$		4	2	0	0

element currently in the basis that is replaced. For example, if Y were produced, each unit of Y "uses up" one unit of S_1 and two units of S_2. The profits from these variables would be lost if Y were produced.

The C_j represent the increase in the value of the objective function that comes from introduction of a unit of a given variable. In the case of Y, two units are added to profit for each unit of Y introduced. The difference $C_j - Z_j$ thus represents the net addition to the objective function from making positive a unit of each variable now zero and nonzero also.

The Z_j for Table 8.1 are calculated below:

$$Z_{P_0} = 9(0) + 10(0) = 0$$
$$Z_X = 3(0) + 2(0) = 0$$
$$Z_Y = 1(0) + 2(0) = 0$$
$$Z_{S_1} = 1(0) + 0(0) = 0$$
$$Z_{S_2} = 0(0) + 1(0) = 0.$$

Thus, for Table 8.1, the $(C_j - Z_j)$ are all equal to the respective C_j's.

The variable that adds most to the objective function for each unit introduced is product X. This variable, the one with the largest value for $C_j - Z_j$, is introduced into the program. If one unit of X should be introduced, then it is profitable to continue to introduce X until a constraint is reached. The first constraint to be reached is determined by the first variable to be reduced to zero by the introduction of X. As mentioned in the discussion of the algebraic solution to this problem, three units of X reduces S_1 to zero, whereas it would take five units to reduce S_2 to zero. The S_1 constraint is reached first (as shown graphically in Figure 8.1). This is seen by dividing the element in the P_0 column by the element in the corresponding row for the column of the entering variable:

for S_1, $\frac{9}{3} = 3$;   for S_2, $\frac{10}{2} = 5$.

Therefore choose S_1. If an element in any row for an entering variable is negative, this row should be ignored.

The element in the column of the entering variable and in the row of the variable being replaced is known as the pivot element. It is the element circled in Table 8.1. The introduction of the new variable is achieved by following two rules.

1. *For the pivot row:* Divide all elements now in the row by the pivot element. This step is identical with calculating the coefficients for

equation (8.18) from equation (8.16). The elements in the S_1 row are thus 3, 1, $\frac{1}{3}$, $\frac{1}{3}$, and 0, as tabulated in Table 8.2.

Table 8.2

Second Simplex Tableau

C_j			4	2	0	0
	Basis	P_0	X	Y	S_1	S_2
4	X	3	1	$\frac{1}{3}$	$\frac{1}{3}$	0
0	S_2	4	0	$\frac{4}{3}$	$-\frac{2}{3}$	1
	Z_j	12	4	$\frac{4}{3}$	$\frac{4}{3}$	0
	$C_j - Z_j$		0	$\frac{2}{3}$	$-\frac{4}{3}$	0

2. *For all other rows:* Subtract from each element in a given row the product of the element in the pivot column for that row and the element in that column for the new pivot row calculated in the preceding step.

The calculation of the new elements for Table 8.2, the second simplex tableau, is given in Table 8.3 for the new S_2 row. The numbers in Table 8.3 are the same as

Table 8.3

Calculation of Values of Nonpivot Row Elements for Tableau Two

	Minus Product of		
Old Element a_j of Old Row i	Element of Row i in Pivot Column	Element a_j of New Pivot Row	New Element a_j for Row i
10	(2)	(3) =	4
2	(2)	(1) =	0
2	(2)	$(\frac{1}{3})$ =	$\frac{4}{3}$
0	(2)	$(\frac{1}{3})$ =	$-\frac{2}{3}$
1	(2)	(0) =	1

those in equation (8.17) (with the signs reversed for the coefficients of variables currently zero) and are tabulated in Table 8.2.

The variable X is now included in the basis in place of S_1, and its value C_j is placed in this row. The new Z_j's are calculated as before:

$$\begin{aligned}
Z_{P_0} &= 3(4) + 4(0) = 12 \\
Z_X &= 1(4) + 0(0) = 4 \\
Z_Y &= \tfrac{1}{3}(4) + \tfrac{4}{3}(0) = \tfrac{4}{3} \\
Z_{S_1} &= \tfrac{1}{3}(4) + -\tfrac{2}{3}(0) = \tfrac{4}{3} \\
Z_{S_2} &= 0(4) + 1(0) = 0.
\end{aligned}$$

The values $C_j - Z_j$ are determined by subtracting the value Z_j in each column from the C_j at the head of that column. The numbers $(C_j - Z_j)$ are the same as the coefficients in equation (8.18) with the same sign.

Introduction of variable Y adds $\frac{2}{3}$ dollars per unit to the objective function, so it is made positive in the next iteration. S_2 is reduced to zero first:

$$\text{for } X, \quad \frac{3}{1/3} = 9; \quad \text{for } S_2, \quad \frac{4}{4/3} = 3;$$

therefore S_2 is driven out and the element $\frac{4}{3}$ is identified as the pivot.

The pivot row is 3, 0, 1, $-\frac{1}{2}$, and $\frac{3}{4}$, respectively (see equation (8.19)), and the other row(s) are developed as shown in Table 8.4. Compare these values with equation (8.20). The third simplex tableau is that shown in Table 8.5.

Table 8.4

Calculation of Values of Nonpivot Row Elements for Tableau Three

Old Element a_i of Old Row i	Minus Product of		New Element a_j for Row i
	Element of Row i in Pivot Column	Element a_j of New Pivot Row	
3	$(\frac{1}{3})$	(3) =	2
1	$(\frac{1}{3})$	(0) =	1
$\frac{1}{3}$	$(\frac{1}{3})$	(1) =	0
$\frac{1}{3}$	$(\frac{1}{3})$	$(-\frac{1}{2})$ =	$\frac{1}{2}$
0	$(\frac{1}{3})$	$(\frac{3}{4})$ =	$-\frac{1}{4}$

Table 8.5

Third Simplex Tableau

C_j			4	2	0	0
	Basis	P_0	X	Y	S_1	S_2
4	X	2	1	0	$\frac{1}{2}$	$-\frac{1}{4}$
2	Y	3	0	1	$-\frac{1}{2}$	$\frac{3}{4}$
	Z_j	14	4	2	1	$\frac{1}{2}$
	$(C_j - Z_j)$		0	0	-1	$-\frac{1}{2}$

The new Z_j's are

$$\begin{aligned}
Z_{P_0} &= 2(4) + 3(2) = 14 \\
Z_X &= 1(4) + 0(2) = 4 \\
Z_Y &= 0(4) + 1(2) = 2 \\
Z_{S_1} &= \tfrac{1}{2}(4) + -\tfrac{1}{2}(2) = 1 \\
Z_{S_2} &= -\tfrac{1}{4}(4) + \tfrac{3}{4}(2) = \tfrac{1}{2}.
\end{aligned}$$

The third tableau is optimal. There is no variable whose introduction would increase the value of the objective function of the program. Compare the $(C_j - Z_j)$ row with the coefficients in equation (8.23).

Throughout these calculations, the coefficients in the columns for the variables included in the basis are consistently of the form

$$\begin{matrix} 1 & 0 \\ 0 & 1. \end{matrix}$$

The calculations generate this pattern for each new nonzero variable. The significance of this is explained in Chapter 9.

8.5 The Dual Program

From the statement of the program with its three parts in the system (8.1)—with the objective function, the capacity constraints, and the nonnegativities, it is possible to develop almost immediately a second program known as the dual program. The development of the dual program is first presented as a routine procedure in accordance with a series of simple rules. After this has been done, the reasoning underlying the procedure is suggested.

First the key relationships are stated. Associated with any maximization program is a dual program that is a minimization problem. For each ordinary variable in the primal problem, there is a constraint in the dual program. For each capacity constraint in the primal program, there is an ordinary variable in the dual program. The direction of each inequality in the primal program is reversed in the dual to the linear program, except in the case of the nonnegativity requirements. (That is, given less than or equal to signs in the primal program, the dual has greater than or equal to signs for each of the constraints.)[2] The coefficients in the objective function of the primal program become the values on the right-hand side (RHS) in the dual program, and the values on the RHS of the constraints in the primal program become the coefficients of the objective function in the dual program. These relationships are followed in writing out the dual program for system (8.1). First, variables V_i (in parentheses) are associated with each constraint:

Objective function Max $\quad \pi = 4X + 2Y$
Subject to the $\quad\quad\quad\quad 3X + Y \leq 9 \quad (V_1)$ (8.1)
contraints $\quad\quad\quad\quad\quad 2X + 2Y \leq 10 \quad (V_2)$
With nonnegativities $\quad\quad X \geq 0, Y \geq 0$

For each of the ordinary variables, a constraint is constructed by using the column of coefficients of that variable, beginning with the first constraint. That is, the coefficient for X in each of the constraints is now associated with the new variables

---

[2] For a maximization problem all the constraint inequalities are normally in the $\leq$ direction. If a particular constraint happens to be in the $\geq$ direction, the direction of that inequality can be reversed by multiplying it through by (-1). For a minimization problem, all the constraint inequalities are normally in the $\geq$ direction. If a given constraint happens to be stated as an equality, the equality can be replaced by two inequalities, one in each direction. Then the direction of the appropriate inequality can be reversed by multiplying it through by (-1).

V_1 and V_2; thus, the first constraint becomes

$$3V_1 + 2V_2 \geq 4. \tag{8.26}$$

As stated in the system of relationships just given, the direction of the inequality has been reversed so that it is now greater than or equal to zero, and the coefficient of X from the objective function of the primal is now the (RHS) value of this constraint. Similarly, it is possible to associate with the variable Y a new constraint made up of the coefficients for Y from the original program and the variables V_1 and V_2 with the result:

$$V_1 + 2V_2 \geq 2. \tag{8.27}$$

Inequalities (8.26) and (8.27) together form the new constraints for the dual program. There is one for each of the ordinary variables in the primal program. The objective function involving V_1 and V_2 variables is also developed, with the coefficients of each of the variables in the objective function coming from the values of the constraints in the primal program:

$$\gamma = 9V_1 + 10V_2.$$

This program must be a minimization program. Since the primal program involved maximization of profit, the dual is treated as a minimization of cost

$$\text{Min } (\gamma = 9V_1 + 10V_2). \tag{8.28}$$

The main factor that remains unchanged in the dual program is that all the variables in the dual program must be nonnegative. This is not at all surprising, for a problem involving real-world factors that cannot take on negative value. Assembling the various parts gives the dual program as follows:

$$\begin{aligned}
\text{Min} \quad & \gamma = 9V_1 + 10V_2 \\
\text{Subject to} \quad & 3V_1 + 2V_2 \geq 4 \quad (X) \\
& V_1 + 2V_2 \geq 2 \quad (Y) \\
& V_1 \geq 0, V_2 \geq 0.
\end{aligned} \tag{8.29}$$

The process can be reversed by treating the program just developed as the primal program and writing its dual. First, a variable is associated with each of the constraints in expression (8.29), the variable X with the first constraint and the variable Y with the second constraint. The coefficients of V_1 are used to develop the first constraint with the sign of the inequality reversed [in relation to (8.29)] as follows:

$$3X + Y \leq 9. \tag{8.30}$$

The value of the RHS of the constraint is equal to the coefficient of V_1 in the objective function. The second constraint is developed from the coefficients associated with the ordinary variable V_2 as follows:

$$2X + 2Y \leq 10. \tag{8.31}$$

Again, the sign of the inequality is reversed and the value of the RHS of the constraint is equal to the coefficient of V_2 from the objective function. Next, the objective function itself is written with coefficients of the variables X and Y taken

Sec. 8.5] THE DUAL PROGRAM

from the values of the RHS of the constraints as indicated in equation (8.32)

$$\text{Max } (\pi = 4X + 2Y). \tag{8.32}$$

The objective function must involve maximization since the program being treated as the primal program is a minimization problem. Again, each of the new variables introduced must be positive. The dual to the dual is now written out in full as follows:

$$\begin{aligned}
\text{Max} \quad & \pi = 4X + 2Y \\
\text{Subject to} \quad & 3X + Y \leq 9 \\
& 2X + 2Y \leq 10 \\
& X \geq 0, Y \geq 0.
\end{aligned} \tag{8.33}$$

It is identical with the program shown as (8.1). The dual program shown in expression (8.29) is shown graphically in Figure 8.2. (Note that this is the dual program, not the dual to the dual.)

Figure 8.2

Graph of the Dual Program.

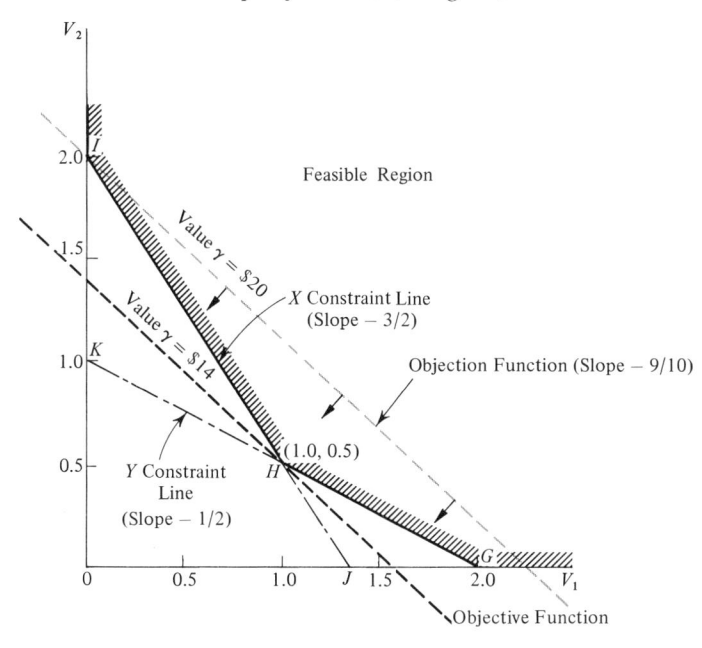

Again, there are two constraints: one associated with the variable X and another with the variable Y. The X constraint runs from the point $(0, 2)$ to the point $(\frac{4}{3}, 0)$. The Y constraint runs from the point $(0, 1)$ to the point $(2, 0)$. The feasible region now is the cross-hatched region on the right-hand side of each of these constraints because they are greater than or equal to constraints.

It is possible in the case of the dual to determine the optimal program solution by sweeping the extreme points of the feasible region with the objective function downward and to the left to identify that corner representing the minimum cost. The objective function for the dual program and the direction of its sweep are shown in Figure 8.2. The objective function is a constant cost line. Thus, the lower the cost, the more desirable the result. At the point (0, 2), point *I* in Figure 8.2, the value of the objective function is 20: V_2 has a value of 2, V_1 a value of 0, for a value of the objective function of $20 = 10(2)$. The right-most dashed line shown in the graph is the 20-unit objective function. At the extreme point J the objective function has a value of 18, $V_1 = 2$, and $V_2 = 0$. It is possible to continue further through the feasible region and achieve a lower value still at the point *H*. This appears to be the optimal (lowest) value, and since the value of the objective function for the primal program and that for the dual program are identical in an optimal solution, then the value of the objective function of the dual program would be expected to be equal to 14. With a value of 1 for V_1 and a value of $\frac{1}{2}$ for V_2, the objective function at the point *H* takes on a value of 14; that is, at (1, 0.5), the point *H*,

$$9(1) + 10(.5) = 14. \qquad (8.34)$$

It is desirable to continue the discussion through a few additional steps. The dual program of (8.29) may be restated for simplex solution by the conversion of the inequalities to equalities through the insertion of slack variables designated as L_1 for the *X* constraint and L_2 for the *Y* constraint. Since the inequalities are in the greater than direction, slack variables are added to the RHS of these constraints, or subtracted from the left-hand side (LHS). Subtraction is adopted here, with the following result:

$$\begin{aligned}
\text{Min} \quad & \gamma = 9V_1 + 10V_2 + 0L_1 + 0L_2 \\
\text{Subject to} \quad & 3V_1 + 2V_2 - L_1 + 0L_2 = 4 \\
& V_1 + 2V_2 + 0L_1 - L_2 = 2 \\
\text{And} \quad & V_1 \geq 0, V_2 \geq 0, L_1 \geq 0, L_2 \geq 0. \qquad (8.35)
\end{aligned}$$

The constraints can be rearranged to facilitate their interpretation. A simplified version of the constraints is

$$\begin{aligned}
(3V_1 + 2V_2) - L_1 &= 4 \\
(V_1 + 2V_2) - L_2 &= 2.
\end{aligned}$$

Transposing the ordinary variables to the RHS gives

$$\begin{aligned}
-L_1 &= 4 - (3V_1 + 2V_2) \\
-L_2 &= 2 - (V_1 + 2V_2)
\end{aligned}$$

and changing signs gives:

$$\begin{aligned}
L_1 &= (3V_1 + 2V_2) - 4 \\
L_2 &= (V_1 + 2V_2) - 2. \qquad (8.36)
\end{aligned}$$

Just what are the ordinary variables V_1 and V_2 of the dual program? The variables V_1 and V_2 are associated with each of the constraints of the original program. They represent an imputed cost of each of the scarce resources in the original program; that is, the ability of the firm to obtain a contribution to overhead and profit is limited by the scarcity of these two inputs. If a larger amount of each of the constrained resources were available, it would be possible to attain more profits in this program. The degree to which profits are constrained by the scarceness of each resource is given by the values of the ordinary variables of the dual program; that is, the value of V_1 is an indication of the rate at which profits would increase if the grinding machine constraint were relaxed at the margin—if additional grinding machine time were made available. Another way of stating this would be to say that if the grinding machine were to break down, the profits available to the firm, as expressed in the linear program, would be reduced at the rate of V_1 per hour. The imputed cost V_2 is interpreted similarly. This is a measure of the profits foregone because the warehouse resource is limited. These values, V_1 and V_2, are called the *shadow prices* on the scarce resources. They are measures of the profit lost because of bottlenecks in the production process.

The shadow prices can be read from the simplex tableau once the optimal solution has been achieved. Table 8.5, the optimal tableau, is repeated below.

Table 8.5

Third Simplex Tableau

C_j			4	2	0	0
	Basis	P_0	X	Y	S_1	S_2
4	X	2	1	0	$\frac{1}{2}$	$-\frac{1}{4}$
2	Y	3	0	1	$-\frac{1}{2}$	$\frac{3}{4}$
	Z_j	14	4	2	1	$\frac{1}{2}$
	$(C_j - Z_j)$		0	0	-1	$-\frac{1}{2}$

Recall that the shadow price represents the value per unit of an increment in each of the resources that is scarce in the optimal program. In the $C_j - Z_j$ row and the columns for the zero-valued variables S_1 and S_2 in Table 8.5, the numbers -1 and $-\frac{1}{2}$ appear. These are the values of the shadow prices with signs reversed. The reversed signs can be understood since the $C_j - Z_j$ row represents the change in profit resulting from a unit change in basis variable. That is, if S_1 were made positive and slack introduced on the grinding machine—perhaps through a breakdown of the machine—then profits would be reduced by $1 for each hour the machine is down, within the limits over which the $1 shadow price applies. Similarly, if one square foot of warehouse space were left unused, profits would be reduced by $0.50. Positive values of the shadow prices are associated with increases in the availability of each scarce resource.

In the dual program interpreted as in equation system (8.36), the dual slack variables presented become very significant also. The slack variable L_1 is associated with the constraint generated from the coefficients of the X variable in the original program, and is equal to a term in parenthesis involving V_1 and V_2 minus an additional term, the number 4, representing the contribution to overhead and profit from the production of one unit of variable X

$$L_1 = (3V_1 + 2V_2) - 4. \tag{8.36a}$$

Similarly, the slack variable L_2 is associated with the constraint developed from the requirements of the production of a run of product Y

$$L_2 = (V_1 + 2V_2) - 2. \tag{8.36b}$$

In the equation for L_1, the factor inside the parenthesis represents the total of the imputed costs of the resources used in the production of product X. These are opportunity costs. They are the values imputed to the scarce resources as derived from the profit that could be achieved from use of these resources to produce other products included in the program. The figure outside of the brackets, -4, represents the contribution made by sale of one unit of product X. The equation shows the relationship between the costs and benefits from the production of product X. If it were possible for benefits to exceed costs, the value of L_1 would be negative and production of product X would be highly desirable. However, the original form of the constraint required that these inputs have a value *greater than* or equal to the contribution from the sale of X; that is, $(3V_1 + 2V_2)$ of necessity is greater than, or at best equal to, 4. Thus, a negative value of L_1 is not possible, and this is in accord with the nonnegativity requirements for variables in any linear program. Thus, L_1 will be either 0 or positive. A positive value of L_1 suggests that the imputed costs of producing product X exceed the contribution made from sale of product X. The inputs can produce higher profits when used to produce *other* products in the program. Under such conditions, it is not desirable to produce product X. Thus, if the slack variable in the dual program is positive, the product whose production process is associated with the development of this slack variable will not be produced.

In the equation for L_2, again, the values in the brackets represent the total imputed cost of resources utilized in the production of product Y and 2 represents the contribution per unit of output of product Y. The variable L_2 can be either 0 or positive. If zero, the costs and benefits are matched and there is no opportunity cost from producing Y; it is therefore worthwhile to produce product Y. If L_2 is positive, the costs of using the scarce resources to produce Y exceed the benefits from such production; therefore, it is not desirable to produce Y.

It has been stated that V_1 and V_2 are imputed costs associated with the utilization of the scarce resources of the firm. These costs are very real, although they do not appear in the accounting statements of the firm. The objective function of the dual program requires the minimization of the total of the values (costs) imputed multiplied by the number of units of each scarce resource available. There are nine grinder hours available and there are 10,000 square feet in the warehouse; thus, the total value associated with these scarce resources is $9V_1 + 10V_2$. The grand total of the costs imputed to these resources is minimized in the optimal solution to the program which automatically requires that the value of the slack variable in the dual program be, at most, equal to zero for any product produced in the program.

There are two basic rules that summarize the relationships between the primal and dual programs in a linear program. The first is

$$\text{Max } \pi = \text{Min } \gamma. \tag{8.37}$$

The maximum profit in the primal program is equal to the minimum cost in the dual program. At the optimal solution, the value of the objective function in the primal and the dual program is identical. The relationship between the primal ordinary variables and the dual slack variables has been discussed. There is a similar relationship between the primal slack variables and the dual ordinary variables. This relationship might have been anticipated from the reciprocal relationships between the primal and dual programs. A formal statement of this relationship is

$$\begin{aligned}[Q_j L_i] &= 0 \\ [S_i V_j] &= 0\end{aligned} \quad \text{for } i = j. \tag{8.38}$$

Where Q_j is a positive value of output of any of the products of the firm, the relationship implies that, given a positive Q_j the value of the dual slack variable L_i (where $j = i$) must be zero, such that the factor $Q_j \cdot L_i$ is equal to zero. Similarly, given a positive L_i, then Q_j must, of necessity, be zero. If there is a loss in contribution to overhead and profit from producing any product of the firm, then that product should not be produced; that is, its value should be zero.

Similarly, the relationship between the primal slack variable S_i and the dual ordinary variable V_j (for $i = j$) can be rationalized. If one slack variable S_i is positive, that scarce resource is not used to capacity. The V_j (for $j = i$) associated with the constraint equation for that scarce resource will be equal to zero. If the resource is not used to capacity, it does not restrict the ability of the firm to achieve profits. No profits are lost if more of this resource is used since there is more than enough of the resource available for current output levels. Similarly, a positive value of V_j occurs only when a resource is utilized to capacity; it is really scarce. If a resource is really scarce, an increment in its use to produce one product cannot be achieved *except* at the expense of not producing some other product or products. When these other products are not produced, there is a definite cost to the firm from not producing them—the cost of the profits that must be foregone because products not produced cannot be sold. Thus, for V_j to be positive, S_i must be zero. The equations (8.37) and (8.38) are a succinct summary of the preceding discussion.

8.6 Sensitivity Analysis

It is possible to demonstrate graphically the sensitivity of the program to errors or changes in important coefficients and values, such as the amounts of the resources available and the coefficients of the objective function. First, the impact of a change in a single factor (with all others held constant), the amount of time available on the grinding machine, is investigated.

In the original program there is a maximum of nine hours of grinding machine time. If an additional six hours were added, what would be the impact on the solution to the program? The new total of 15 grinding hours requires the movement of the grinding machine constraint AE in Figure 8.1 to the left until the lower extremity of the line passes through the point D. It then would be possible to produce five units of product X (see Figure 8.3).

Figure 8.3

Linear Program with Machine Constraint Relaxed to 15 Hours.

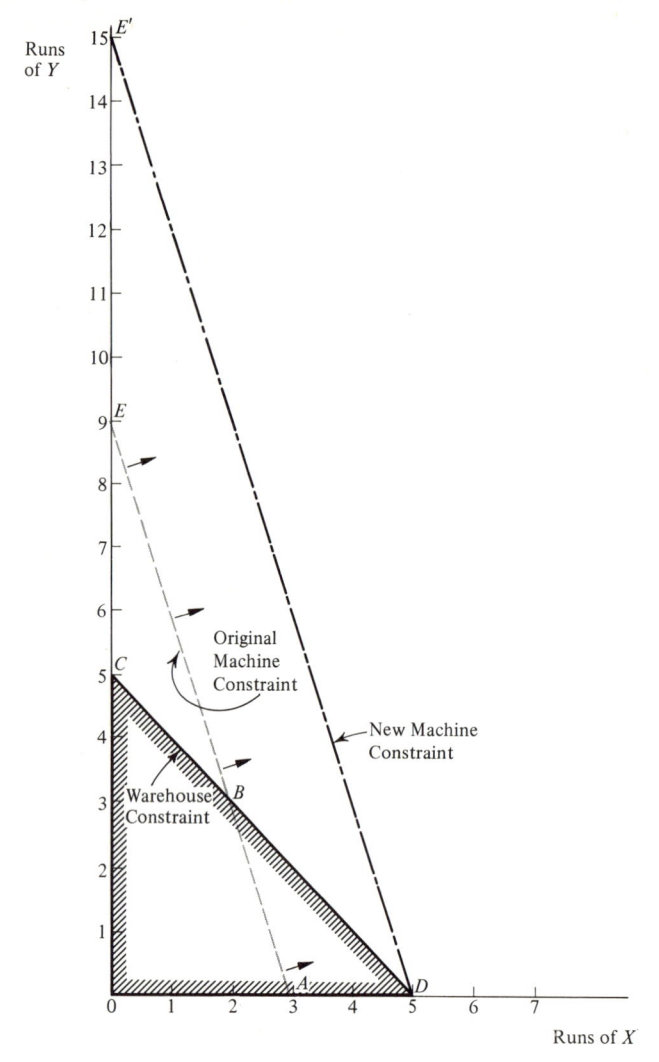

If the new units of machine time were added gradually, the solution in the primal program would involve decreasing amounts of product Y and increasing amounts of X until at point D only X is produced. That is, the production of Y would be changing successively from three units output, point B, down to none. The motivation for substituting X for Y with increasing availability of machine time can be seen by considering the objective function. Each unit of product X leads to $4 contribution to overhead and profit, each unit of Y to only $2. Product Y was produced in the first place because of the scarcity of grinding machine time. Profit per unit of grinding machine time was important then. Just beyond point D this constraint is

Sec. 8.6]   SENSITIVITY ANALYSIS

no longer binding; this resource is no longer scarce. Given an increased amount of machine time, it became less and less necessary to produce product Y, the product that utilizes only a small amount of machine time. It became increasingly possible to produce more of product X that uses a relatively large amount of machine time but also provides more profit per unit produced.

The new machine constraint is that as shown in Figure 8.3. Five units of product X and no units of Y are produced. The relationships stated at the end of Section 8.5 are useful for anticipating changes in the dual program that are the result of changes such as that just introduced in the amount of grinding time. Under the original machine constraint at the point D in Figure 8.3 there now is slack in the warehouse constraint, and S_1 is positive. By equation (8.38),

$$S_i V_j = 0$$

and the shadow price for this resource must be zero. Also since product X continues to be produced, by equation (8.38),

$$XL_j = 0$$

the slack in the first dual constraint is necessarily zero. Similarly, the slack in the second dual constraint must be positive, since the value of Y is zero; product Y is not produced.

The slopes of each of the lines involved in the optimal solution to the dual program are shown in Figure 8.4. The slope of the first constraint, associated with

Figure 8.4

Dual to Program with Increase in Machine Hours.

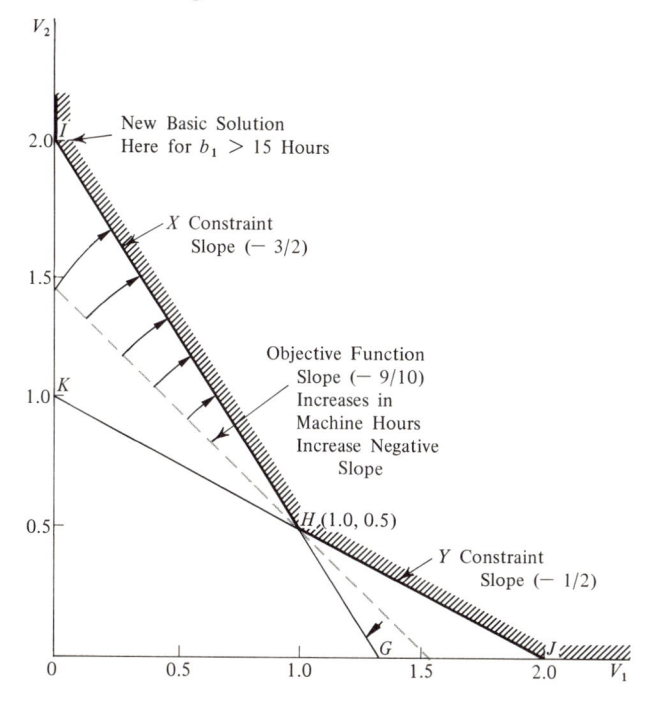

the output of product X, is $-\frac{3}{2}$. The slope of the objective function is $-\frac{9}{10}$; the slope of the Y constraint line is $-\frac{1}{2}$. Because the slope of the objective function falls partway between the slopes of the two constraint lines, the optimal solution is found at the corner point H. The slope of the objective function of the dual program is the ratio of the amounts of the scarce resources. The numerator in that slope is the number of units of the scarce resource, grinding hours. The increase in the number of grinding hours then increases the numerator of the slope of the objective function of the dual causing it to shift from $-\frac{9}{10}$ to $-\frac{15}{10}$ or $-\frac{3}{2}$. Note that when the constraint in the graph of the primal program reaches a point such as point D, Figure 8.3, the slope of the objective function in Figure 8.4 coincides with the slope of the X constraint; thus, the optimal solution is no longer unique. Any value for the scarce resource inputs at any point along the line IH gives the same result. The dual objective function coincides with the Y constraint line over its full length.

When the constraint, such as the grinding machine constraint, reaches a point such as D, the primal program is said to be degenerate; that is, there is no slack in either of the primal constraints and the value of one of the product variables is 0. Thus, the solution violates the definition of a basic solution. There are three zero variables and only one nonzero variable.

Definition—Degenerate Solution: *A feasible solution to a linear program that occurs at an extreme point for which the number of nonzero variables is fewer than the number of constraints is a degenerate solution.*

The degenerate solution in the primal program occurs when the objective function in the dual program coincides with one of the constraint lines. This occurs at a single point: a single value of the input, a single slope for the objective function. Continued change beyond this point means a change in the solution itself. All variables take on new values in the new solution. In the case illustrated, once a new solution is reached, no further increase in machine time, no matter how great or how slight, is of any value in this program. From point D and beyond, the *warehouse* constraint is binding. Machine time is no longer economically scarce. Similarly, in the graph of the dual program, further increases in the amount of machine time lead to a further increase in the slope of the objective function, but do not shift the optimal solution from the corner point I. At the corner point I only resource two, the warehouse, has a positive value imputed to it, only $V_2 > 0$. Resource one has no value, $V_1 = 0$. Further changes do not shift the point of the optimal basis. The warehouse resource is used to capacity, and it is the only constraint on profits. All profits are imputed, therefore, to the warehouse constraint.

The imputed value of variable V_2 in Figure 8.4 is seen to be $2 (dollars) in the new solution. Since there are only 10 units of warehouse space in the program, the value of the dual objective function must be $2 \times 10 = 20$ dollars. The value of the primal objective function with only 5 units of product X is $5 \times \$4 = \$20$. This verifies the relationship indicated in equation (8.37):

$$\text{Max } \pi = \text{Min } \gamma \qquad (8.37)$$

for the new solution to the program. This analysis is now carried through on

Figure 8.5

Program with Machine Constraint Decreased to Six Hours.

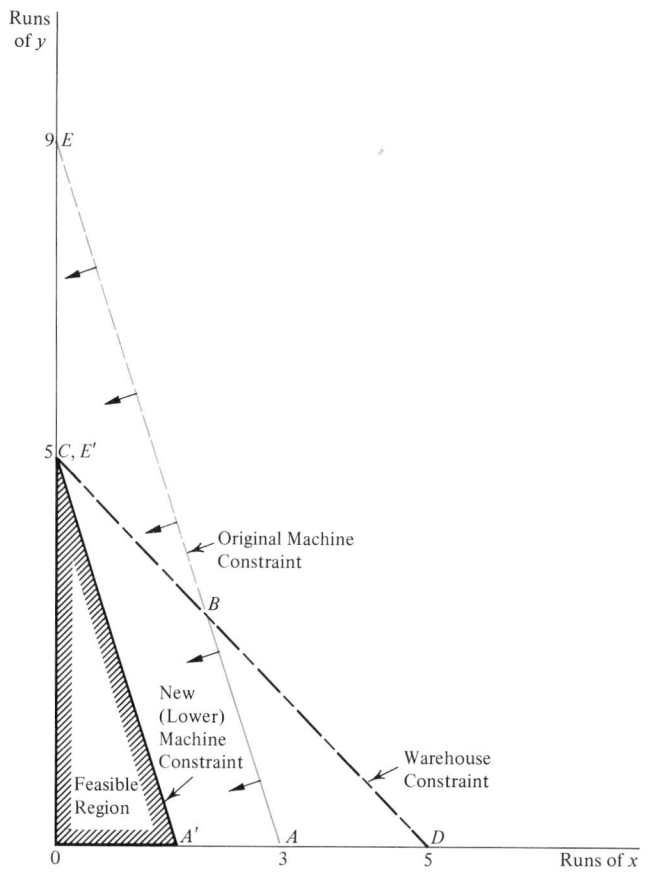

Figures 8.5 and 8.4 for the lower limit on the interval over which the shadow price $V_1 = 1.0$ applies for the machine constraint.

If four fewer hours of machine time were available, the machine constraint would move inward in the primal program until its upper extremity reached point C. The gradual change from nine hours of machine time to five hours leaves the same variables in the basis, but leads to gradual change in the amount of X and Y produced, product Y is gradually substituted for product X, as shown graphically by movement of the extreme point (corner) of the feasible region from point B along the line BC to C. At point C, only Y is produced, and the primal program becomes degenerate. This signals the coincidence of the objective function of the dual program with a constraint, the Y constraint. In the dual program, the Y constraint is binding, and the value of the Y constraint slack variable L_2 is thus zero, as suggested in Figure 8.4.

The coincidence of the dual objective function with the Y constraint implies the equality of their slopes. That of the Y constraint is $-\frac{1}{2}$; that of the objective function is

$$-\frac{9-4}{10} = -\frac{5}{10} = -\frac{1}{2}$$

also. Further decreases in the amount of machine time available do not change the new basis. Only Y will be produced if anything is produced. The value of V_1 is 2 at point J, and further decrease in machine hours will have no effect on this shadow price. The range of machine hours over which the original shadow price applies is thus $5 \leq b_1 \leq 15$.

The value of the objective function for a production of five units of Y at the lower limit point is $5(2) = 10$ for both the primal and the dual programs. The similar range for value of the warehouse b_2 constraint is

$$6 \leq b_2 \leq 18$$

The verification of this is left to the reader.

Essentially the same analysis can be carried out for shifts in the coefficients in the primal objective function that cause it to coincide with the constraints in the primal program, thus signalling a change in the solution to the program. Changes in the coefficients of the primal objective function correspond to shifts in the locations of the constraints of the dual program. The same degeneracy signals are useful and important here.

Modern computer programs for solution of linear programming problems include the solution of the dual program and a great deal of information about the sensitivity of particular solutions to changes in the variables. A limitation of the type of sensitivity analysis described is that it presents the impact of a change in a single parameter at a time and does not suggest the interrelationships from simultaneous changes in more than one variable. Other computer routines can be used to present sensitivity information for combinations of changes important for particular organizations. The most powerful form of sensitivity analysis, however, is the ability to rerun a program with several changes built into the rerun. The formal sensitivity analysis of each separate run also provides important information for the selection of the most significant changes to be tried in successive reruns.

8.7 Parametric Programming

Sensitivity analysis can be carried a few steps further by a procedure known as *parametric programming*. The essence of this procedure is the determination of the full range of values that the shadow prices for the scarce resources take on as the amount of *one* of the inputs is varied from zero to infinity. For example, if the machine time were varied from zero hours to an infinite number of hours, the shadow prices for machine time and warehouse space would each go through a series of changes. From the dual solution and sensitivity analysis one shadow price for each resource is known and the range of machine time availability over which that price applies is also known. The end points of that range identify the limits beyond which a new basic solution, and thus new shadow prices, apply for each scarce resource.

Sec. 8.7]   PARAMETRIC PROGRAMMING

Figure 8.6

Changes in Shadow Price V_1 with Changes in Available Machine Time (for Warehouse Capacity Constant at 10,000 Square Feet)

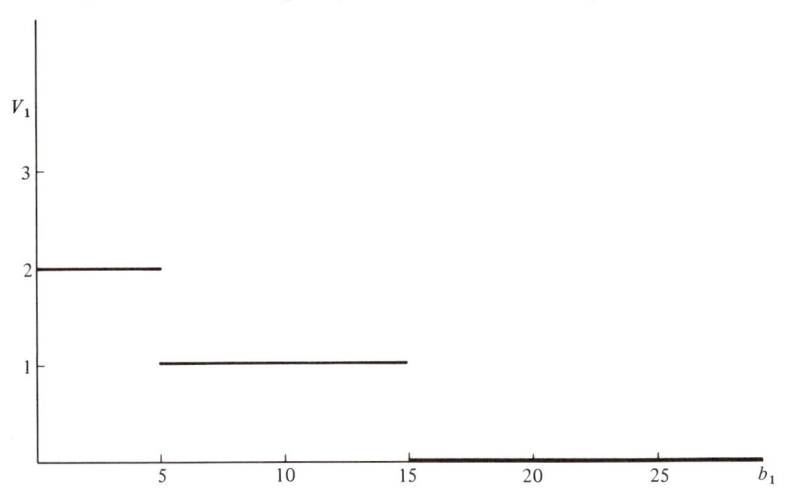

Parametric programming will be illustrated by carrying through the analysis for the resource of machine time. The one dollar shadow price V_1 applies for machine time between five and fifteen hours of capacity. It was noted from Figure 8.3 that adding machine hours in excess of fifteen hours (in fact, from fifteen to infinity) is worth nothing, the shadow price drops to and remains at zero since the machine constraint is no longer binding. This value can be read from the graph of the dual program in Figure 8.4 at point I; the value of the warehouse space shadow price V_2 of $2.00 can also be read at that point. At the other end point of the range, five machine hours, the basis also changes. The value of V_1 for fewer than five machine hours, at point J of Figure 8.4 is seen to be $2.00 (it is only coincidence that this value is identical to the value that V_2 takes on at point I) and the value of V_2 drops to zero. This information is recorded in a simple diagram, Figure 8.6. Similar information for shadow price V_2 is recorded in Figure 8.7.

It is also possible to identify from the dual program those products that will be produced within each range of scarce resource. As long as the solution to the program remains at point J in Figure 8.4 for values $0 < b_1 < 5$, only product Y will be produced—only the Y constraint is binding. When the solution is at point H for values $5 < b_1 < 15$, both X and Y will be produced—both the X and Y constraints are binding. When the solution falls at point I, for values $15 < v_1 < \infty$, only X will be produced—only the X constraint is binding.

The information in Figures 8.6 and 8.7 can readily be combined, and a statement of the products produced within each range of shadow price added at an appropriate place at the base of the graph to provide a compact summary of the parametric program data. It is left to the reader to carry through a parametric program solution for changes in the availability of warehouse floorspace and prepare such a figure for himself.

Figure 8.7

Changes in Shadow Price V_2 with Changes in Available Machine Time (for Warehouse Capacity Constant at 10,000 Square Feet)

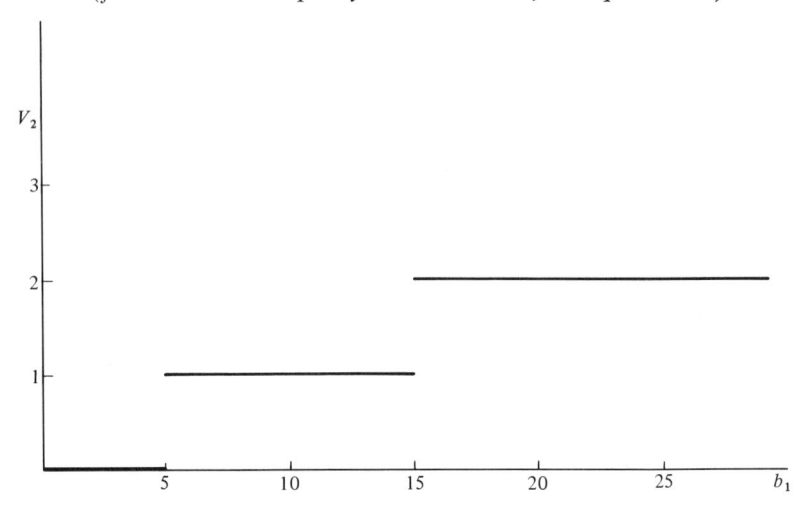

8.8 Feasibility Variables

For a standard maximization problem with constraints that are less than or equal to some values b_i, an initial basic, feasible solution is easily arrived at by setting the values of the ordinary variables at zero and, thus, the slack variables equal to the RHS of the constraints b_i. When the program is a standard minimization problem, this is not possible since the value of the variables would be negative; that is,

$$-L_i = c_i$$

and thus the solution would not be feasible. However, if new artificial variables (W_i) are introduced into each constraint, a basic solution can be found that is also feasible.

$$W_i - L_i = c_i.$$

For L_i assumed to be zero, this equation becomes the solution

$$W_i = c_i$$

that is feasible by definition. The variables W_i are called feasibility variables. But now new means must be found to eliminate them from the solution, so that the optimal solution will not be distorted. If each of these variables is assigned a large positive coefficient in the minimization objective function, then they will be eliminated from the optimal solution. The simplex procedure for a minimization is essentially the same as before, with the exception that variables in columns with the largest negative values of $C_j - Z_j$ are those introduced at each iteration.

8.9 Summary

A linear program is made up of three parts: the objective function, the resource constraint inequalities, and the nonnegativity requirements. The intersection of the point sets determined by the constraint inequalities plus the nonnegativities is known as the feasible region—the set of points which satisfies all the constraints in the system. This set must be convex and is known as a simplex. A solution to a linear program occurs at an extreme point, a corner, of the feasible region (the simplex). The program can more easily be dealt with mathematically if the constraint inequalities are restated as equalities. This is accomplished by introducing slack variables into each constraint. For a maximization problem with constraints stated as less-than-or-equal-to inequalities, a basic solution easily arrived at is one in which only the slack variables are nonzero, all ordinary variables are zero. When such a solution is not feasible, the method of feasibility variables presented in Section 8.7 can be used. A solution in which the slack variables are nonzero is a *basic solution*, may be a *feasible solution* if all nonzero variables are positive; but it is not an *optimal solution*. In a maximization problem an optimal solution is a basic, feasible solution for which the objective function takes on its maximum value. The simplex method of solving a linear program is a systematic procedure for moving from one extreme point of the simplex to another with higher value for the objective function such that the optimal solution is reached in a finite number of steps. The method is described and illustrated in Section 8.4.

Every linear program (known as the primal program) has associated with it another program called its *dual* as described in Section 8.5. The dual to the dual is the primal. The optimal solution to the primal is identical with the optimal solution to the dual program. There are a number of useful insights that can be gained from an economic interpretation of a dual program to a linear program.

Some important relationships between the primal and the dual problems can be summarized by the equation Max π = Min γ, which states that the *optimal* value of the primal and the dual programs are identical; and by

$$\begin{aligned}[Q_j L_i] &= 0 \\ [S_i V_j] &= 0\end{aligned} \quad \text{for } i = j \tag{8.38}$$

where Q_j is the product shown in column j in the primal program, L_i is the slack variable in the dual constraint i whose coefficients were derived from those in column j in the primal, S_i is the slack variable in primal constraint i, and V_j is the dual shadow price, the dual variable associated with scarce resource i in the primal. The coefficients of V in column j of the dual are identical to those for the ordinary variables in row i of the primal. The relationship implied by equation (8.38) is that if the value of the dual slack variable L_i is positive, the resources used in producing the product Q_j can be used more effectively in the production of products other than Q_j; an opportunity loss would result from producing Q_j so Q_j should not be produced. Only if there is no opportunity loss, only if $L_i = 0$, should Q_j be positive. The second equation of (8.38) implies that resource i has a shadow price V_i only if S_i is zero. The shadow price is a measure of the degree to which the profits of a firm are restricted by the scarcity of a resource. If a resource is not scarce, it does not restrict profitability and therefore the shadow price associated with that resource is zero. When a resource is scarce, its slack variable is zero (it is used to capacity) and the shadow price is positive.

Through sensitivity analysis it is possible to determine the range of values of various variables over which their shadow prices apply. This very useful information for a manager is discussed in Section 8.6, and extended through parametric programming in Section 8.7. Feasibility variables required when an initial, basic solution to a linear program is not feasible are discussed in Section 8.8.

Key Concepts

Concept	Reference Section
Objective Function	8.1
Capacity Constraints	8.1
Nonnegativities	8.1
Feasible Region	8.2
Extreme Points	8.2
Slack Variables	8.3
Basic Solution	8.3
Feasible Solution	8.3
Optimal Solution	8.3
Simplex Method	8.4
A Simplex	8.4
Simplex Tableau	8.4
Pivot Element	8.4
Dual Program	8.5
The Dual to the Dual	8.5
Shadow Prices	8.5
Max π = Min γ	8.5
$(Q_j L_i) = 0$	8.5
$(S_i V_j) = 0$	8.5
Sensitivity Analysis	8.6
Degenerate Solution	8.6
Parametric Programming	8.7
Feasibility Variables	8.8

Linear Programming Problems

I. Graphical Solution

1. The Ajax pretzel factory manufactures two kinds of pretzels: soft and hard Dutch pretzels. These pretzels are automatically produced in continuous process by Ajax's single pretzel-making machine, the Saline Twist'n Bake 500. Ajax management is interested in optimizing its contribution to fixed costs and profits from the sale of its two products. However, in the short run, its capability to produce is restricted by the amount of material, labor, equipment and other resources available during the period of time over which the decision will be effective.

LINEAR PROGRAMMING PROBLEMS

On a given day, the management determines the following availability of productive resources:

Pretzel mix (including salt) = 120 lbs.
Yeast = 32 oz.
Twist'n Bake 500 capability = 120 dozen of either hard or soft pretzels
Labor = 15 hours

From the bill of materials (recipe) for each product and their operations sheets, management determines that each dozen of the pretzels requires inputs as follows:

	Soft	Hard
Pretzel mix/dozen	1.0 lb.	0.6 lb.
Yeast/dozen	0.4 oz.	—
Labor/dozen	0.15 hrs.	0.10 hrs.

At the moment, the market is quite favorable and management expects to be able to sell at prevailing prices all the company produces. The selling price for soft pretzels is currently 70¢ per dozen and, for hard pretzels, 60¢ per dozen. The direct material and labor costs associated with the production of each is 50¢ per dozen and 45¢ per dozen, respectively.

(a) Formulate this problem as a linear programming problem.

Let X_H = quantity (in dozens) of hard pretzels produced.
X_S = quantity (in dozens) of soft pretzels produced.

(b) Solve the problem graphically.
(c) How does the restriction on pretzel mix affect the optimal solution?

2. A candy manufacturer is attempting to find the optimal mixture for a box of chocolates. There are two kinds of candy, fruit and nuts, that he wants in the box. He lists the important characteristics of each kind of candy and the characteristics of the box.

	Fruit	Nuts	Box
Number of pieces	X_F	X_N	35
Weight per piece (oz.)	1.6	0.8	32
Space per piece (sq. in.)	2.0	1.0	65
Cost per piece ($)	0.02	0.01	0.60

(a) Can the manufacturer get everything he wants? Prove why or why not.
(b) Suppose that we restate the problem in terms of inequalities (that is, total number of pieces ≥ 35, weight per box ≥ 32 oz., 40 sq. in. $\leq$ space available by using a variable number of fillers ≤ 65 sq. in., and cost per box $\leq \$0.60$). Set up this problem graphically and find the combination(s) of fruit and nuts that yields minimum cost. Show all your calculations.
(c) Suppose that after examining your solution (b), the manufacturer is unhappy about the number of pieces of fruit candy in the box and states that there must be at least 10 pieces in every box. Show graphically what this restriction means in terms of cost. (That is, what is the new optimum combination and how much more does it cost?)

3. The following represent additional linear programming problems which can be solved graphically:

(a) Max $Z = x + 1.5y$
Subject to:
$$2x + 3y \leq 6$$
$$x + 4y \leq 4$$
$$x, y \geq 0$$

(b) Min $g = 32w + 84v$
Subject to:
$$4w + 7v \geq 11$$
$$2w + 6v \geq 4$$
$$w, v \geq 0$$

(c) Min $g = 6w + 4v$
Subject to:
$$2w + v \geq 1$$
$$3w + 4v \geq 1.5$$
$$w, v \geq 0$$

(d) Max $z = 11x + 4y$
Subject to:
$$4x + 2y \leq 32$$
$$7x + 6y \leq 84$$
$$x, y \geq 0$$

II. Simplex Solution

4. Solve problem 1 by the simplex method.

5. Solve problems 3(a) and (d) by the simplex method.

III. Economics of Duality

6. (a) Formulate the dual to problem 1. (Neglect the redundant constraint.)
(b) What is the solution to the dual problem?
(c) What is the economic meaning of the solution to the dual in terms of problem 1?

7. (a) Formulate the dual to problem 2(b). (Neglect any redundant constraints.)
(b) Solve the dual problem using the simplex method (or use the computer).
(c) What is the economic interpretation of your dual solution for the situation in problem 2?
(d) What generality can you draw about using the simplex procedures described in Chapter 8 for the solution of minimization problems (for example, problem 2)?
(e) Formulate the duals to the parts of problem 3. What interrelation exists between the four parts of problem 3?

IV. Sensitivity Analysis

8. Suppose, in problem 1, that the pretzel manufacturer had been uncertain about the contribution he could expect from the sale of soft and hard pretzels (either because his variable costs and/or his selling prices were uncertain).

(a) If he could obtain contributions of 40¢ per dozen soft and 30¢ per dozen hard pretzels would his optimal production quantities (that is, his solution) change? Use graphical analysis to answer this or use the computer.
(b) What about 4¢ and 3¢ (respectively)? Can you generalize the above results to all linear programming problems?
(c) Suppose that the contributions were as follows, what would be the optimal X_S and X_H? Use graphical analysis.

Contributions from	Soft	Hard
	25¢	15¢
	$22\frac{1}{2}$¢	15¢
	15¢	15¢
	14¢	15¢

(d) What generalization can you draw from the above results about the circumstances under which the solution will change?

9. Suppose, owing to severe shortage of pretzel mix, that the manufacturer in problem 1 was unable to obtain more than 98 lbs. of mix.

(a) Set up graphically this new situation. What is the new optimal production X_S and x_H? What is the new maximum contribution?

(b) What is the dual opportunity cost associated with the mix constraint? *Hint:* You may find it easier to solve part (c) first as the value for the dual opportunity cost is discontinuous at 98 lbs. of mix (that is, it assumes two values—one for upward changes in the amount of mix and one for downward changes).

(c) Suppose that the shortage continues and at some later point the manufacturer is able to obtain only 80 lbs. of pretzel mix. What is the dual opportunity cost associated with the mix constraint? (*Hint:* Since now the yeast and labor constraints are redundant, they can be ignored in formulating the dual.) What is the dual opportunity cost at 70 lbs. of mix?

(d) Plot the dual opportunity cost for pretzel mix against the amount (pounds) of pretzel mix available.

V. Problem Formulation

Formulate the following problems as linear programs (do not solve them).

10. The Internal Revenue Service operates a number of regional computer centers to process federal tax returns. Every location in the country sends tax returns to one of the regional centers. Initially the allocation of reporting locations to centers was done strictly on a straight-line mileage basis; i.e., for any given location, the appropriate center was the center which was the closest in distance (measured by straight-line mileage) to it. Now suppose that in order to speed up collections the IRS decides to pair up centers to locations so as to minimize the total time required (in days) for the U.S. mails to carry the tax returns to the regional centers. In a simplified example suppose there are three regional IRS Centers and four major cities with tax returns to be processed. The following information is provided:

Mailing Time (Days) from City i to Center j

City \ Center	1	2	3
1	2.5	2.0	1.0
2	3.0	3.5	1.5
3	2.0	1.5	2.5
4	1.0	1.2	1.0

Center Processing Capacity (Millions of Returns)

Center 1	30
Center 2	60
Center 3	20

Location of Returns to be Processed (Millions of Returns)

City 1	15
City 2	25
City 3	35
City 4	15

How should centers and cities be matched up so that the total "return days" of delay in mailing is minimized?

11. A detailed study of alternative means of providing certain required tasks in a medical clinic has been performed. The study explored the use of various grades of labor in performing the required tasks, and recorded both the cost of each grade and the speed of each grade in performing the services. As a simplified example, suppose there are three tasks, A, B and C which must be performed; and four types of labor (I, II, III and IV) which can be used to perform the tasks. The use of lower-training grades of labor requires double-checking to maintain the quality of operation, and the additional time required to use these lower labor grades is reflected in the following data:

Task Requirements per Day

Task A	40
Task B	200
Task C	480

Hours Required for Labor Type i to Perform One Task of Type j

Labor Type \ Tasks	A	B	C
I	1	2	3
II	3	4	6
III	4	6	7
IV	5	8	6

Cost of Labor Grades per Hour		Maximum Availability of Labor Grades	
Labor Type I	$10.00	I	20
II	5.00	II	50
III	2.00	III	200
IV	1.60	IV	800

How should labor types be selected and assigned to minimize the total cost of performing the required tasks?

12. The Department of Defense (or HEW or The Acme Company) must award contracts for the procurement of 30,000 units of some product. The product will be used in three different locations, and shipping costs as well as bidding costs are important (the bidders are located in three different cities). Suppose the following bids have been obtained:

Bidder 1	Up to 20,000 units at $1.50/unit
Bidder 2	Up to 5,000 units at $1.20/unit
Bidder 3	Up to 15,000 units at $1.60/unit

Unit transportation costs from bidders to locations are:

Bidder \ Location	1	2	3
1	.10	.30	.25
2	.60	.30	.45
3	.25	.35	.30

Requirements by location are 10,000, 12,000, and 8,000 respectively, for locations 1, 2 and 3. How should bids be accepted to minimize total cost?

The A & B Company

The A & B Company, a firm located in Louisville, Kentucky, produces two major products, axes and bats, in its modern plant located on the outskirts of the city. In recent months two of A & B's production facilities have become bottlenecks. After the firm meets the requirements of its long-term contracts with two leading mailorder houses, it has only six hours per day available on its lathes, and six hours on its grinding machines. All other facilities are available in amounts equal to foreseeable needs.

Mr. S. L. Ugger, the production manager, plans to schedule the plant so as to maximize the firm's profits.

Prior to the regular Thursday scheduling meeting, the industrial engineers and cost analysts provided Mr. Ugger with the following data: Axes require one hour of lathe time and two hours of grinder time and contribute $5 to overhead and profit per production run; bats require two hours of lathe time and one hour of grinder time and contribute $2 to overhead and profit per production run. Mr. Ugger wrote a memorandum requesting that the industrial engineers analyze the situation for him. He wished answers to several questions:

> Which product should he produce daily and in what amounts?
>
> What would the cost implications be for a breakdown on the lathe? On the grinder?
>
> If he could subcontract grinder or lathe time, what is the maximum price that he could pay for it? For how many units would this apply?
>
> If the sales manager gave a $2 discount on axes, what difference would this make to the scheduling of the plant? To the cost of a breakdown on the lathe or the grinder? To the price he could afford to pay for subcontracting?

Mr. Jones, the chief industrial engineer, called in his assistant, a recent business school graduate, and gave him instructions to

1. Formulate the questions given to Mr. Jones as a linear programming problem.
2. Solve the problem by total enumeration of the solution.
3. Solve the problem graphically.
4. Solve the problem algebraically.
5. Formulate the dual to the problem.
6. Solve the dual graphically for $5 contribution for axes.
7. Carry out the sensitivity analysis on the program.
8. Resolve the dual for a $3 contribution to overhead and profit for axes and carry out the sensitivity analysis.
9. Complete the parametric programming analysis of the problem for the $3 contribution for axes.
10. Repeat the parametric programming solution for the $5 contribution for axes.

Acme Lift Corporation (A)

The Acme Lift Corporation had developed a standard line of fork-lift trucks, pallet elevators, traveling cranes, jib cranes, and similar materials-handling equipment. No inventory of completed equipment was carried because the demand for various types was too unpredictable and because customers often wanted minor modifications of design in the equipment ordered. During 1966, demand for its products was heavy and the company had lost some business because of *lack of capacity in its forge shop and its consequent inability* to meet required delivery schedules.

Company cost studies had indicated that *indirect variable overhead costs* of two dollars were incurred for every dollar of standard *direct labor expense*, and that this relationship held generally throughout its plant and for all levels of production over 25 per cent of capacity. *Standard direct labor expense* was computed at $2.00 per hour. The company attempted to price its products so as to earn a margin for profits and fixed overhead of 25 per cent of the selling price. However, in the case of its fork-lift trucks, competitive pressures had forced the company to accept a 15 per cent margin.

Early in July, 1966, the company received orders for travel cranes from a steel warehousing concern, and for 40 fork lift trucks from a western railroad. Both orders required delivery dates early in the fall of 1966. Before accepting either order, the general manager of the company discussed the situation with his sales manager and factory superintendent.

The factory superintendent reported that in order to meet the required delivery dates both jobs would have to be put through the forge shop sometime during the month of August, but there was room in the forge shop schedule for only one of them. There were only 1,200 *standard labor* hours of forge-shop capacity uncommitted during the month, and it was estimated that each travel crane would require approximately 367 man hours of forge-shop work, while a fork-lift truck would take about $27\frac{1}{2}$ hours per truck. It was not possible to complete both orders by the required delivery date utilizing the company's present forge-shop facilities, but it would be possible to handle either one of the orders with the company's present facilities and, *if the forge-shop work were given to another company for one of the orders, it would be possible for the company to accept both orders and complete them on time*. The superintendent believed that if Acme provided materials, he could get the forge-shop work done by another company on either one of the orders during August for $15,000. He also presented data from the accounting department indicating that under normal conditions the company would expect to *earn a margin available for fixed overhead expense and profit of* $750 on the sale of one fork-lift truck and $6000 on the sale of one traveling crane.

The sales manager indicated that both customers were insistent on stated delivery dates. Both firms were new customers for the Acme Lift Corporation, and both seemed equally *good prospects for repeat business*. The only basis for choice between the two orders, as the sales manager saw the situation, was that the company had a lower profit margin on fork-lift trucks than on travel cranes.

Questions:

1. Assuming that the company wants to maximize its *short-run profits:*
    (a) If it accepts only one of the two orders, which one should it accept and why?
    (b) If it subcontracts the forge work on one of the orders, how much can the firm afford to pay for the forge job before it loses money on that order?
2. Are there other considerations that should bear on the decision in this case?

Acme Lift Corporation (B)

While the Acme Lift Corporation was still considering what to do about the potential bottleneck in its projected forge-shop schedule during August, it learned of a new machine which, by speeding up work in the forge shop, could add the equivalent of 400 standard labor hours to the existing capacity of 1600 hours per month, when used with the same labor and other equipment now available. (This increase in efficiency would be linear over the entire range of production.)

If ordered immediately, the machine could be installed by August 1. It had a life expectancy of 24 months and the direct operating costs of the machine would be negligible. *The after-tax cost of the machine, including an allowance for the present value of the tax savings from depreciation, would be* $36,000. The company uses an after-tax discount rate of 12.7 per cent per year—1 per cent per month—to evaluate investment proposals. (Assume a 50 per cent tax rate for Acme.)

Compare the advantages and disadvantages of purchasing this machine versus subcontracting forge-shop work. (The cost estimate of subcontracting given in Acme Lift Corporation (A) is relevant for this problem.) Quantify your answer to the extent possible.

Acme Lift Corporation (C)

In August 1966 the factory superintendent of Acme Lift was planning his production schedule for the month of September. He knew that there were sufficient orders on hand for all products to more than use the existing capacity of the plant. In this situation he was concerned with two questions:

(a) What products should he produce to make the most profitable use of his available plant capacity?
(b) How much could he afford to pay to have some of the necessary work done outside?

The superintendent was impressed with the fact that the opportunity cost concept had been helpful to Acme in the previous month in deciding to produce fork trucks or travel cranes, and in deciding how much to pay for outside forge work. Therefore he decided to try to use these concepts in connection with the more complicated question facing him this month.

As a beginning, he decided to concentrate on two of the company's products, fork-lift trucks and loading shovels, and on two of the company's shops, the forge shop and the press shop.

The fork trucks required 25 man hours per unit in the forge shop (the new machinery had been installed there), 15 hours per unit in the press shop, and made a contribution to overhead of $750 per unit. The shovels required 20 man hours per unit in the forge shop, 25 man hours per unit in the press shop, and contributed $1000 per unit to overhead.

The superintendent was tentatively willing to assign a total of 200 hours of forge-shop time and 225 hours of press-shop time to these two products.

Assume that the company could sell as many fork trucks or shovels as it could produce.

1. What is the optimal quantity of each product to produce, if no work was done outside the plant? (Fractional units are acceptable, since the remaining work could be completed the next month.)

2. What is the maximum amount the firm could pay to have an increment of forge-shop work done outside the plant? (It would supply the materials.)

3. What is the maximum amount the firm could pay to have an increment of press-shop work done outside the plant? (It would supply the materials.)

4. If an additional 100 man hours of capacity were available in each shop, how would the optimal production quantities change, as compared to the answer to question (1) above?

5. Could the firm afford to pay the maximum amounts given in parts (2) and (3) above for *both* 100 additional hours of press-shop capacity and 100 additional hours of forge-shop capacity? Explain why or why not.

6. Carry out a full parametric programming solution to the problem.

Acme Lift Corporation (D)

Since his experience with the September production scheduling of two products was successful, the plant superintendent of Acme decided to try to schedule his entire plant's production during October using the methods he had been developing. Table 1 gives a list of all of the products produced by Acme, the requirements of each product in each shop in the plant, and the capacity of each shop.

Table 1

Shop Requirements and Contribution for Each Product (Man Hours of Work per Unit)

Shops and Man Hours Available in Each	Travel Crane	Fork Lift	Shovel Loader	Pallet Elevator	Model 1100 Crane	Total Man Hours Available
Forge Shop	330	25	20	10	30	2000
Press Shop	600	15	25	20	100	2500
Stamping Shop	200	40	100	12	50	4000
Machine Shop	150	75	30	25	40	3000
Assembly Shop	120	20	50	50	60	5000
Contribution	6000	750	1000	500	2000	

After checking with the sales manager, he learned that the company could sell up to four travel cranes, up to 30 fork lifts, up to ten shovel loaders, up to 20 pallet elevators, and up to 10 Model 1100 cranes. At least two travel cranes and ten fork lifts had to be produced, to satisfy orders on which the sales manager had promised firm delivery dates. Subject to these constraints the superintendent was free to decide what product mix would be most profitable.

1. Can you suggest a method of arriving at a detailed production plan for this plant?

2. Can you suggest a method of deciding how much Acme could afford to pay for outside work in case one or more shops in the plant were operating at capacity? (Assembly-shop work cannot be done outside.)

3. Can you suggest a method for guiding the sales manager as to what types of products are most profitable for Acme?

Red Brand Canners[3]

On Monday, September 13, 1965, Mr. Mitchell Gordon, Vice-President of Operations, asked the Controller, the Sales Manager, and the Production Manager to meet with him to discuss the amount of tomato products to pack that season. The tomato crop, which had been purchased at planting, was beginning to arrive at the cannery, and packing operations would have to be started by the following Monday. Red Brand Canners was a

---

[3] © 1965 by the Board of Trustees of the Leland Stanford Junior University. Distributed by the Intercollegiate Case Clearing House, Soldiers Field, Boston, Mass. 02163. All rights reserved to the contributors. Printed in USA.

medium-size company that canned and distributed a variety of fruit and vegetable products under private brands in the western states.

Mr. William Cooper, the Controller, and Mr. Charles Myers, the Sales Manager, were the first to arrive in Mr. Gordon's office. Dan Tucker, the Production Manager, came in a few minutes later and said that he had picked up Produce Inspection's latest estimate of the quality of the incoming tomatoes. According to their report, about 20% of the crop was Grade "A" quality and the remaining portion of the 3,000,000-pound crop was Grade "B".

Gordon asked Myers about the demand for tomato products for the coming year. Myers replied that they could sell all of the whole canned tomatoes they could produce. The expected demand for tomato juice and tomato paste, on the other hand, was limited. The Sales Manager then passed around the latest demand forecast, which is shown in Exhibit 1. He reminded the group that the selling prices had been set in light of the long-term marketing strategy of the Company, and potential sales had been forecasted at these prices.

Bill Cooper, after looking at Myer's estimates of demand, said that it looked like the company, "should do quite well (on the tomato crop) this year." With the new accounting system that had been set up, he had been able to compute the contribution for each product, and according to his analysis the incremental profit on the whole tomatoes was greater than for any other tomato product. In May, after Red Brand had signed contracts agreeing to purchase the grower's production at an average delivered price of 6 cents per pound Cooper had computed the tomato products contributions (see Exhibit 2).

Dan Tucker brought to Cooper's attention that, although there was ample production capacity, it was impossible to produce all whole tomatoes, as too small a portion of the tomato crop was "A" quality. Red Brand used a numerical scale to record the quality of both raw produce and prepared products. This scale ran from zero to ten, the higher number representing better quality. Rating tomatoes according to this scale, "A" tomatoes averaged nine points per pound and "B" tomatoes averaged five points per pound. Tucker noted that the minimum average input quality for canned whole tomatoes was eight and for juice it was six points per pound. Paste could be made entirely from "B" grade tomatoes. This meant that whole tomato production was limited to 800,000 pounds.

Gordon stated that this was not a real limitation. He had been recently solicited to purchase 80,000 pounds of Grade "A" tomatoes at $8\frac{1}{2}$¢ per pound and at that time had turned down the offer. He felt, however, that the tomatoes were still available.

Myers, who had been doing some calculations, said that although he agreed that the Company "should do quite well this year," it would not be by canning whole tomatoes. It seemed to him that the tomato cost should be allocated on the basis of quality and quantity rather than by quantity only as Cooper had done. Therefore, he had recomputed the marginal profit on this basis (see Exhibit 3), and from his results, Red Brand should use 2,000,000 pounds of the "B" tomatoes for paste, and the remaining 400,000 pounds of "B" tomatoes and all of the "A" tomatoes for juice. If the demand expectations were realized, a contribution of $48,000 would be made on this year's tomato crop.

Exhibit 1

Demand Forecasts

Product	Selling Price per Case	Demand Forecast (Cases)
24—$2\frac{1}{2}$ whole tomatoes	$4.00	800,000
24—$2\frac{1}{2}$ choice peach halves	5.40	10,000
24—$2\frac{1}{2}$ peach necture	4.60	5,000
24—$2\frac{1}{2}$ tomato juice	4.50	50,000
24—$2\frac{1}{2}$ cooking apples	4.90	15,000
24—$2\frac{1}{2}$ tomato paste	3.80	80,000

Exhibit 2
Product Item Profitability

Product	24—2½ Whole Tomatoes	24—2½ Choice Peach Halves	24—2½ Peach Necture	24—2½ Tomato Juice	24—2½ Cooking Apples	24—2½ Tomato Paste
Selling price	$4.00	$5.40	$4.60	$4.50	$4.90	$3.80
Variable costs:						
Direct labor	1.18	1.40	1.27	1.32	.70	.54
Variable OHD	.24	.32	.23	.36	.22	.26
Variable selling	.40	.30	.40	.85	.28	.38
Packaging mat'l.	.70	.56	.60	.65	.70	.77
Fruit[1]	1.08	1.80	1.70	1.20	.90	1.50
Total variance costs	3.60	4.38	4.20	4.20	2.80	3.45
Contribution	.40	1.02	.40	.12	1.10	.35
Less allocated OHD	.28	.70	.52	.21	.75	.23
Net profit	.12	.32	(.12)	(.09)	.35	.12

[1] Product usage is as given below

Product	Pounds per Case
Whole Tomatoes	18
Peach Halves	18
Peach Necture	17
Tomato Juice	20
Cooking Apples	27
Tomato Paste	25

Exhibit 3
Marginal Analysis of Tomato Products

Z = Cost per pound of A tomatoes in cents
Y = Cost per pound of B tomatoes in cents

(1) $(600{,}000 \text{ pounds} \times Z) + (2{,}400{,}000 \text{ pounds} \times Y) = (3{,}000{,}000 \text{ pounds} \times 6)$

(2) $\dfrac{Z}{9} = \dfrac{Y}{5}$

$Z = 9.32¢$ per pound
$Y = 5.18¢$ per pound

Product	Canned Whole Tomatoes	Tomato Juice	Tomato Paste
Selling price	$4.00	$4.50	$3.80
Variable cost (excl. tomato costs)	2.52	3.18	1.95
	$1.48	$1.32	$1.85
Tomato cost	1.49	1.24	1.30
Marginal profit	($ .01)	$ .08	$ .55

Analyze the problem.

Refrigerated Provisioner's Case

The Refrigerated Provisioners' Company, a distributor of frozen vegetables, serves over 300 retail grocery stores in a large metropolitan area. Refrigerated distributes 15 different vegetables that are packed in cartons with 12 ($5\frac{1}{4}$ inches $\times$ 4 inches $\times$ $1\frac{5}{8}$ inches) packages to a carton.

The company replenishes its supply of frozen foods only once each week. With the exception of the summer months when fresh vegetables are available to consumers, the company rarely has any inventory remaining in the warehouse at the end of a week. An entire week's supply of frozen vegetables arrives each Monday morning at the company's cold-storage warehouse. The warehouse is capable of holding 22,786 cartons.[4]

Refrigerated's supplier of frozen foods extends to the company a line of credit amounting to $40,000 per week. That is, the company's weekly credit purchases from its supplier may not exceed $40,000. Additional purchases must be financed through short-term bank loans since the company feels that it must maintain its present cash balances in order to preserve its liquidity position. The extent to which the company relies on bank loans depends on the prevailing interest rates and the demand conditions for frozen vegetables.

The following table represents the predicted sales (in twelve-package cartons) for the company's fifteen products during a forthcoming week. This forecast is expressed in terms of the maximum and minimum expected sales for the particular week. The minimum sales figure is based on a contractual agreement that Refrigerated has with a large retail grocery chain to provide certain minimum quantities of certain designated products each week. The maximum figure represents an estimate of the expected number of cartons of each vegetable that can be sold during the week in question.

	Minimum	*Maximum*
Whipped potatoes	300	1500
Creamed corn	400	2000
Black-eyed peas	250	900
Artichokes	—	150
Carrots and peas	300	1200
Succotash	200	800
Okra	150	600
Cauliflower	100	300
Green peas	750	3500
Spinach	400	2000
Lima beans	500	3300
Brussels sprouts	100	500
Green beans	500	3200
Squash	100	500
Broccoli	400	2500

The cost and selling price per carton of each of the products in the company's product line are listed on the next page.

[4] This figure was arrived at from the following data: Although the warehouse contains 16,000 cubic feet of space (40 feet $\times$ 40 feet $\times$ 10 feet), frozen vegetables may be stacked only to a maximum height of six feet. This restriction is imposed by overhead refrigeration equipment and the possibility of damaging the frozen food packages by the excessive weight placed on those cartons that are on the bottom of the stack. The warehouse also contains two five-foot aisles that are parallel to the length of the building and two five-foot aisles that are parallel to the width of the building. The aisles must remain clear at all times for efficient materials handling.

	Cost Unit	Selling Price/Unit
Whipped potatoes	$2.15	$2.27
Creamed corn	2.20	2.48
Black-eye peas	2.40	2.70
Artichokes	4.80	5.62
Carrots and peas	2.60	2.92
Succotash	2.30	2.48
Okra	2.35	2.20
Cauliflower	2.85	3.13
Green peas	2.25	2.48
Spinach	2.10	2.27
Lima beans	2.80	3.13
Brussels sprouts	3.00	3.18
Green beans	2.60	2.92
Squash	2.50	2.70
Broccoli	2.90	3.13

1. Assuming the company does not utilize its present cash balances or a short-term bank loan to finance its frozen food purchases, determine by linear programming the optimal quantity of each product that the company should order for the forthcoming week.
2. What is the maximum interest rate that Refrigerated should consider paying in order to increase its purchases beyond the $40,000 limitation imposed by the supplier? If the company can obtain a short-term note for less than this maximum interest rate, what effect will this have on the solution?
3. Suppose the company can lease additional cold storage space near its present location. Using the assumptions in Question (1) and then Question (2), determine the maximum rental fee (per cubic foot of cold storage capacity) that the company can afford to pay each week for increased storage capacity.
4. Carry out a parametric program solution for the resources that are currently scarce.

CHAPTER 9

Vectors and Matrices

The usual procedure in the book thus far has been to present conceptual and theoretical material first and then to introduce applications. With Chapters 8 and 9 this procedure has been reversed. In Chapter 8 an important application of linear algebra, linear programming, was presented using basic algebra, graphic analysis, and the simplex method.

The conceptual and theoretical basis of linear programming is given in the discussion of vectors and matrices in this chapter. Vectors and matrices are discussed to clarify the linear programming procedures and determinants are used in the solution of systems of simultaneous equations.

9.1 Vectors

To define a vector, it is necessary to understand and define a field. This section defines both terms, then gives illustrations of vectors and some basic rules for their manipulation.

A *field* is the most basic concept in linear algebra, and can be defined as follows:

> ***Definition—Field:*** *A field is a set of numbers that includes the sum, difference, product, and quotient of every two numbers in the set, and for which the numbers are commutative and associative in addition and multiplication as well as distributive in multiplication.*

A field has a number of important properties, among which are the following:

1. The field must have a unique element zero (0) such that $a + 0 = a$ for all a's in the field.
2. For each element a in the field, there is a unique element $-a$ such that $a + (-a) = 0$.
3. The field contains a unique nonzero element one (1) such that for all elements a of the field, $1 \cdot a = a$.
4. To every nonzero element a in the field, there corresponds a unique element $1/a$ in the field such that $a \cdot (1/a) = 1$.

Consider the set T, where

$$T = \{1, 2, 7, \ldots, 10\}.$$

This set of numbers does not fit the definition of a field. However, by repeated application of addition, subtraction, multiplication, and division, a field could be developed from this set of numbers. The smallest possible field is the field of *rational* numbers. The next larger field is the field of real numbers. The elements of a field are known as *scalars* or numbers. An ordered subset of elements of a field is called a *vector*.[1]

> ***Definition—Vector:*** *A vector is a n-tuple of elements from a field or a subset of a field.*

Generally the field of real numbers is used in defining vectors. For example, a

---

[1] It is not essential that a vector be composed of ordered *numbers;* however, this more restricted definition is useful in this exposition.

column vector **v** is designated as

$$\mathbf{v} = \begin{bmatrix} a_1 \\ a_2 \\ \vdots \\ a_n \end{bmatrix}$$

where the a_i constitute an *n*-tuple of real numbers. The numbers a_i, when referred to individually, are known as the *components* of the vector. The vector

$$\mathbf{v}' = [a_1, a_2, \ldots, a_n]$$

is called a *row vector*. The symbol **v**' is used here to designate such row vectors. The components of vectors are enclosed within brackets [] to distinguish them from ordinary sets. When a vector is not specified as a row vector or column vector, it is assumed to be a column vector.

A vector of real numbers, whether a row or column vector, can also be expressed graphically. A single-component row vector, $\mathbf{v}'_1$ can be expressed as

$$\mathbf{v}'_1 = [1]$$

or as a line segment of given magnitude and oriented in a particular direction, which would be graphed as shown in Figure 9.1. The head of an arrow is used to

Figure 9.1

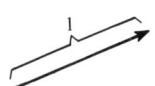

indicate the direction of the vector with the base of the arrow set at its origin (the zero point in the field of definition). This vector has direction, northeast, and one (arbitrary) unit of length. Such a vector is known as a unit vector (in the direction shown). Every vector has two key characteristics, direction and length.

A row vector $\mathbf{v}'_2$ with base at point 0 and proceeding horizontally for one unit can be expressed graphically as shown in Figure 9.2. A vector can be multiplied by

Figure 9.2

Horizontal Vector $\mathbf{v}'_2$ of Unit Length.

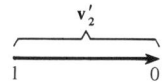

a real number k, a scalar from the field of definition. Graphically, this implies making the vector k times as long as it was previously. The vector $3\mathbf{v}'_2$ is shown in Figure 9.3 with length three times that of $\mathbf{v}'_2$. The procedure of multiplying a vector by a scalar is called scalar multiplication. Because scalars are numbers taken from

Figure 9.3

Horizontal Vector $\mathbf{v}'_2$ of Unit Length and $3\mathbf{v}'_2$ of 3 Units Length

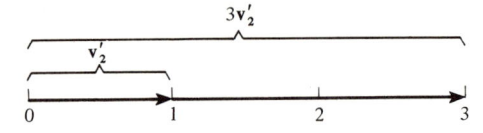

the field of real numbers, they may be either positive or negative. The meaning of a negative scalar $-k$ times a vector is shown in Figure 9.4, where $-k = -1$ for vector $\mathbf{v}'_2$.

Figure 9.4

Horizontal Vectors $\mathbf{v}'_2$ and $-1\mathbf{v}'_2$, Each of Unit Length.

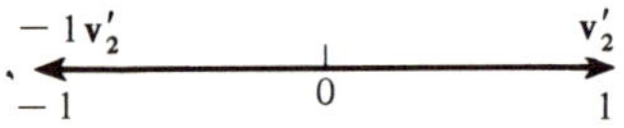

Given any vector, such as $\mathbf{v}'_2$, it is possible to develop any other vector in the same direction; that is, any other vector along the same line or a line *parallel* to that line (since the key characteristics of a vector are its *direction* and *length*). The set of vectors that can be generated by scalar multiplication of a given vector are known as *collinear* vectors. For any vector, there is, therefore, an infinite set of collinear vectors consisting of the original vector plus the product of that vector times every scalar from the field of real numbers. From the vector $\mathbf{v}'_2$, for example, the set of all possible collinear vectors generates a horizontal line that may be designated the x axis.

If another vector, $\mathbf{v}'_3$, also originating at point 0 and having the same unit length, but *vertical* direction, is introduced into Figure 9.4, the result is the graph shown in Figure 9.5. The vectors collinear with $\mathbf{v}'_2$ constitute the horizontal x axis, and the

Figure 9.5

Horizontal Vectors $\mathbf{v}'_2$ and $-1\mathbf{v}'_2$ and Vertical Vector $\mathbf{v}'_3$, All with Unit Length.

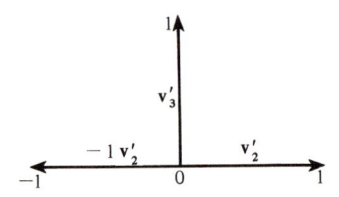

vectors collinear with $\mathbf{v}'_3$ constitute a vertical line that can be designated the y axis. However, there is *no* scalar multiple of $\mathbf{v}'_2$ that can be equated to $\mathbf{v}'_3$, and none of $\mathbf{v}'_3$ that can produce $\mathbf{v}'_2$; the vectors $\mathbf{v}'_2$ and $\mathbf{v}'_3$ are noncollinear—they do *not* lie along the same line, although they do lie in the same plane.

Vectors may be added to each other and subtracted from each other. To add

Sec. 9.1] VECTORS

vector v'_2 to v'_3 graphically, the base of vector v'_2 is placed at the head of v'_3, as shown in Figure 9.6. The new vector, $v'_4 = v'_3 + v'_2$, is that drawn from the base of v'_3 to the head of v'_2.

Figure 9.6

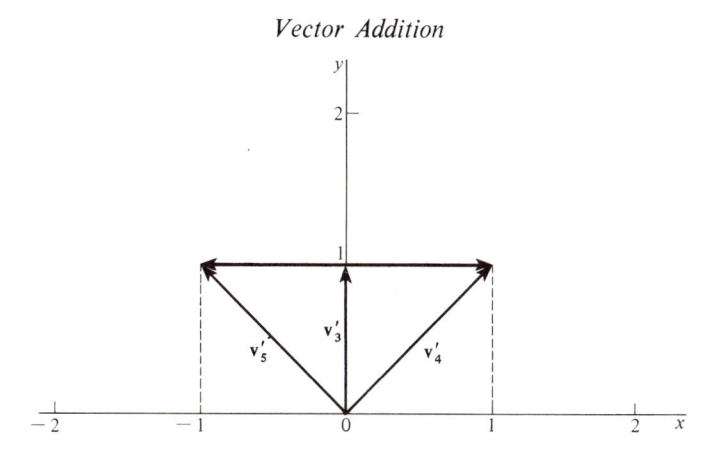

Vector Addition

To subtract vector v'_2 from v'_3, the base of $-v'_2$ is placed at the head of v'_3 and the new vector $v'_5 = v'_3 - v'_2$ drawn from the base of v'_3 to the head of $-v'_2$. The vector v'_4 is known as the *resultant* of vector addition and v'_5 as the resultant of vector subtraction.

Given that the vectors v'_2 and v'_3 have the following components,

$$v'_2 = [1, 0]$$
$$v'_3 = [0, 1]$$

and given the following definition of *vector addition:*

Definition—Vector Addition: The vector addition of $v_i + v_j$ is defined to be the sum of the respective components of the vectors.

then adding v'_2 and v'_3 implies

$$v'_2 + v'_3 = [(0, 1) + (1, 0)] = [(0 + 1, 1 + 0)] = [1, 1] = v'_4.$$

The components of the new vector are

$$v'_4 = [1, 1].$$

Also, given the following definition of *vector subtraction:*

Definition—Vector Subtraction: The vector subtraction of $v_i - v_j$ is defined as the components of the first vector minus the components of the second vector.

subtracting v'_2 from v'_3 implies

$$v'_3 - v'_2 = [(0, 1) - (1, 0)] = [(0 - 1, 1 - 0)] = [-1, 1] = v'_5.$$

The components of the new vector $\mathbf{v}'_5$ are

$$\mathbf{v}'_5 = [-1, 1].$$

The subtraction of $\mathbf{v}'_2$ from $\mathbf{v}'_3$ is equivalent to multiplying $\mathbf{v}'_2$ by the scalar -1 and adding the product to $\mathbf{v}'_3$

$$(-1)\mathbf{v}'_2 = -1[1, 0] = [-1, 0]$$
$$\mathbf{v}'_3 + (-1)\mathbf{v}'_2 = [0, 1] + [-1, 0] = [(0 - 1, 1 - 0)] = [-1, 1].$$

Given an arbitrary point such as point P in Figure 9.7, a vector $\mathbf{v}'_n$ can be generated from appropriate scalar multiples of $\mathbf{v}'_2$ and $\mathbf{v}'_3$ to reach the point P. That is, $\mathbf{v}'_n$ is the resultant of a scalar k_1 times $\mathbf{v}'_2$ plus another scalar k_2 times $\mathbf{v}'_3$:

$$\begin{aligned}\mathbf{v}'_n &= (k_1)\mathbf{v}'_2 + (k_2)\mathbf{v}'_3 \\ &= k_1[1, 0] + k_2[0, 1] \\ &= [k_1, k_2].\end{aligned}$$

The particular point chosen for P in Figure 9.7 is the point with coordinates $(1, 2)$; that is, $k_1 = 1, k_2 = 2$. This is the linear combination of the vectors $\mathbf{v}'_2$ and

Figure 9.7

Linear Combination of $\mathbf{v}'_2$ and $\mathbf{v}'_3$ to Generate Arbitrary Vector $\mathbf{v}'_n$.

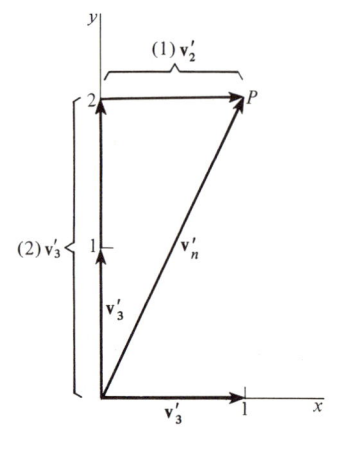

$\mathbf{v}'_3$ required to generate the vector $[1, 2]$. The unit vectors generate the x and y axes. Scalar multiples of each vector represent the coordinates of points on the plane determined by the axes. The unit vectors are noncollinear; there is no scalar multiple of either vector that can generate the other vector. The unit vectors are *coplanar;* that is, they lie in the same plane and can together be used to generate any other vector in the plane.

Similarly, it is possible to take any two noncollinear vectors generated by $\mathbf{v}'_2$ and $\mathbf{v}'_3$ and regenerate these original unit vectors (and also any of the other vectors in the plane, since it is possible to use $\mathbf{v}'_2$ and $\mathbf{v}'_3$ to generate any other vector). For example, in Figure 9.8, if appropriate multiples of vectors **OM** and **OM'**,[2] the

---
[2] Vectors indicated by letters are shown boldface.

Figure 9.8

Vectors **OM** *and* **OM'** *Used to Generate Vector* v'_3

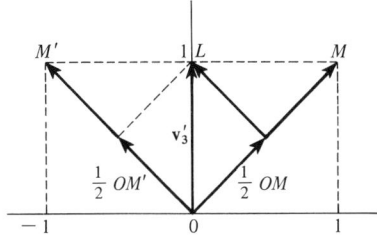

vectors [1, 1] and [−1, 1] are combined, they could generate v'_3 as shown graphically. The same is shown algebraically in the following development:

$$\mathbf{OM} = [1, 1]$$
$$\mathbf{OM'} = [-1, 1]$$
$$k_1\mathbf{OM} + k_2\mathbf{OM'} = \mathbf{v}'_3$$
$$k_1[1, 1] + k_2[-1, 1] = [0, 1].$$

The addition rule gives

$$[k_1 - k_2, k_1 + k_2] = [0, 1]$$

which implies that

$$k_1 - k_2 = 0$$

or

$$k_1 = k_2$$

and

$$k_1 + k_2 = 1.$$

Substituting for k_2 its equivalent, k_1 gives:

$$k_1 + k_1 = 1$$
$$2k_1 = 1$$
$$k_1 = \tfrac{1}{2}$$

and

$$k_2 = \tfrac{1}{2}.$$

9.2 Vector Spaces

The preceding discussion has shown that a single vector may generate any vector collinear with itself, and that two noncollinear, but coplanar[3] vectors can generate any other vector in a given plane. It has also illustrated that *orthogonal* unit vectors (that is, vectors at right angles to each other such as $\mathbf{v}_2$ and $\mathbf{v}_3$) can be generated by any pair of coplanar, noncollinear vectors; or to state it in another way, two noncollinear, coplanar vectors may be reduced to orthogonal unit vectors by scalar

---
[3] Note that two noncollinear vectors *determine* a plane; therefore any pair of such vectors are necessarily coplanar.

multiplication and vector addition and subtraction. These relationships are now generalized and presented more formally.

Definitions—Linear Dependence, Linear Independence: A set of vectors $\mathbf{v}_1, \mathbf{v}_2, \ldots \mathbf{v}_m$ is said to be linearly dependent *if there is a set of m scalars* $k_1, k_2, \ldots, k_m$, *not all zero, such that*

$$k_1\mathbf{v}_1 + k_2\mathbf{v}_2 + \cdots + k_m\mathbf{v}_m = 0;$$

where $0 =$ *the zero vector* $[0, 0, \ldots, 0]$ *with m components, all zero. If all of the scalars must be zero in order for this equation to hold, then the vectors are* linearly independent.

This definition implies that if any scalar is nonzero, the set is linearly dependent. Testing the vectors $\mathbf{v}_2'$ and $\mathbf{v}_3'$ for linear dependence under this definition gives:

$$\mathbf{v}_3' = [0, 1]$$
$$\mathbf{v}_2' = [1, 0]$$
$$k_1\mathbf{v}_3' + k_2\mathbf{v}_2' = [0, 0]$$
$$k_1[0, 1] + k_2[1, 0] = [0, 0]$$
$$[0, k_1] + [k_2, 0] = [0, 0]$$
$$0 + k_2 = 0$$
$$0 + k_1 = 0$$

therefore

$$k_2 = k_1 = 0.$$

Because the required scalars must both be zero, then $\mathbf{v}_2'$ and $\mathbf{v}_3'$ are linearly independent.

Next $\mathbf{v}_2'$, **OM**, and $\mathbf{v}_3'$ are tested for linear dependence:

$$k_1(\mathbf{v}_2') + k_2(\mathbf{OM}) + k_3(\mathbf{v}_3') = [0, 0]$$
$$k_1[1, 0] + k_2[1, 1] + k_3[0, 1] = [0, 0]$$
$$k_1 + k_2 = 0$$
$$k_2 + k_3 = 0$$
$$k_1 = -k_2$$
$$k_3 = -k_2.$$

Any scalar can be used for both k_1 and k_3, as long as the negative of that scalar is used for k_2 to satisfy the requirements for linear dependence. This is reasonable since **OM** was generated by a linear combination of $\mathbf{v}_2'$ and $\mathbf{v}_3'$. Note that the set of vectors $\{[1, 0, 0, 0], [0, 1, 0, 0], [0, 0, 1, 0], [0, 0, 0, 1], [0, 0, 0, 2]\}$ is linearly dependent, even though four of the vectors are linearly independent.

The unit vectors can be extended to the three-dimensional case as shown in Figure 9.9. These vectors are linearly independent, since only for the scalars $k_i = 0$ will the sum equal zero:

$$k_1\mathbf{J} + k_2\mathbf{K} + k_3\mathbf{L} = [0, 0, 0].$$

This result can be stated generally for n dimensions: orthogonal unit vectors in n

Sec. 9.2] *VECTOR SPACES* 311

Figure 9.9

Orthogonal Unit Vectors in Three Dimensions.

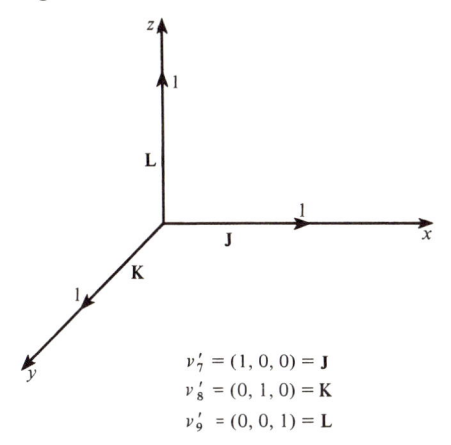

$v'_7 = (1, 0, 0) = \mathbf{J}$
$v'_8 = (0, 1, 0) = \mathbf{K}$
$v'_9 = (0, 0, 1) = \mathbf{L}$

dimensions are linearly independent. The concept of a *vector space* and the *basis* for a vector space are useful in establishing this.

Definition—Vector Space: *Given the field F and the vectors* $\mathbf{v}_1, \mathbf{v}_2, \ldots, \mathbf{v}_n$ *defined over F, a* vector space *is the set* $\mathbf{V}$, *of vectors for which the sum of every two vectors, plus the product k$\mathbf{v}$ of every vector $\mathbf{v}$ with a scalar k from F is included in* $\mathbf{V}$.

The sum is an operation which associates with any two vectors $\mathbf{v}_1$ and $\mathbf{v}_2$ in $\mathbf{V}$ a vector $\mathbf{v}_1 + \mathbf{v}_2$ also in $\mathbf{V}$ such that:

1. $\mathbf{v}_1 + \mathbf{v}_2 = \mathbf{v}_2 + \mathbf{v}_1$
2. $\mathbf{v}_1 + (\mathbf{v}_2 + \mathbf{v}_3) = (\mathbf{v}_1 + \mathbf{v}_2) + \mathbf{v}_3$
3. $\mathbf{V}$ has a unique vector 0, the zero vector, such that $\mathbf{v}_i + 0 = \mathbf{v}_i$ for all $\mathbf{v}_i$ in $\mathbf{V}$.
4. To each $\mathbf{v}_i$ there corresponds a unique vector $-\mathbf{v}_i$ such that $\mathbf{v}_i + (-\mathbf{v}_i) = 0$.

The scalar product is an operation that associates with any scalar k in the field and any vector $\mathbf{v}$ in $\mathbf{V}$, a vector $k\mathbf{v}$ also in $\mathbf{V}$, such that:

1. $1\mathbf{v}_i = \mathbf{v}_i$ for all $\mathbf{v}_i$ in $\mathbf{V}$.
2. $[k_i k_j]\mathbf{v}_i = k_i[k_j \mathbf{v}_i]$.
3. $k_i[\mathbf{v}_i + \mathbf{v}_j] = k_i \mathbf{v}_i + k_i \mathbf{v}_j$.
4. $[k_i + k_j]\mathbf{v}_i = k_i \mathbf{v}_i + k_j \mathbf{v}_i$.

The x, y plane meets the definition of a vector space for an n of two; the y axis meets the definition of a vector space for an n of one. It is now possible to define a *basis* for a vector space.

Definition—Basis: *A basis for a vector space* $\mathbf{V}_n$ *is a set of n linearly independent vectors* $\mathbf{v}_1, \mathbf{v}_2, \ldots, \mathbf{v}_n$, *from which any other vector in the space* $\mathbf{V}_n$ *can be generated.*

The vectors $\mathbf{v}'_2$ and $\mathbf{v}'_3$ fit this definition of a basis for the vector space $\mathbf{V}_2$. The vectors $\mathbf{v}'_2$ and $\mathbf{v}'_3$ were used to generate any (arbitrary) vector in the plane $\mathbf{V}_2$; the vector $\mathbf{v}'_2$ is the basis for $\mathbf{V}_1$, the x axis.

Up to this point the word *dimension* has been used without definition; for example, "in the three-dimensional case." A rigorous definition of the word dimension is now given.

Definition—Dimension of a Vector Space: *The dimension of a vector space $\mathbf{V}_m$ is the maximum number, m, of linearly independent vectors in $\mathbf{V}_m$, which can be a basis for $\mathbf{V}_m$.*

Thus, because only one vector $\mathbf{v}'_2$ provided a basis for a space $\mathbf{V}_1$, then $\mathbf{V}_1$ is a *one-dimensional* vector space. Any vector (except the zero vector) can be used as the basis for a particular one-dimensional vector space. The xy plane is a two-dimensional vector space $\mathbf{V}_2$. There is an infinite number of pairs of linearly independent vectors in $\mathbf{V}_2$. Each forms a basis for the xy plane; it is possible to generate any other vector in the plane from such pairs of vectors.

A basis for a n-dimensional vector space has n elements. Thus two different bases for a given vector space have the same number of elements. The following two implications of the preceding discussion are important:

1. If V is a vector space, and $(\mathbf{v}_1, \ldots, \mathbf{v}_n)$ is a maximal set of linearly independent elements of $\mathbf{V}$, then $(\mathbf{v}_1, \ldots, \mathbf{v}_n)$ is a basis for $\mathbf{V}$.
2. If $\mathbf{V}$ is a vector space with dimension n, and $\mathbf{v}_1, \ldots, \mathbf{v}_n$ are linearly independent elements of $\mathbf{V}$, then $\mathbf{v}_1, \ldots, \mathbf{v}_n$ is a basis for $\mathbf{V}$.

The combination of any pair of basis vectors for $\mathbf{V}_2$ with a third vector was seen (in Figure 9.8, for example) to produce a linearly *dependent* set of vectors, confirming that the maximum number of linearly independent vectors in the xy plane is two. Similarly, linear combinations of the unit orthogonal vectors $\mathbf{J}$, $\mathbf{K}$, and $\mathbf{L}$ shown in Figure 9.9 can be used to generate any other vector in the three-dimensional space that they span. The vectors $\mathbf{J}$, $\mathbf{K}$, and $\mathbf{L}$ may be oriented in space in an infinite number of ways; one way is with $\mathbf{J}$ and $\mathbf{K}$ determining the horizontal plane and a vertical axis $\mathbf{L}$ raised in the plane to determine the Cartesian coordinate system.

The three orthogonal unit vectors are linearly dependent with any other vector in space $\mathbf{V}_3$. In other words, three is the maximum number of linearly independent vectors for spanning the space $\mathbf{V}_3$.

To emphasize the fact that the familiar Cartesian coordinate system is just a particular basis for a vector space, orthogonal unit vectors have been used as an illustrative first basis for each space discussed. However, it must be recognized that by definition *any* two (m) noncollinear (linearly independent) vectors can be used as a coordinate system, that is, as a basis for two (m) space; this is the *meaning* of the word basis. Any vector in the vector space can be generated from linear combinations of the basis vectors. In the discussion of linear programming in Chapter 8, movement from corner to corner of the feasible region can now be viewed as the process of changing the basis vectors for spanning the space. The old basis is replaced by a new basis of linearly independent vectors at each iteration. The coordinates or scalar multiples of the particular vectors in the basis (that is, the coefficients in the simplex tableau) change accordingly. The vectors used as the basis were labeled the *basis vectors* in the simplex tableau in Chapter 8.

Note that, given any two (*m*) basis vectors, the orthogonal unit vectors for that space can be generated, and these unit vectors constitute the simplest set of basis vectors for the space. Also the process of multiplying a vector by a scalar and then adding (or subtracting) one vector to (or from) another represents the process of linearly combining vectors. Moving from one basis for a vector space to another basis requires only a linear transformation of the original vectors. Any basis for an *n*-dimensional space has *n* vectors. Thus for the vector space V_n, any set of vectors including more than *n* vectors must be linearly dependent, and no set having less than *n* vectors can span V_n. An *ordered* basis is a basis with a particular order for its elements. For example, in the *xy* plane a vector [1, 2] is generally regarded as one with $x = 1$, $y = 2$; that is, the ordered basis consists of ordered pairs (*x*, *y*). The reverse order could equally well have been specified, but the convention is that the first component specifies the scalar used to multiply the unit vector in the *x* direction.

9.3 Matrices

This section introduces matrices, and the operations of matrix addition and subtraction, scalar multiplication, and matrix multiplication are defined. A matrix is a rectangular array of elements. Therefore, a vector can be considered a matrix with only one column (or row). Initially, matrices were introduced to simplify notation, but they were soon found to possess very useful properties. For example, given the system of equations shown in equation (9.1):

$$2x_1 + 4x_2 + 7x_3 = 5 \\ x_1 + 3x_2 + 7x_3 = 6 \tag{9.1}$$

it is possible to express the coefficients of the variables x_i as a matrix, as indicated in equation (9.2):

$$A \equiv \begin{bmatrix} 2 & 4 & 7 \\ 1 & 3 & 7 \end{bmatrix}_{(2 \times 3)}. \tag{9.2}$$

Matrix *A* is called the matrix of coefficients of the equation system; it has two rows and three columns and is a (2 × 3) matrix. The number of rows and columns of the matrix are known as the order of the matrix. The general notation for the order of a matrix is (*m* × *n*), where *m* designates rows and *n* columns.

The matrix of coefficients may be enlarged as indicated in equation (9.3) into what is known as the augmented matrix of coefficients by including the two constants on the right-hand side of the equations, as shown in A_1; A_1 is

$$A_1 \equiv \begin{bmatrix} 2 & 4 & 7 & 5 \\ 1 & 3 & 7 & 6 \end{bmatrix}_{(2 \times 4)} \tag{9.3}$$

a (2 × 4) matrix.

Matrix addition and subtraction is defined analogously to vector addition and subtraction. These operations can be carried out only for matrices of the same order; the number of rows and columns of two matrices must be the same if they are to be added to each other or subtracted from each other. Given the matrix B, another (2×4) matrix indicated in equation (9.4),

$$B \equiv \begin{bmatrix} 1 & 2 & 1 & 2 \\ 2 & 1 & 2 & 1 \end{bmatrix}_{(2 \times 4)} \tag{9.4}$$

it is possible to add it to the matrix A_1. This is done by adding to each element of A_1 the corresponding element of B. The result is the matrix C as shown in equation (9.5):

$$A_1 + B = C = \begin{bmatrix} (2+1)(4+2)(7+1)(5+2) \\ (1+2)(3+1)(7+2)(6+1) \end{bmatrix} = \begin{bmatrix} 3 & 6 & 8 & 7 \\ 3 & 4 & 9 & 7 \end{bmatrix}. \tag{9.5}$$

Each element of the matrix C is the sum of an element of A_1 plus the corresponding element of B. For example, the first element in the first row of matrix C is equal to the sum of the first elements in the first row of matrices A_1 and B, $3 = 2 + 1$. In general notation, given two $m \times n$ matrices A_1 and B with elements a_{ij}, b_{ij}, their sum is also an $m \times n$ matrix C with elements $c_{ij} = a_{ij} + b_{ij}$.

The same procedure is followed in the subtraction of matrices. Again, each individual element in the matrix B is subtracted from the corresponding element in matrix A_1 as shown in equation (9.6):

$$A_1 - B = D = \begin{bmatrix} (2-1)(4-2)(7-1)(5-2) \\ (1-2)(3-1)(7-2)(6-1) \end{bmatrix} = \begin{bmatrix} 1 & 2 & 6 & 3 \\ -1 & 2 & 5 & 5 \end{bmatrix}. \tag{9.6}$$

The operation of scalar multiplication is defined as multiplication of each element of the matrix by a scalar. Again this is analogous to scalar multiplication of vectors. Thus the scalar product of the scalar k and the matrix A_1 is a matrix D with elements $d_{ij} = ka_{ij}$. To illustrate:

$$3A_1 = 3 \begin{bmatrix} 2 & 4 & 7 & 5 \\ 1 & 3 & 7 & 6 \end{bmatrix} = \begin{bmatrix} 6 & 12 & 21 & 15 \\ 3 & 9 & 21 & 18 \end{bmatrix}. \tag{9.7}$$

The operation of *matrix* multiplication is defined as follows. The matrix product of the $m \times n$ matrix A and the $n \times l$ matrix B with elements a_{ij} and b_{jh} is the $m \times l$ matrix E with elements $e_{ih} = \sum_{k=1}^{n} a_{ik} b_{kh}$. As with matrix addition and subtraction, matrix multiplication is defined only for certain matrices; it can only be carried out for particular matrices. The number of columns of the first matrix must equal the number of rows of the second matrix; if this is not the case, multiplication is not defined. In matrix A and E in equation (9.8),

$$A \equiv \begin{bmatrix} 2 & 4 & 7 \\ 1 & 3 & 7 \end{bmatrix}_{(2 \times 3)} \qquad E \equiv \begin{bmatrix} 4 & 2 \\ 2 & 1 \\ 3 & 2 \end{bmatrix}_{(3 \times 2)} \tag{9.8}$$

it can be seen that matrix A is a (2×3) matrix and matrix E a (3×2) matrix.

Thus the number of elements in a row of *A* is the same as the number of elements in a column of *E*, as indicated by the notation for the order of these matrices. Multiplication is thus defined for these matrices; the two can be multiplied. Another way to state this is the number of columns of a premultiplying matrix *A* (the matrix that is stated first in the multiplication) must be the same as the number of rows of *E*, the postmultiplying matrix.[4]

The matrix that results from the postmultiplication of *A* by *E* will have two rows and two columns, again as suggested in the notation for the order of the matrices indicated in equation (9.9):

$$(2 \times \underline{3})(\underline{3} \times 2) = (2 \times 2) \tag{9.9}$$

(Reversal of the order of multiplication would result in a (3 × 3) matrix as will be shown in a moment.) The postmultiplication of *A* by *E* is shown in equation (9.10). Each element in the first row of *A* is multiplied by each of the elements in the first column of *E*:

$$G = A \times E$$
$$= \begin{bmatrix} \{(2 \times 4) + (4 \times 2) + (7 \times 3)\}, & \{(2 \times 2) + (4 \times 1) + (7 \times 2)\} \\ \{(1 \times 4) + (3 \times 2) + (7 \times 3)\}, & \{(1 \times 2) + (3 \times 1) + (7 \times 2)\} \end{bmatrix}$$
$$\tag{9.10}$$

and the sum of these products constitutes the first element of the first row of the new matrix *G*. The second element of the first row is also determined by the elements in the first row of *A*, but this time they are multiplied by individual elements in the second column of *E*. As a general statement, then, in determining the elements of a product of two matrices, a particular element is determined by the row of the premultiplying matrix and the appropriate column of the postmultiplying matrix. The element of the first row, second column of *G* is determined by the elements in the first row of matrix *A* and the second column of matrix *E*. The matrix *G*, then, is as indicated in equation (9.11):

$$G = \begin{bmatrix} 37 & 22 \\ 31 & 19 \end{bmatrix}_{(2 \times 2)} \tag{9.11}$$

The matrix *G* resulting from the postmultiplication of *A* by *E* is a square matrix. All earlier matrices have been rectangular in form; that is, *m* has not been equal to *n*. There are many special properties of square matrices that make them unusually interesting to deal with, as will be shown later in this chapter.

The matrix *G* resulted from a postmultiplication of *A* by *E*. Now, it is desirable to investigate the product of the matrix *A* premultiplied by matrix *E*. It can be seen from the notation of the order of these matrices that, as before, a square matrix will result from this multiplication; but this time it will be a (3 × 3) matrix:

$$(3 \times \underline{2})(\underline{2} \times 3) = (3 \times 3) \tag{9.12}$$

The multiplication is possible because matrix *E* has two columns and matrix *A* two

---

[4] The matrix multiplication $A \times E$ can be referred to as the premultiplying of *E* by *A*, or the postmultiplying of *A* by *E*.

rows. The multiplication is indicated in equation (9.13):

$$H = E \times A = \begin{bmatrix} 4 & 2 \\ 2 & 1 \\ 3 & 2 \end{bmatrix} \begin{bmatrix} 2 & 4 & 7 \\ 1 & 3 & 7 \end{bmatrix}$$

$$= \begin{bmatrix} \{(4 \times 2) + (2 \times 1)\}, & \{(4 \times 4) + (2 \times 3)\}, & \{(4 \times 7) + (2 \times 7)\} \\ \{(2 \times 2) + (1 \times 1)\}, & \{(2 \times 4) + (1 \times 3)\}, & \{(2 \times 7) + (1 \times 7)\} \\ \{(3 \times 2) + (2 \times 1)\}, & \{(3 \times 4) + (2 \times 3)\}, & \{(3 \times 7) + (2 \times 7)\} \end{bmatrix} \quad (9.13)$$

and the overall result is indicated in equation (9.14):

$$E \times A = H = \begin{bmatrix} 10 & 22 & 42 \\ 5 & 11 & 21 \\ 8 & 18 & 35 \end{bmatrix}_{(3 \times 3)} \quad (9.14)$$

The matrix H is a (3×3) matrix and is in no way similar to G, a (2×2) matrix. The fact that the results of these two matrix multiplications are square matrices is a coincidence. The order of other matrices that could be multiplied to result in a rectangular matrix are suggested in equations (9.15):

$$\begin{aligned} (1 \times 3)(3 \times 2) &= (1 \times 2) \\ (2 \times 3)(3 \times 1) &= (2 \times 1) \\ (7 \times 3)(3 \times 3) &= (7 \times 3) \end{aligned} \quad (9.15)$$

Matrix multiplication does not necessarily result in square matrices.

Since vectors are themselves matrices, $(1 \times n)$ or $(n \times 1)$ (they have a single row or single column), they may also be multiplied. The matrix product of the $(1 \times m)$ matrix A and the $(m \times 1)$ matrix B with elements a_{1j}, b_{j1} is the (1×1) matrix P

$$P = [3 \ 1 \ 5] \begin{bmatrix} 2 \\ 3 \\ 1 \end{bmatrix} = 6 + 3 + 5 = 14,$$

a scalar. The element $p_{11} = \sum_{j=1}^{k} a_{1j} b_{j1}$; on the other hand, the matrix product of the $(m \times 1)$ matrix B and the $(1 \times m)$ matrix A is the $(m \times m)$ matrix Q:

$$Q = \begin{bmatrix} 2 \\ 3 \\ 1 \end{bmatrix} [3 \ 1 \ 5] = \begin{bmatrix} 6 & 2 & 10 \\ 9 & 3 & 15 \\ 3 & 1 & 5 \end{bmatrix}$$

with elements $q_{ij} = \sum_{n=1}^{1} b_{in} a_{nj} = b_{i1} a_{1j}$.

9.4 Special Matrices

Certain matrices are especially interesting. Two such matrices are the zero matrix and the identity matrix. Each is a square matrix of any order, and each has properties that approximate, for matrix algebra, useful properties of scalars in scalar

Sec. 9.4] SPECIAL MATRICES

algebra. The *zero matrix* is indicated in equation (9.16):

$$\begin{bmatrix} 0 & 0 & 0 \\ 0 & 0 & 0 \\ 0 & 0 & 0 \end{bmatrix}_{(3\times 3)} \equiv 0. \qquad (9.16)$$

Pre- or postmultiplication by the zero matrix results in the zero matrix; that is, pre- or postmultiplication of a given matrix H by the zero matrix results in the zero matrix, as indicated in equation (9.17):

$$0 \cdot H = H \cdot 0 = 0. \qquad (9.17)$$

The properties of the zero matrix are similar to those for the number 0 in scalar algebra.

The identity matrix, an example of which is as shown in equation (9.18),

$$\begin{bmatrix} 1 & 0 & 0 \\ 0 & 1 & 0 \\ 0 & 0 & 1 \end{bmatrix}_{(3\times 3)} \equiv I \qquad (9.18)$$

has properties similar to those of the number 1 in scalar algebra. Pre- or postmultiplication of a square matrix by the identity matrix results in the original matrix; that is, multiplying the matrix H by the identity matrix leaves it unchanged, as suggested in equation (9.19):

$$I \cdot H = H \cdot I = H. \qquad (9.19)$$

Another special matrix is the *symmetric* matrix.

Definition—Symmetric Matrix: *A symmetric matrix is a $(n \times n)$ matrix A such that $a_{ij} = a_{ji}$ for all pairs ij; that is, its elements are symmetric with respect to the main diagonal.*

An example of a symmetric matrix is

$$\begin{bmatrix} 1 & 4 & 5 \\ 4 & 2 & 6 \\ 5 & 6 & 3 \end{bmatrix}.$$

A special kind of symmetric matrix is the *diagonal* matrix.

Definition—Diagonal Matrix: *A diagonal matrix is a $(n \times n)$ matrix A such that $a_{ii} \neq 0$ for all i, and $a_{ij} = 0$ for $i \neq j$; that is, nonzero elements along the main diagonal, and zero elements elsewhere.*

The following is an example of a diagonal matrix:

$$\begin{bmatrix} 1 & 0 & 0 \\ 0 & 2 & 0 \\ 0 & 0 & 3 \end{bmatrix}.$$

The identity matrix is a special case of a diagonal matrix.

Another important concept is the *transpose* of a matrix.

Definition—Transpose: *The transpose of an (m × n) matrix A with elements a_{ij} is an (n × m) matrix A^T such that $a_{ij}^T = a_{ji}$ for all pairs ij; that is, the rows and columns of the original matrix are interchanged.*

If

$$A = \begin{bmatrix} 1 & 2 & 3 \\ 4 & 5 & 6 \end{bmatrix}, \quad \text{then } A^T = \begin{bmatrix} 1 & 4 \\ 2 & 5 \\ 3 & 6 \end{bmatrix}.$$

9.5 Matrix Inversion

A scalar equation such as equation (9.20) can be solved readily:

$$ax = b. \tag{9.20}$$

Solving for x, both sides are multiplied by $1/a$, the reciprocal of a (if $a \neq 0$):

$$\frac{1}{a}(a)x = \frac{1}{a}(b)$$

or

$$(1)x = \frac{1}{a}(b).$$

There is an analogous procedure for solving matrix equations. The system of equations

$$\begin{aligned} 3x_1 + x_2 &= 9 \\ 2x_1 + 2x_2 &= 10 \end{aligned} \tag{9.21}$$

may be rewritten in vector form as

$$AX = b \tag{9.22}$$

where

$$A \equiv \begin{bmatrix} 3 & 1 \\ 2 & 2 \end{bmatrix}, \quad X \equiv \begin{bmatrix} x_1 \\ x_2 \end{bmatrix}, \quad \text{and} \quad b \equiv \begin{bmatrix} 9 \\ 10 \end{bmatrix}.$$

A is the matrix $[a_{ij}]$ of coefficients of x in the system, **X** the vector $[x_i]$, and **b** the vector $[b_i]$. Given the matrix equation (9.22), it would appear desirable to investigate the solution of the matrix equation in a manner analogous to the solution of the scalar equation (9.20).

It has been noted that premultiplication by the identity matrix leaves any given matrix unchanged. Thus, if it were possible to multiply the left-hand side of equation (9.22) by a matrix, such that the identity matrix would result, leaving on the left-hand side only the matrix **X**, then a solution to the matrix equation would be obtained. This, in fact, can be done. Just as in scalar algebra, the scalar a was multiplied by its reciprocal $1/a$, it is possible to multiply the matrix A by another matrix to produce the identity matrix. The matrix that makes possible this result is defined to be the *inverse* matrix of the matrix A, shown in symbols as A^{-1} (and sometimes called the reciprocal of A). The inverse is only defined for square matrices.[5] For a square matrix A, if $LA = I$, then L is called a left-hand inverse, and if $AR = I$, then R is called a right-hand inverse. A matrix A that has a left and a right inverse is said to be *invertible*. If A ($n \times n$) has a left inverse L, and a right inverse R, then $L = R$. *Proof:* Suppose $LA = I$ and $AR = I$, then since multiplying a given matrix by the identity matrix I leaves that matrix unchanged,

$$L = LI$$

and since $AR = I$

$$L = LI = L(AR).$$

Since matrices are associative in multiplication

$$LI = L(AR) = (LA)R,$$

but $LA = I$ so

$$(LA)R = IR,$$

and

$$LI = (LA)R = IR$$

so

$$L = R.$$

Therefore, a square matrix with *either* a left or a right inverse is invertible; its inverse exists. A matrix for which the inverse exists is known as a *nonsingular* matrix.

Definition—Nonsingular Matrix: *Given a square matrix A of order ($n \times n$) such that its product with another matrix of order ($n \times n$) gives the identity matrix I*

$$A^{-1}A = I = AA^{-1} \tag{9.23}$$

then A is said to be nonsingular *and the matrix A^{-1} is known as the* inverse matrix *of A.*

This definition implies that not all square matrices have an inverse. The inverse has the following properties:

1. If A has an inverse, so does A^{-1}, and $(A^{-1})^{-1} = A$.
2. The product of invertible matrices is invertible.
3. The identity matrix is its own inverse, since by definition $I^{-1}I = I$, which gives $I^{-1} = I$.
4. The zero matrix has no inverse.

The definition also suggests a method of determining the inverse matrix. Given any matrix A, it is necessary to determine the elements of a matrix such that, when it is multiplied by A, the identity matrix results. All the elements of the matrix A are known; all the elements of the identity matrix are known. It is possible, then, to establish a system of equations whose solution determines the inverse of the

---

[5] Even if a matrix P could be found, such that either PA or AP equalled I for A not a square matrix, an inverse is defined only for square matrices. If A were a rectangular matrix, P would be known as a left-hand inverse or a right-hand inverse and its dimension would be different in each case.

matrix A. This is shown in the system of equations (9.24) for a simple matrix:

$$A^{-1} \times \begin{bmatrix} 3 & 1 \\ 2 & 2 \end{bmatrix} = \begin{bmatrix} 1 & 0 \\ 0 & 1 \end{bmatrix}. \tag{9.24}$$

The order of the inverse can be determined by examining the notation symbols for the matrices used in the product.

$$(y \times 2) \times (2 \times 2) = (2 \times 2).$$

Therefore y must also equal 2. The objective is to find a (2×2) matrix A^{-1}, such that when A is premultiplied by it, the (2×2) identity matrix results; that is,

$$\begin{bmatrix} e & f \\ g & h \end{bmatrix} \times \begin{bmatrix} 3 & 1 \\ 2 & 2 \end{bmatrix} = \begin{bmatrix} 1 & 0 \\ 0 & 1 \end{bmatrix}, \tag{9.25}$$

or carrying out the multiplication and equating terms

$$3e + 2f = 1 \tag{9.26}$$
$$e + 2f = 0 \tag{9.27}$$
$$3g + 2h = 0 \tag{9.28}$$
$$g + 2h = 1. \tag{9.29}$$

Solving equations (9.27) and (9.28) for the unknowns in terms of each other gives

$$e = -2f \tag{9.30}$$
$$g = -\tfrac{2}{3}h. \tag{9.31}$$

Substituting these values in equations (9.26) and (9.29) gives

$$3e - e = 1, \quad e = \tfrac{1}{2}$$
$$-\tfrac{2}{3}h + 2h = 1, \quad h = \tfrac{3}{4}$$

so from equations (9.30) and (9.31)

$$f = -\tfrac{1}{4} \quad \text{and} \quad g = -\tfrac{1}{2}.$$

Therefore the inverse of A is:

$$A^{-1} = \begin{bmatrix} \tfrac{1}{2} & -\tfrac{1}{4} \\ -\tfrac{1}{2} & \tfrac{3}{4} \end{bmatrix}. \tag{9.32}$$

Testing this to see if, in fact, $A^{-1}A = I$, gives

$$\begin{bmatrix} \tfrac{1}{2} & -\tfrac{1}{4} \\ -\tfrac{1}{2} & \tfrac{3}{4} \end{bmatrix} \begin{bmatrix} 3 & 1 \\ 2 & 2 \end{bmatrix} = \begin{bmatrix} (\tfrac{1}{2})3 - \tfrac{1}{4}(2), & \tfrac{1}{2}(1) - \tfrac{1}{4}(2) \\ (-\tfrac{1}{2})3 + \tfrac{3}{4}(2), & -\tfrac{1}{2}(1) + \tfrac{3}{4}(2) \end{bmatrix} = \begin{bmatrix} 1 & 0 \\ 0 & 1 \end{bmatrix}. \tag{9.33a}$$

Also testing to see if postmultiplication by A^{-1} leads to the same result gives

$$\begin{bmatrix} 3 & 1 \\ 2 & 2 \end{bmatrix} \begin{bmatrix} \tfrac{1}{2} & -\tfrac{1}{4} \\ -\tfrac{1}{2} & \tfrac{3}{4} \end{bmatrix} = \begin{bmatrix} 3(\tfrac{1}{2}) - \tfrac{1}{2}, & 3(-\tfrac{1}{4}) + \tfrac{3}{4} \\ 2(\tfrac{1}{2}) + 2(-\tfrac{1}{2}), & 2(-\tfrac{1}{4}) + 2(\tfrac{3}{4}) \end{bmatrix} = \begin{bmatrix} 1 & 0 \\ 0 & 1 \end{bmatrix} \tag{9.33b}$$

which demonstrates that the predicted result holds.

This procedure for inverting the matrix is cumbersome and more expeditious methods of inversion are available. Two are explored here: (1) inversion of the matrix through elementary row and column operations, and (2) inversion through use of determinants.

9.6 Row and Column Operations

Just as it is possible in scalar algebra to manipulate a system of linear equations by particular linear transformations of the equations without altering the solution to the system, so it is possible to carry out linear transformations in matrix algebra without changing the solution to the system. For example, it is possible in scalar algebra to exchange the locations of equations in the system, to multiply any given equation by any nonzero scalar and to subtract such a product from another equation, yet still obtain the original solution to the system. Similarly, in matrix algebra it is possible to interchange the rows of a matrix, to multiply any given row by a nonzero scalar, and to replace one row by the difference between its (original) elements and the product of a scalar times the elements of another row in the matrix. These three transformations of matrices are known as elementary row operations. In matrix algebra, it is also possible to carry out any one of these elementary operations in relation to the columns of the matrix, so that previous statements could be read with the word "column" exchanged for "row."

Elementary row and column operations can often be used to transform a given matrix into the identity matrix. It is stated without proof, and can be readily verified through example, that the results of row and column operations can be reversed merely by reversing the order of their application. Also when a given matrix A is transformed into the identity matrix I by row and column operations, these same operations performed on the identity matrix I lead to the matrix A^{-1}, the inverse of A, as stated in Theorem 9.1:

> **Theorem 9.1:** *Given the matrix A, the series of row and column operations required to transform the matrix to the identity matrix I, when applied to the identity matrix, will convert the identity matrix to A^{-1}, the inverse matrix of A.*

Thus, given a matrix A such as the one inverted by row and column operations in Section 9.5, it is possible to transform the matrix into the identity matrix by row and column operations. The sequence of elementary operations that will transform the given matrix into the identity matrix is as follows:

1. Divide the first row of the matrix by the component a_{11} in the first column; the result is the number 1 in the first row and first column. (If a_{11} is zero, exchange row one for a row with a first element nonzero.)
2. Subtract scalar multiples of this row from all other rows to obtain zeroes in the first column of all other rows in the matrix.
3. Divide the second row by the element in the second column; this provides the scalar 1 at this point in the diagonal of the matrix. (If a_{22} is zero, exchange row two for a lower row with nonzero second element.)
4. Subtract scalar multiples of this row from all other rows to obtain zeroes in the second column of all other rows in the matrix.
5. Continue the process for the rest of the rows in the matrix.

If the inverse of an $(n \times n)$ matrix exists, then there are n linearly independent row vectors that make up the matrix and the basis for the matrix is n dimensional.

The *rank* of a matrix is determined by the number of linearly independent row vectors that it contains. Therefore the rank of an ($n \times n$) nonsingular matrix is n.

Generally the row and column operations are carried out simultaneously on both the matrix A and the matrix I, but for clarity here the procedure is illustrated first on A and then on I in Example 9.1.

Example 9.1

Given the matrix A

$$A \equiv \begin{bmatrix} 3 & 1 \\ 2 & 2 \end{bmatrix} \tag{9.34}$$

use elementary row and column operations to obtain the identity matrix $I_{(2 \times 2)}$; then repeat these operations on the matrix I to obtain A^{-1}.

1. Dividing the first row by 3 gives

$$\begin{bmatrix} 1 & \frac{1}{3} \\ 2 & 2 \end{bmatrix}. \tag{9.35}$$

2. Using the product of the scalar 2, times the elements in the first, and then subtracting that product from each lower row to reduce all other elements in the first column to zero (but leaving row 1 itself unchanged in the matrices) gives

$$\begin{bmatrix} 1 & \frac{1}{3} \\ 0 & \frac{4}{3} \end{bmatrix}. \tag{9.36}$$

3. Dividing the second row through by $\frac{4}{3}$ gives

$$\begin{bmatrix} 1 & \frac{1}{3} \\ 0 & 1 \end{bmatrix}. \tag{9.37}$$

4. Using the product of the scalar $\frac{1}{3}$, times the elements in (the second column of) the new second row to reduce all other elements in that column to zero gives

$$\begin{bmatrix} 1 & 0 \\ 0 & 1 \end{bmatrix}. \tag{9.38}$$

5. These steps are now repeated (without comment) for the identity matrix:

$$I \equiv \begin{bmatrix} 1 & 0 \\ 0 & 1 \end{bmatrix}.$$

The results of the various row operations are:

$$\begin{bmatrix} \frac{1}{3} & 0 \\ 0 & 1 \end{bmatrix} \tag{9.39}$$

$$\begin{bmatrix} \frac{1}{3} & 0 \\ -\frac{2}{3} & 1 \end{bmatrix} \tag{9.40}$$

$$\begin{bmatrix} \frac{1}{3} & 0 \\ -\frac{1}{2} & \frac{3}{4} \end{bmatrix} \tag{9.41}$$

$$\begin{bmatrix} \frac{1}{2} & -\frac{1}{4} \\ -\frac{1}{2} & \frac{3}{4} \end{bmatrix} \tag{9.42}$$

and this is the inverse of A identical to equation (9.32) which is in accord with Theorem 9.1.

Beginning with the equation (9.34), it is now possible to trace through the development of the changing matrices in this example by comparing these results of successive transformations with those achieved by the simplex method in simplex tableaus 1, 2, and 3 in Chapter 8. In tableau 1 (Section 8.4) the columns under x and y are the same as in equation (9.34); tableau 2 (Section 8.4) is identical with matrix (9.36). The two additional row operations transform matrix (9.36) into matrix (9.38), the form shown in tableau 3 (Section 8.4). Similarly, for the transformation of equation (9.38), the elements in the successive matrices can be identified in the three tableaus under the columns for S_1 and S_2. Matrix (9.38) is the form shown in tableau 1; after two of the row operations, the matrix is transformed into matrix (9.40), the form shown in tableau 2; after the final two row operations have been carried out, matrix (9.40) is transformed into matrix (9.42), the form shown in tableau 3 (Section 8.4). In effect, then, the simplex method is a specific system of row and column operations used to shift the basis for spanning the vector space in the program. A complete change in basis from the first basis to one involving entirely new vectors involves essentially the same procedure as inverting the matrix of components of the original basis vectors.

The general form of the linear program can be stated in vector notation. The linear program in equations (8.1) in Chapter 8 is stated in terms of matrix notation as shown in equation system (9.43):

$$\begin{array}{ll} \text{Max} & \mathbf{c'X} \\ \text{Subject to} & A\mathbf{X} \leq \mathbf{b} \\ & \mathbf{X} \geq \mathbf{0}. \end{array} \qquad (9.43)$$

The objective function is stated as a maximization of the product of the transpose of the column vector **c** multiplied by the vector of variables. The program is subject to the constraint of the product of the matrix of coefficients of the system postmultiplied by the vector of variables **X**, which must be less than or equal to the vector **b**. All of the variables in the system must be greater than or equal to 0. It was seen that the dual program could be obtained from the primal, and this can also be readily presented in matrix notation as shown in equation (9.44):

$$\begin{array}{ll} \text{Min} & \mathbf{b'V} \\ \text{Subject to} & A'\mathbf{V} \geq \mathbf{c} \\ & \mathbf{V} \geq \mathbf{0}. \end{array} \qquad (9.44)$$

In the dual, the matrix of coefficients is transposed and postmultiplied by the new dual variables, and these are represented as the column vector **V**. The objective function is changed from a maximization to a minimization, and the coefficients of the dual variables in the objective function are the transpose of the right-hand side of the primal problem as shown in equation (9.44). Similarly, the right-hand side of the dual program is the transpose of the row vector of coefficients in the objective function of the primal, thus returning this vector, the vector **c**, to the normal column form for vectors. Once again, all of the variables in the system must be positive.

It is possible to trace through the simplex solution to a linear program in general matrix notation focusing the discussion on the constraints. The constraints, as initially stated in a linear program, can be stated as matrices of the following form

$$A_{(m \times n)} X_{(n \times 1)} \leq b_{(m \times 1)}$$

where A is the matrix of coefficients of the system of inequalities, X the matrix (vector) of variables and b the matrix (vector) of values of the RHS of the inequalities. When stated for simplex solution, a slack variable is added for each of the m inequalities with appropriate coefficients added to the coefficient matrix A to give:

$$A_{[m \times (n+m)]} X_{[(n+m) \times 1]} = b_{(m \times 1)}$$

The coefficient matrix consists of the original matrix of coefficients plus an identity matrix (which is necessarily square) for the coefficients of the slack variables. The identity matrix is made up of mutually orthogonal vectors and is its own inverse ($I \times I = I$). The first simplex solution is given by setting all ordinary variables equal to zero so that the product of the necessarily linearly independent (and thus invertible) matrix of slack variable coefficients multiplied by the vector of slack variables is equal to the values on the right-hand side of the equations. The slack variables thus form a basis for the vector space. In the simplex method of solution to the program, new variables (new columns) which consistently improve the value of the objective function are substituted for variables currently in the basis. The coefficients of the basis variables make up a matrix designated B which remains $(m \times m)$. Row and column operations are used to accomplish the substitution of one variable for another in the basis. The same row and column operations are simultaneously carried on for the variables in the initial basis as is done in Example 9.2 below. For the B matrix determined at each iteration the inverse of B appears in the columns of the variables in the *initial* basis.[6]

The maximum number of simplex iterations is determined by the number of "corners" of the simplex, some of which may not be feasible. At each corner m variables are nonzero so the upper bound for the number of iterations is given by the formula for the number of combinations of $(m + n)$ items taken m at a time.

The procedures for inverting a matrix by row and column operations are now illustrated and confirmed in Examples 9.2 and 9.3 for a 3×3 matrix C.

Example 9.2

Given the matrix,

$$C \equiv \begin{bmatrix} 4 & 0 & 2 \\ 2 & 10 & 2 \\ 3 & 9 & 1 \end{bmatrix} \qquad (9.45)$$

find its inverse by row and column operations.

This example is carried through by placing the original matrix C and the identity matrix side by side. The row and column operations are carried out on both

---

[6] This is true for the initial basis also since $I^{-1} = I$. Only when the solution becomes degenerate does this not hold, and then there is no inverse; the basis is singular.

Sec. 9.6]   ROW AND COLUMN OPERATIONS   325

matrices simultaneously. It is thus possible to trace out the impact of these operations on the original identity matrix from which the inverse is generated. The two matrices are:

$$\begin{bmatrix} 4 & 0 & 2 \\ 2 & 10 & 2 \\ 3 & 9 & 1 \end{bmatrix} \begin{bmatrix} 1 & 0 & 0 \\ 0 & 1 & 0 \\ 0 & 0 & 1 \end{bmatrix}. \tag{9.46}$$

1. Dividing row 1 by the scalar 4 gives

$$\begin{bmatrix} 1 & 0 & \frac{1}{2} \\ 2 & 10 & 2 \\ 3 & 9 & 1 \end{bmatrix} \begin{bmatrix} \frac{1}{4} & 0 & 0 \\ 0 & 1 & 0 \\ 0 & 0 & 1 \end{bmatrix}. \tag{9.47}$$

2. Multiplying row 1 by the scalar 2 and subtracting the product from row 2 (but leaving row 1 itself unchanged in the matrices) gives

$$\begin{bmatrix} 1 & 0 & \frac{1}{2} \\ 0 & 10 & 1 \\ 3 & 9 & 1 \end{bmatrix} \begin{bmatrix} \frac{1}{4} & 0 & 0 \\ -\frac{1}{2} & 1 & 0 \\ 0 & 0 & 1 \end{bmatrix}. \tag{9.48}$$

Multiplying row 1 by the scalar 3 and subtracting the product from row 3 gives

$$\begin{bmatrix} 1 & 0 & \frac{1}{2} \\ 0 & 10 & 1 \\ 0 & 9 & -\frac{1}{2} \end{bmatrix} \begin{bmatrix} \frac{1}{4} & 0 & 0 \\ -\frac{1}{2} & 1 & 0 \\ -\frac{3}{4} & 0 & 1 \end{bmatrix}. \tag{9.49}$$

3. Dividing row 2 by 10 then gives

$$\begin{bmatrix} 1 & 0 & \frac{1}{2} \\ 0 & 1 & \frac{1}{10} \\ 0 & 9 & -\frac{1}{2} \end{bmatrix} \begin{bmatrix} \frac{1}{4} & 0 & 0 \\ -\frac{1}{20} & \frac{1}{10} & 0 \\ -\frac{3}{4} & 0 & 1 \end{bmatrix}. \tag{9.50}$$

Now multiplying row 2 by 9 and subtracting it from row 3 gives

$$\begin{bmatrix} 1 & 0 & \frac{1}{2} \\ 0 & 1 & \frac{1}{10} \\ 0 & 0 & -\frac{14}{10} \end{bmatrix} \begin{bmatrix} \frac{1}{4} & 0 & 0 \\ -\frac{1}{20} & \frac{1}{10} & 0 \\ -\frac{3}{10} & -\frac{9}{10} & 1 \end{bmatrix}. \tag{9.51}$$

Dividing row 3 through by $-\frac{14}{10}$ gives

$$\begin{bmatrix} 1 & 0 & \frac{1}{2} \\ 0 & 1 & \frac{1}{10} \\ 0 & 0 & 1 \end{bmatrix} \begin{bmatrix} \frac{1}{4} & 0 & 0 \\ -\frac{1}{20} & \frac{1}{10} & 0 \\ \frac{3}{14} & \frac{9}{14} & -\frac{10}{14} \end{bmatrix}. \tag{9.52}$$

Now multiplying row 3 by $\frac{1}{2}$ and subtracting the product from row 1 gives

$$\begin{bmatrix} 1 & 0 & 0 \\ 0 & 1 & \frac{1}{10} \\ 0 & 0 & 1 \end{bmatrix} \begin{bmatrix} \frac{1}{7} & -\frac{9}{28} & \frac{5}{14} \\ -\frac{1}{20} & \frac{1}{10} & 0 \\ \frac{3}{14} & \frac{9}{14} & -\frac{10}{14} \end{bmatrix}. \tag{9.53}$$

Multiplying row 3 by $\frac{1}{10}$ and subtracting the product from row 2 gives

$$\begin{bmatrix} 1 & 0 & 0 \\ 0 & 1 & 0 \\ 0 & 0 & 1 \end{bmatrix} \begin{bmatrix} \frac{1}{7} & -\frac{9}{28} & \frac{5}{14} \\ -\frac{1}{14} & \frac{1}{28} & \frac{1}{14} \\ \frac{3}{14} & \frac{9}{14} & -\frac{5}{7} \end{bmatrix}. \tag{9.54}$$

Example 9.3

Premultiply the matrix C by its inverse C^{-1} as found in Example 9.2, to confirm that

$$C^{-1} \times C = I.$$

$$\begin{bmatrix} \frac{1}{7} & -\frac{9}{28} & \frac{5}{14} \\ -\frac{1}{14} & \frac{1}{28} & \frac{1}{14} \\ \frac{3}{14} & \frac{9}{14} & -\frac{5}{7} \end{bmatrix} \begin{bmatrix} 4 & 0 & 2 \\ 2 & 10 & 2 \\ 3 & 9 & 1 \end{bmatrix} = ? \qquad (9.55)$$

Carrying out the matrix multiplication results in the following matrix:

$$\begin{bmatrix} \frac{4}{7} + (-\frac{9}{14}) + \frac{15}{14}, & 0 + (-\frac{45}{14}) + \frac{45}{14}, & \frac{2}{7} + (-\frac{9}{14}) + \frac{5}{14} \\ -\frac{4}{14} + \frac{1}{14} + \frac{3}{14}, & 0 + \frac{5}{14} + \frac{9}{14}, & -\frac{1}{7} + \frac{1}{14} + \frac{1}{14} \\ \frac{6}{7} + \frac{9}{7} - \frac{15}{7}, & 0 + \frac{45}{7} - \frac{45}{7}, & \frac{3}{7} + \frac{9}{7} - \frac{5}{7} \end{bmatrix} \qquad (9.56)$$

which reduces to

$$\begin{bmatrix} 1 & 0 & 0 \\ 0 & 1 & 0 \\ 0 & 0 & 1 \end{bmatrix} \equiv I_{(3 \times 3)} \qquad (9.57)$$

confirming that the inverse has been found.

Given a matrix such as C or C^{-1}, it is possible to transform it by elementary operations into the identity matrix. Since the effects of elementary operations can be reversed merely by reversing the order of their application to a given matrix, it is possible to generate C or C^{-1} from an identity matrix; in fact, this has just been done for C^{-1}. When two matrices can be generated from the same matrix by elementary operations (and thus from each other), they are called *equivalent matrices*:

> ***Definition—Equivalent Matrices:*** *Given two matrices A and B, each of order $(m \times n)$, if the one can be generated from the other as a result of elementary matrix operations, then A and B are called* equivalent *matrices.*

Thus any matrix and its inverse are equivalent, and both are equivalent to the identity matrix of their order. Any matrix of a given order that is invertible is equivalent to any other invertible matrix of the same order.

An important concept is that of *elementary matrix*.

> ***Definition—Elementary Matrix:*** *An elementary matrix is a matrix that can be obtained from an identity matrix by a single elementary (row or column) operation.*

An elementary row operation on any matrix A can be obtained by premultiplying A by the elementary matrix P, obtained by performing the same row operation on the corresponding identity matrix. A sequence of elementary row operations on a matrix A can be achieved by premultiplying A by the corresponding sequence of elementary matrices. (Elementary matrices can be developed and used for column operations and postmultiplication also.) If the sequence of elementary row operations results in A being transformed into an identity matrix, then $P_k P_{k-1} \ldots P_1 A = I$. Since elementary matrices are invertible (by definition), then $A = P_1^{-1} \ldots P_{k-1}^{-1} P_k^{-1}$, which shows that every invertible matrix can be written as a product of elementary matrices.

9.7 Determinants

The equation system (9.58), with two equations and two unknowns, can be solved through the normal procedures for scalar algebra:

$$2x + y = 6$$
$$2x - 2y = 3. \tag{9.58}$$

The equations are first restated in general terms as shown in equations (9.59) and (9.60) and then solved for x and y:

$$a_{11}x + a_{12}y = b_1 \tag{9.59}$$

$$a_{21}x + a_{22}y = b_2. \tag{9.60}$$

Solving equation (9.59) in terms of x gives

$$x = \frac{b_1 - a_{12}y}{a_{11}}, \quad \text{for } a_{11} \neq 0. \tag{9.61}$$

Solving equation (9.60) in terms of y gives

$$y = \frac{b_2 - a_{21}x}{a_{22}}, \quad \text{for } a_{22} \neq 0. \tag{9.62}$$

Substituting for y in equation (9.59) gives

$$x = \frac{b_1 - a_{12}\left(\dfrac{b_2 - a_{21}x}{a_{22}}\right)}{a_{11}} \tag{9.63}$$

and simplifying gives

$$a_{11}x = b_1 - \frac{a_{12}b_2 + a_{12}a_{21}x}{a_{22}}$$
$$a_{11}a_{22}x = a_{22}b_1 - a_{12}b_2 + a_{12}a_{21}x$$
$$a_{11}a_{22}x - a_{12}a_{21}x = a_{22}b_1 - a_{12}b_2.$$

Thus the value of x in terms of the coefficients in the original equation system is given by

$$x = \frac{a_{22}b_1 - a_{12}b_2}{a_{11}a_{22} - a_{12}a_{21}}, \quad a_{11}a_{22} - a_{12}a_{21} \neq 0. \tag{9.64}$$

Similarly, it is possible to solve for y, with the following result:

$$y = \frac{a_{11}b_2 - a_{21}b_1}{a_{11}a_{22} - a_{12}a_{21}}, \quad a_{11}a_{22} - a_{12}a_{21} \neq 0. \tag{9.65}$$

The denominators in equations (9.64) and (9.65) are identical. Also, the denominator represents the *determinant* for this equation system:

Definition—A Determinant: *Given a (2 × 2) equation system with coefficient matrix $[a_{ij}]_{(2\times 2)}$, the determinant of the system is a scalar defined to be the difference of the cross products:*

$$\begin{vmatrix} a_{11} & a_{12} \\ a_{21} & a_{22} \end{vmatrix} = a_{11}a_{22} - a_{12}a_{21}. \qquad (9.66)$$

The symbol for the determinant is the vertical straight lines enclosing its elements.[7]

A determinant is therefore a number associated with a square matrix with value for the (2 × 2) case equal to the difference between the cross products determined by the arrows in equation (9.66). This method of indicating the evaluation of a determinant is known as the method of arrows. The determinant indicated in equation (9.66) is that for a (2 × 2) square matrix.

The denominator in equations (9.64) and (9.65) is the determinant of the system of equations. It can be noted also that the numerator in each case is also a cross product of determinant form. The value of x can be restated in terms of determinants as shown in equation (9.67):

$$x = \frac{\begin{vmatrix} b_1 & a_{12} \\ b_2 & a_{22} \end{vmatrix}}{\begin{vmatrix} a_{11} & a_{12} \\ a_{21} & a_{22} \end{vmatrix}}. \qquad (9.67)$$

Similarly, the value for y can be restated in terms of determinants as shown in equation (9.68):

$$y = \frac{\begin{vmatrix} a_{11} & b_1 \\ a_{21} & b_2 \end{vmatrix}}{\begin{vmatrix} a_{11} & a_{12} \\ a_{21} & a_{22} \end{vmatrix}}. \qquad (9.68)$$

The pattern of the determinants in equations (9.67) and (9.68) is an interesting one. The denominator is the determinant for the overall system; the numerator in each case is the determinant for the system with the vector for the right-hand side (RHS), $[b_i]$ substituted in the determinant matrix for the column vector of coefficients of the variable whose value is being sought. That is, b_1 and b_2 have been substituted for a_{11} and a_{21} in the solution for the value of x, the variable in column 1 of the equation system; and b_1 and b_2 have been substituted for a_{12} and a_{22} in the solution for the value of y, the variable in column 2 of the equation system.

[7] Another symbol for the determinant of a matrix A is det $[a_{ij}]$; this symbol is used in several of the problems for this chapter.

Sec. 9.7]   DETERMINANTS

It is possible to determine immediately whether the system of equations has a solution. If the system determinant—the number in the denominator of equations (9.67) and (9.68)—does not equal zero, then there is a solution. The original equation system (9.58) is now solved for the values of x and y by use of determinants. Given,

$$\begin{aligned} 2x + y &= 6 \\ 2x - 2y &= 3. \end{aligned} \quad (9.58)$$

The system determinant can be seen to be

$$|A| = \begin{vmatrix} 2 & 1 \\ 2 & -2 \end{vmatrix} \quad (9.69)$$

which, evaluated in accordance with the arrows, is

$$-4 - (2) = -6. \quad (9.70)$$

To solve for x, the vector of the scalars of the RHS is substituted for the first vector in the determinant for the system to find the determinant used in the numerator in equation (9.67). The value for x is shown and evaluated in equation (9.71):

$$x = \frac{\begin{vmatrix} 6 & 1 \\ 3 & -2 \end{vmatrix}}{-6} = \frac{-12 - (3)}{-6} = \frac{-15}{-6} = \frac{5}{2}. \quad (9.71)$$

The value of y can also be found in equation (9.72).

$$y = \frac{\begin{vmatrix} 2 & 6 \\ 2 & 3 \end{vmatrix}}{-6} = \frac{6 - 12}{-6} = \frac{-6}{-6} = 1. \quad (9.72)$$

The key determinant for applications is the determinant of the matrix of coefficients of the system. The method for solving a system of equations just described is known as Cramer's rule.

Equation (9.73) shows an alternative form for a solution of a (2 × 2) system of equations through the use of determinants:

$$\frac{x}{\begin{vmatrix} b_1 & a_{12} \\ b_2 & a_{22} \end{vmatrix}} = \frac{y}{\begin{vmatrix} a_{11} & b_1 \\ a_{21} & b_2 \end{vmatrix}} = \frac{1}{\begin{vmatrix} a_{11} & a_{12} \\ a_{21} & a_{22} \end{vmatrix}}. \quad (9.73)$$

Essentially it is the same as that shown earlier in equations (9.67) and (9.68), except that the form of this statement can more easily be extended to further dimensions. The use of equation (9.73) is illustrated in Example 9.4.

Example 9.4

Given the system of linear equations used in Section 9.7 and the values of the appropriate determinants expressed in equations (9.70) and (9.71), the values of the variables for this system of equations are calculated as:

$$\frac{x}{\begin{vmatrix} 6 & 1 \\ 3 & -2 \end{vmatrix}} = \frac{y}{\begin{vmatrix} 2 & 6 \\ 2 & 3 \end{vmatrix}} = \frac{1}{\begin{vmatrix} 2 & 1 \\ 2 & -2 \end{vmatrix}}$$

$$\frac{x}{-15} = \frac{y}{-6} = \frac{1}{-6}$$

$$x = \tfrac{5}{2}$$
$$y = 1.$$

Determinants have several useful properties:

1. If the rows and columns of a given matrix are transposed, the value of the determinant remains unchanged:

$$\begin{vmatrix} 6 & 1 \\ 3 & -2 \end{vmatrix} = -12 - (3) = -15$$

$$\begin{vmatrix} 6 & 3 \\ 1 & -2 \end{vmatrix} = -12 - (3) = -15.$$

2. If two rows (columns) are interchanged, the numerical value of the determinant remains unchanged, but the sign of the determinant is changed:

$$\begin{vmatrix} 6 & 1 \\ 3 & -2 \end{vmatrix} = -12 - (3) = -15$$

$$\begin{vmatrix} 1 & 6 \\ -2 & 3 \end{vmatrix} = 3 - (-12) = 15$$

$$\begin{vmatrix} 3 & -2 \\ 6 & 1 \end{vmatrix} = 3 - (-12) = 15.$$

3. If each element in a row (column) is multiplied by a scalar, the value of the determinant is multiplied by this same scalar:

$$\begin{vmatrix} 6 & 1 \\ 3 & -2 \end{vmatrix} = -12 - (3) = -15$$

$$\begin{vmatrix} 12 & 2 \\ 3 & -2 \end{vmatrix} = -24 - (6) = -30 = 2(-15)$$

$$\begin{vmatrix} 6 & 2 \\ 3 & -4 \end{vmatrix} = -24 - (6) = -30 = 2(-15).$$

4. If the elements of one row (column) are equal to the elements of another row, or a scalar multiple of that row (column), then the value of the determinant is 0:

$$\begin{vmatrix} 1 & 1 \\ -2 & -2 \end{vmatrix} = -2 - (-2) = 0$$

and

$$\begin{vmatrix} 3 & 1 \\ -6 & -2 \end{vmatrix} = -6 - (-6) = 0.$$

Sec. 9.7] DETERMINANTS

(This second illustration relates property 3 to property 4 also.)

5. If the elements of a row (column) of a given matrix are each the sum of two parts, then the determinant is equal to the sum of two determinants:

$$\begin{vmatrix} (6+3)(1-2) \\ 3 \quad -2 \end{vmatrix} = -2(6+3) - 3(1-2) = -15$$
$$= -2(6) - 2(3) - 3(1) - 3(-2)$$
$$= \begin{vmatrix} 6 & 1 \\ 3 & -2 \end{vmatrix} + \begin{vmatrix} 3 & -2 \\ 3 & -2 \end{vmatrix}$$
$$= -12 - 3 + -6 + (-6) = -15.$$

6. If a scalar multiple of the elements in one row (column) is added to the corresponding elements of another row (column), then the value of the determinant remains unchanged:

$$\begin{vmatrix} 6+(3)1 & 1+(-2)1 \\ 3 & -2 \end{vmatrix} = \begin{vmatrix} 9 & -1 \\ 3 & -2 \end{vmatrix} = -18 - (-3) = -15.$$

The system of equations can be restated in matrix form, as shown in equation (9.74):

$$A_{(m \times n)} X_{(n \times 1)} = B_{(n \times 1)} \qquad (9.74)$$

and stated in terms of the particular equation, this gives

$$[a]_{(2 \times 2)} \cdot [x]_{(2 \times 1)} = [b]_{(2 \times 1)}. \qquad (9.75)$$

This system of equations can be solved if the matrix of coefficients is premultiplied by its inverse to reduce the left-hand side of equation (9.75) to the identity matrix I times the matrix of variables X. To maintain equality, it is necessary also to premultiply the matrix B by the inverse of A as well. When this is done, the solution to the system of equations is obtained.

In the next examples the inverse to a matrix is first determined by scalar algebra, then by matrix algebra using determinants. When determinants are used, the system of equations is solved and the results compared with the solutions to the system in Section 9.7 and in Example 9.4. The same system of equations from Section 9.7 is used:

$$\begin{aligned} 2x + y &= 6 \\ 2x - 2y &= 3. \end{aligned} \qquad (9.58)$$

Example 9.5

Given the matrix

$$A \equiv \begin{bmatrix} 2 & 1 \\ 2 & -2 \end{bmatrix},$$

determine the inverse to the matrix by use of simultaneous equations, assuming premultiplication of A by its inverse to achieve the identity matrix:

$$\begin{bmatrix} c_{11} & c_{12} \\ c_{21} & c_{22} \end{bmatrix} \begin{bmatrix} 2 & 1 \\ 2 & -2 \end{bmatrix} = \begin{bmatrix} 1 & 0 \\ 0 & 1 \end{bmatrix}. \qquad (9.76)$$

Carrying out the matrix multiplication and setting the elements of the respective matrices equal to each other, the following system of four equations results:

$$2c_{11} + 2c_{12} = 1 \tag{9.77}$$

$$c_{11} - 2c_{12} = 0 \tag{9.78}$$

$$2c_{21} + 2c_{22} = 0 \tag{9.79}$$

$$c_{21} - 2c_{22} = 1. \tag{9.80}$$

Equations (9.78) and (9.79) can be solved for c_{11} in terms of c_{12}, and for c_{21} in terms of c_{22} as shown:

$$c_{11} = 2c_{12} \tag{9.81}$$

$$c_{21} = -c_{22}. \tag{9.82}$$

Inserting the results of equation (9.81) into equation (9.77) gives

$$6c_{12} = 1$$
$$c_{12} = \tfrac{1}{6}. \tag{9.83a}$$

Therefore

$$c_{11} = \tfrac{1}{3}. \tag{9.83b}$$

Similarly, substituting for c_{21} in equation (9.80) gives

$$-3c_{22} = 1$$
$$c_{22} = -\tfrac{1}{3}. \tag{9.84a}$$

Therefore

$$c_{21} = \tfrac{1}{3}. \tag{9.84b}$$

Thus the inverse of the matrix A is

$$A^{-1} = \begin{bmatrix} \tfrac{1}{3} & \tfrac{1}{6} \\ \tfrac{1}{3} & -\tfrac{1}{3} \end{bmatrix}. \tag{9.85}$$

The usual test can be made to determine if, in fact, this is accurate:

$$\begin{bmatrix} \tfrac{1}{3} & \tfrac{1}{6} \\ \tfrac{1}{3} & -\tfrac{1}{3} \end{bmatrix} \begin{bmatrix} 2 & 1 \\ 2 & -2 \end{bmatrix} = \begin{bmatrix} \tfrac{2}{3} + \tfrac{2}{6} & \tfrac{1}{3} - \tfrac{2}{6} \\ \tfrac{2}{3} - \tfrac{2}{3} & \tfrac{1}{3} + \tfrac{2}{3} \end{bmatrix} = \begin{bmatrix} 1 & 0 \\ 0 & 1 \end{bmatrix} \tag{9.86}$$

confirming that the matrix A^{-1} is the inverse of the matrix A.

Now the problem is repeated in Example 9.6 to determine the inverse through use of determinants.

Example 9.6

Given the matrix

$$A = \begin{bmatrix} 2 & 1 \\ 2 & -2 \end{bmatrix},$$

determine its inverse by use of determinants:

$$\overset{A^{-1}}{\begin{bmatrix} c_{11} & c_{12} \\ c_{21} & c_{22} \end{bmatrix}} \overset{A}{\begin{bmatrix} a_{11} & a_{12} \\ a_{21} & a_{22} \end{bmatrix}} = \overset{I}{\begin{bmatrix} 1 & 0 \\ 0 & 1 \end{bmatrix}}. \tag{9.87}$$

Sec. 9.7] DETERMINANTS 333

Carrying out the matrix multiplication gives the system of four linear equations:

$$a_{11}c_{11} + a_{21}c_{12} = 1 \tag{9.88}$$

$$a_{12}c_{11} + a_{22}c_{12} = 0 \tag{9.89}$$

$$a_{11}c_{21} + a_{21}c_{22} = 0 \tag{9.90}$$

$$a_{12}c_{21} + a_{22}c_{22} = 1. \tag{9.91}$$

It can be noted that the first two of these equations, equations (9.88) and (9.89), involve the elements of the matrix of coefficients from A plus two unknowns c_{11} and c_{12}; it is a system of two linear equations with two unknowns. The same is true of the next two equations, equations (9.90) and (9.91), with the exception that here the unknowns are c_{21} and c_{22}. It was noted in Section 9.7 that any system of two linear equations can be solved readily through the use of determinants. The system determinant is used as the value in the denominator of a system of equations to give the values of the unknowns. It so happens that in the (2×2) case shown the system determinant for each of the two pairs of simultaneous equations is identical with the determinant of the matrix A:

$$\begin{bmatrix} a_{11} & a_{12} \\ a_{21} & a_{22} \end{bmatrix} = a_{11}a_{22} - a_{12}a_{21}$$

for A, and

$$\begin{bmatrix} a_{11} & a_{21} \\ a_{12} & a_{22} \end{bmatrix} = a_{11}a_{22} - a_{12}a_{21}$$

for equations (9.88) and (9.89) and also (9.90) and (9.91). By Cramer's rule the solution to each of these two systems of linear equations is as indicated:

$$c_{11} = \frac{\begin{vmatrix} 1 & a_{21} \\ 0 & a_{22} \end{vmatrix}}{\begin{vmatrix} a_{11} & a_{21} \\ a_{12} & a_{22} \end{vmatrix}} \tag{9.92}$$

$$c_{12} = \frac{\begin{vmatrix} a_{11} & 1 \\ a_{12} & 0 \end{vmatrix}}{\begin{vmatrix} a_{11} & a_{21} \\ a_{12} & a_{22} \end{vmatrix}} \tag{9.93}$$

for c_{11} and c_{12}, and

$$c_{21} = \frac{\begin{vmatrix} 0 & a_{21} \\ 1 & a_{22} \end{vmatrix}}{\begin{vmatrix} a_{11} & a_{21} \\ a_{12} & a_{22} \end{vmatrix}} \tag{9.94}$$

$$c_{22} = \frac{\begin{vmatrix} a_{11} & 0 \\ a_{12} & 1 \end{vmatrix}}{\begin{vmatrix} a_{11} & a_{21} \\ a_{12} & a_{22} \end{vmatrix}} \tag{9.95}$$

for c_{21} and c_{22}.

It was shown in equation (9.70) above that the value of the system determinant for the matrix A, the denominator of each of these four equations, is -6. Substituting the values for each of the indicated elements a_{ij} from the matrix A into the numerators of the appropriate equations and the value -6 into their denominators, gives

$$c_{11} = \frac{a_{22}}{\begin{vmatrix} a_{11} & a_{21} \\ a_{12} & a_{22} \end{vmatrix}} = \frac{-2}{-6} = \frac{1}{3} \tag{9.92}$$

$$c_{12} = \frac{-a_{12}}{\begin{vmatrix} a_{11} & a_{21} \\ a_{12} & a_{22} \end{vmatrix}} = \frac{-1}{-6} = \frac{1}{6} \tag{9.93}$$

$$c_{21} = \frac{-a_{21}}{\begin{vmatrix} a_{11} & a_{21} \\ a_{12} & a_{22} \end{vmatrix}} = \frac{-2}{-6} = \frac{1}{3} \tag{9.94}$$

$$c_{22} = \frac{a_{11}}{\begin{vmatrix} a_{11} & a_{21} \\ a_{12} & a_{22} \end{vmatrix}} = \frac{2}{-6} = -\frac{1}{3}. \tag{9.95}$$

Thus A^{-1} is

$$A^{-1} = \begin{bmatrix} \frac{1}{3} & \frac{1}{6} \\ \frac{1}{3} & -\frac{1}{3} \end{bmatrix} \tag{9.96}$$

the result achieved through scalar algebra and tested to show it to be the inverse of A.

The solution to the system of equations is found through matrix algebra as follows. The equations

$$2x + y = 6$$
$$2x - 2y = 3$$

are rewritten in general matrix notation as

$$AX = B. \tag{9.97}$$

Premultiplying by the inverse of A gives

$$A^{-1}AX = IX = X = A^{-1}B. \tag{9.98}$$

Substituting the values for A^{-1} and B and carrying out the multiplication on the right-hand side gives

$$X = A^{-1}B = \begin{bmatrix} \frac{1}{3} & \frac{1}{6} \\ \frac{1}{3} & -\frac{1}{3} \end{bmatrix} \begin{bmatrix} 6 \\ 3 \end{bmatrix} = \begin{bmatrix} 2+\frac{1}{2} \\ 2-1 \end{bmatrix} \begin{bmatrix} \frac{5}{2} \\ 1 \end{bmatrix}. \tag{9.99}$$

This is the solution also achieved by scalar algebra.

This procedure is now discussed for the general case (but these paragraphs may be skipped by the beginning reader). Consider the system of simultaneous linear equations $AX = B$, where

$$A = \begin{bmatrix} a_{11} & \cdots & a_{1n} \\ \vdots & & \vdots \\ a_{m1} & & a_{mn} \end{bmatrix}, \quad B = \begin{bmatrix} b_1 \\ \vdots \\ b_m \end{bmatrix}, \quad \text{and} \quad X = \begin{bmatrix} x_1 \\ \vdots \\ x_n \end{bmatrix}.$$

The matrix of coefficients A and the right-hand side B are known, and the problem is to solve for the unknown matrix X. A solution to this system is a particular vector X_0, such that $AX_0 = B$. The elements of the matrix X_0 simultaneously satisfy the n individual equations,

$$a_{i1}x_1 + \cdots + a_{in}x_n = b_i, \quad \text{for } i = 1, \ldots, m.$$

Any system of equations may have either no solution, exactly one solution, or infinitely many solutions. It is customary to distinguish between a homogeneous system, where $B = 0$, and a nonhomogeneous system, where $B \neq 0$.

If $m < n$ (that is, if the number of unknowns exceeds the number of equations), then a homogeneous system will have infinitely many solutions. A nonhomogeneous system may have either infinitely many solutions or no solution. This is generally the case for a linear program, for example, in which different solutions are successively tested to see which is the best. This is because $n - m$ variables can be arbitrarily specified, and the remaining m variables will have a unique solution if the m equations are linearly independent and thus form a basis for the space.

If $m = n$ and if matrix A is invertible, the homogeneous system will have the unique solution $X = 0$, and the nonhomogeneous system, a unique solution $X \neq 0$. If A is not invertible, then the homogeneous system has infinitely many solutions, and the nonhomogeneous system has no solution.

If $m > n$, then both the homogeneous system and the nonhomogeneous system will have either no solution, or a unique solution. Any set of n linearly independent equations will have a unique solution, but all m equations, in general, will have no solution. In the plane, for instance, two equations specify two lines, and two lines will have a unique intersection if they are not parallel. Only if a third noncollinear line intersects at the same point will the three equations in two unknowns have a unique solution; otherwise, there will be no solution. If two, but not three, of the lines are collinear, then there is a unique solution, unless the third is parallel. If all three are collinear, there are infinitely many solutions; if parallel, there are none.

9.8 Cofactors and the Adjoint

The discussion thus far has concentrated on the (2 × 2) system of equations and the (2 × 2) determinant of coefficients. It is also possible to calculate the value of the determinant of an ($n \times n$) matrix as is now illustrated for the (3 × 3) case. Given a general (3 × 3) matrix as shown in equation (9.100):

$$A \equiv \begin{bmatrix} a_{11} & a_{12} & a_{13} \\ a_{21} & a_{22} & a_{23} \\ a_{31} & a_{32} & a_{33} \end{bmatrix} \tag{9.100}$$

the determinant for the system can be established by the method of arrows. The simplest way of presenting this is first to repeat columns 1 and 2 of A at the right of the matrix A and then to calculate the sums of products as indicated in Figure 9.10. The determinant for this system is expressed as shown in equation (9.101):

$$\begin{vmatrix} a_{11} & a_{12} & a_{13} \\ a_{21} & a_{22} & a_{23} \\ a_{31} & a_{32} & a_{33} \end{vmatrix} = \begin{matrix} (a_{11}a_{22}a_{33} + a_{12}a_{23}a_{31} + a_{13}a_{21}a_{32}) - \\ (a_{13}a_{22}a_{31} + a_{11}a_{23}a_{32} + a_{12}a_{21}a_{33}). \end{matrix} \tag{9.101}$$

Figure 9.10

Cross Products for (3 × 3) Matrix in Accordance with the Method of Arrows.

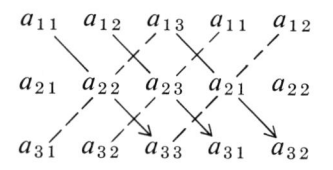

$$\begin{array}{ccccc} a_{11} & a_{12} & a_{13} & a_{11} & a_{12} \\ a_{21} & a_{22} & a_{23} & a_{21} & a_{22} \\ a_{31} & a_{32} & a_{33} & a_{31} & a_{32} \end{array}$$

The cross products formed in accordance with the arrows have been summed. Then the cross products formed in accordance with the dashed arrows have been summed and their sum subtracted from the first sum of cross products.

It is possible to factor out the common elements from the pair of parentheses in equation (9.102):

$$a_{11}(a_{22}a_{33} - a_{23}a_{32}) + a_{12}(a_{23}a_{31} - a_{21}a_{33}) + a_{13}(a_{21}a_{32} - a_{22}a_{31}). \tag{9.102}$$

Each of these factors can then be expressed in terms of a product of an element times a determinant, as illustrated in equation (9.103):

$$a_{11}(a_{22}a_{33} - a_{23}a_{32}) = a_{11} \begin{vmatrix} a_{11} & a_{12} & a_{13} \\ a_{21} & a_{22} & a_{23} \\ a_{31} & a_{32} & a_{33} \end{vmatrix} = a_{11} \begin{vmatrix} a_{22} & a_{23} \\ a_{32} & a_{33} \end{vmatrix}$$

$$a_{12}(a_{23}a_{31} - a_{21}a_{33}) = -a_{12} \begin{vmatrix} a_{11} & a_{12} & a_{13} \\ a_{21} & a_{22} & a_{23} \\ a_{31} & a_{32} & a_{33} \end{vmatrix} = -a_{12} \begin{vmatrix} a_{21} & a_{23} \\ a_{31} & a_{33} \end{vmatrix} \tag{9.103}$$

$$a_{13}(a_{21}a_{32} - a_{22}a_{31}) = a_{13} \begin{vmatrix} a_{11} & a_{12} & a_{13} \\ a_{21} & a_{22} & a_{23} \\ a_{31} & a_{32} & a_{33} \end{vmatrix} = a_{13} \begin{vmatrix} a_{21} & a_{22} \\ a_{31} & a_{32} \end{vmatrix}.$$

Each component of the determinant of the (3 × 3) matrix is the product of an element times a (2 × 2) determinant. The (2 × 2) determinant is made up of elements from the original (3 × 3) matrix less the elements in the row and column for the element that premultiplies the determinant; that is, in the first instance, a_{11} has been multiplied by the determinant formed by deleting row 1 and column 1 from the (3 × 3) matrix as indicated by the shading. Similarly, for a_{12} and a_{13}, the row and column for the element itself has been eliminated, and the resulting (2 × 2) determinant premultiplied by the element. In the case of a_{12}, the factor in parentheses is not the determinant crossproduct, but rather the negative of that crossproduct, which explains the negative sign in front of a_{12}. The (2 × 2) determinants obtained are designated as the *minors* for the original (3 × 3) matrix:

> **Definition—The Minor for an (n × n) Matrix:** Given an (n × n) matrix A, then the determinant of the (n − 1) × (n − 1) matrix obtained by deleting the ith row and the jth column from A is known as the minor *for the element a_{ij}, and is designated by $|N_{ij}|$.*

As with any determinant, the minor is a number. The concept of a minor can be extended further in developing the *cofactor* of a matrix, which is sometimes called the *signed minor* of a particular element of the matrix:

> **Definition—The Cofactor of a Matrix or the Signed Minor of a Matrix:** Given an $(n \times n)$ matrix A, then the scalar which is equal to
> $$C_{ij} = (-1)^{i+j} \cdot |N_{ij}|$$
> is known as the cofactor *or* signed minor *of the element a_{ij} from the matrix A.*

The *cofactor* is also a scalar. The transpose of the matrix of cofactors of a matrix A is also designated in a particular way as the *adjoint* of the matrix A:

> **Definition—The Adjoint of a Matrix:** Given the square matrix A and its cofactors, then the $(n \times n)$ matrix designated
> $$[C_{ij}]^T,$$
> the transpose of the matrix of cofactors, is referred to as the adjoint *of the original matrix A. In symbols, the adjoint of the matrix A is given as adj A or A^{adj}.*

It has just been shown that the determinant of the given matrix can be stated in terms of products of the elements of a given row or column with that of the cofactors of that row and column of the matrix. That is, the product of the elements of the first row of the matrix A multiplied by the corresponding cofactors for that row determines the determinant for the matrix A. This process is known as expansion of cofactors and is expressed in symbols as:

$$|A| = \sum_{j=1}^{n} a_{ij} C_{ij} \qquad (9.104)$$

for any row $i = 1, 2, \ldots, n$; as

$$|A| = a_{11}C_{11} + a_{12}C_{12} + a_{13}C_{13} = \sum_{j=1}^{3} a_{1j}C_{1j} \qquad (9.105)$$

for the first row, or

$$|A| = \sum_{i=1}^{n} a_{ij} C_{ij} \qquad (9.106)$$

for any column $j = 1, 2, \ldots, n$.

It is now possible to express the determinant of any matrix in terms of cofactors. In effect, then, the determinant of a matrix of a given order is being expressed in terms of determinants of matrices of the next lower order.

Note that the sign of the cofactor is determined by the sum of the indexes for the row and column for the given cofactor. If the sum is even, the cofactor is positive; if it is odd, the cofactor is negative. Thus for the cofactor C_{11}, the sum of the rows and columns is equal to $1 + 1 = 2$; therefore this factor is positive. For C_{12}, however, the sum of the indexes for the rows and columns is equal to 3, thus implying a cofactor with negative sign.

Theorem 9.2: *Given a square matrix A, whose determinant is not zero, then the inverse of A is expressed by:*

$$A^{-1} = \frac{1}{|A|} \cdot \text{adj } A. \qquad (9.107)$$

The theorem is not proved, but it is illustrated in Example 9.7. With the earlier matrix [from Example 9.2], it is possible to find the determinant of this matrix, confirm that it is not zero, and then by developing the adjoint of the matrix, determine its inverse.

Example 9.7

$$|E| \equiv \begin{vmatrix} 4 & 0 & 2 \\ 2 & 10 & 2 \\ 3 & 9 & 1 \end{vmatrix}.$$

Expanding by the first row to find $|E|$ gives

$$|E| = 4 \begin{vmatrix} 10 & 2 \\ 9 & 1 \end{vmatrix} + 0 \begin{vmatrix} 2 & 2 \\ 3 & 1 \end{vmatrix} + 2 \begin{vmatrix} 2 & 10 \\ 3 & 9 \end{vmatrix}$$

$$= 4(10 - 18) + 0(2 - 6) + 2(18 - 30) = -56.$$

The system determinant is not zero. The adjoint of E is

$$\begin{array}{lll} C_{11} = +(-8) & C_{21} = -(-18) & C_{31} = +(-20) \\ C_{12} = -(-4) & C_{22} = +(-2) & C_{32} = -(4) \\ C_{13} = +(-12) & C_{23} = -(36) & C_{33} = +(40). \end{array}$$

Thus

$$|C_{ij}| = \begin{vmatrix} -8 & 4 & -12 \\ 18 & -2 & -36 \\ -20 & -4 & 40 \end{vmatrix}$$

$$\text{adj } A = |C_{ij}|^T = \begin{vmatrix} -8 & 18 & -20 \\ 4 & -2 & -4 \\ -12 & -36 & 40 \end{vmatrix}$$

therefore

$$A^{-1} = \frac{1}{|A|} \text{adj } A$$

$$= \frac{1}{-56} \begin{bmatrix} -8 & 18 & -20 \\ 4 & -2 & -4 \\ -12 & -36 & 40 \end{bmatrix} = \begin{bmatrix} \frac{1}{7} & -\frac{9}{28} & \frac{5}{14} \\ -\frac{1}{14} & +\frac{1}{28} & \frac{1}{14} \\ \frac{3}{14} & \frac{9}{14} & -\frac{5}{7} \end{bmatrix}.$$

This means of calculation of the inverse produces the result identical to that achieved in Example 9.2 earlier in the chapter. This means of matrix inversion makes use of cofactors and the adjoint. It should be noted that in this example the expansion of the determinant in developing the cofactors was in terms of the first row in which a 0 existed. The selection of the row for expanding a determinant is a matter of artful choice. Generally the row or column with the largest number of 0's is the one to be selected, so that several of the terms are eliminated. The second most desirable criterion is the number of 1's in a row or column. It should also be noted that in the calculation of the determinant itself, the values of three (n) cofactors were also simultaneously determined, further simplifying the task of calculating the adjoint.

9.9 Summary

Among the concepts discussed in this chapter are the concepts of a field, a vector, a matrix, and a vector space. Vector and matrix operations in an n-dimensional vector space are particularly emphasized. The *basis* for a vector space is a set of linearly independent vectors, from which any other vector in the space can be generated. The maximum number of such vectors (which is also the minimum number for a basis) determines the *dimension* of the space. The operations of addition and subtraction of two vectors are achieved by the algebraic summing of components of the respective vectors. In graphic terms this involves the placing of the base of one vector at the head of a second to determine the *resultant* of the two vectors, that vector connecting the base of the second to the head of the first. Multiplication of a vector by a scalar implies the multiplication of each component of the vector by the scalar.

Addition and subtraction of matrices can be done only for matrices of the same order; that is, the same number of rows and columns. This is so since these operations require the algebraic summing of the corresponding individual elements of each matrix. A matrix may also be multiplied by a scalar by multiplying each element of the matrix by that scalar. Multiplication of matrices generally leads to different results depending upon whether one matrix premultiplies a second or postmultiplies it. The operation of multiplication is defined only for matrices for which the number of columns in the premultiplying matrix is the same as the number of rows in the postmultiplying matrix. The elements of the product matrix are each the sum of the products of the elements in a row of the premultiplying matrix with the corresponding elements in a column of the postmultiplying matrix.

Several special matrices are introduced. The *zero matrix* is a square matrix all of whose elements are zero. Pre- or postmultiplication by the zero matrix results in the zero matrix. The *identity matrix* is a special case of a *diagonal matrix*. A diagonal matrix is a square matrix A such that its diagonal elements, elements a_{ii}, are not zero, while all other elements are. The *identity matrix* is a diagonal matrix with diagonal elements a_{ii} equal to one, $a_{ii} = 1$. Pre- or postmultiplication of a matrix B by the identity matrix results in the matrix B. A diagonal matrix is, in turn, a special case of a symmetric matrix, a square matrix A such that $a_{ij} = a_{ji}$.

Matrix division, an especially important operation, is achieved through matrix inversion plus matrix multiplication. To be *invertible*, a matrix must be square. An *invertible matrix* is known as a *nonsingular matrix* and its determinant is nonzero. Several different approaches to the task of matrix inversion (through the use of simultaneous equations and scalar algebra, of row and column matrix operations, and finally of determinants, cofactors, and the adjoint) are introduced.

Three methods in addition to scalar algebra are described and illustrated for the solution of simultaneous linear equations. Each is based on finding the inverse of the matrix of coefficients of the system of equations. The first requires the use of elementary row and/or column operations. The second and third methods make use of determinants and are called, respectively, Cramer's rule (in two forms), and the classical adjoint method. An ($n \times n$) matrix is shown to be invertible if its determinant is different from zero. When the determinant is zero, then at least two columns (or rows) of the matrix are linearly dependent, that is, its rank is less than n.

Systems of linear equations derive their importance in part from their relation to linear programming. The chapter shows explicitly that procedures introduced as mechanical operations in Chapter 8 on linear programming are, in fact, matrix operations. The typical linear program can be stated in matrix form as

$$\begin{array}{ll} \text{Max} & c'X \\ \text{Subject to} & AX \leq b \\ & X \geq 0. \end{array}$$

For A, an ($m \times n$) matrix, the simplex method of solution proceeds to find an ($m \times m$) matrix B of linearly independent columns chosen from the columns of A as augmented by m columns for coefficients of slack variables. A as augmented becomes an $[m \times (m + n)]$ matrix and X an $[(m + n) \times 1]$ vector. Then by matrix partitioning, $AX = b$ would be rewritten

$$[BN]\begin{bmatrix} X_B \\ X_N \end{bmatrix} = b.$$

Carrying out the multiplication,

$$BX_B + NX_N = b,$$

where X_B is the matrix of basis variables and X_N the matrix of nonbasis variables, and B stands for the matrix of coefficients for basis variables and N for nonbasis ones. The first basis B chosen is the one most easily identified, that including only the slack variables. Since the columns of B are necessarily linearly independent, B^{-1} exists and

$$B^{-1}BX_B + B^{-1}NX_N = B^{-1}b$$

or simplifying and transposing $B^{-1}NX_N$,

$$X_B = B^{-1}b - B^{-1}NX_N.$$

Since X_N is zero by definition, n variables may be chosen to be zero so that the basic solution is $X_B = B^{-1}b$.

The simplex method then proceeds by finding among all basic solutions that one (or one of those) which maximizes the objective function and at the same time is feasible.

Key Concept

Concept	Reference Section
Field	9.1
Vector	9.1
Row and Column Vectors	9.1
Components of a Vector	9.1
Scalar	9.1
Unit Vector	9.1
Directed Line Segment	9.1
Resultant	9.1
Vector Addition, Subtraction	9.1
Scalar Multiplication	9.1
Collinearity	9.1
Vector Space	9.2
Linear Independence, Dependence	9.2
Orthogonal Unit Vectors	9.2
Basis	9.2
Dimension of a Vector Space	9.2
Matrix	9.3
Coefficient Matrix	9.3
Matrix Order	9.3
Matrix Addition, Subtraction	9.3
Scalar Multiplication	9.3
Matrix Multiplication	9.3
Zero Matrix	9.4
Identity Matrix	9.4
Diagonal Matrix	9.4
Symmetric Matrix	9.4
Transpose of a Matrix	9.4
Inverse Matrix	9.5
Singular and Nonsingular Matrix	9.5
Rank of a Matrix	9.5
Elementary Row and Column Operations	9.6
Elementary Matrix	9.6
Linear Program in Matrix Terminology	9.6
Equivalent Matrices	9.6
Determinants	9.7
Cramer's Rule	9.7
Inverting a Matrix through Determinants	9.7
Minor	9.8
Signed Minor	9.8
Cofactor	9.8
Adjoint	9.8
Homogeneous and Nonhomogeneous Linear Equation Systems	9.8
Determinant Evaluation by Method of Arrows	9.8

Problems

1. Show formally that the vextors **x, y, z** [100], [010], [001] are linearly independent.

2. The vectors **x, y, z,** and **ω** are three-dimensional vectors. Could they be linearly independent? Why or why not?

3. In a certain linear program solution

$$\mathbf{x} = \begin{bmatrix} 1 \\ 0 \\ 0 \end{bmatrix}, \quad \mathbf{y} = \begin{bmatrix} 1 \\ 1 \\ 0 \end{bmatrix}, \quad \text{and} \quad \mathbf{z} = \begin{bmatrix} 0 \\ 1 \\ -1 \end{bmatrix}.$$

Are **x, y,** and **z** dependent, or linearly independent?

4. The vectors $\mathbf{x} = [x_1, x_2, x_3]$, $\mathbf{y} = [y_1, y_2, y_3]$, $\mathbf{z} = [z_1, z_2, z_3]$ are such that $\mathbf{z} = \mathbf{x} + \mathbf{y}$. Show that they are linearly dependent.

5. Show that zero (the zero vector) is linearly dependent upon any vector.

6. Let $\mathbf{x} = [1, 0, 0]$, $\mathbf{y} = [0, 1, 0]$, $\mathbf{z} = [1, 1, 0]$, and $\boldsymbol{\omega} = [0, 1, 1]$. (a) Express **z** as a linear combination of **x** and **y**. (b) Show that **ω** cannot be expressed as a linear combination of **x, y,** and **z**. Explain why.

7. (a) Multiply 3**x** where $\mathbf{x} = \begin{bmatrix} \frac{2}{3} \\ -\frac{5}{6} \\ \frac{23}{27} \end{bmatrix}$.

   (b) Take the product of $\mathbf{x} = \begin{bmatrix} 1 \\ 0 \\ -1 \end{bmatrix}$ and $[3, 16, -4] = \mathbf{y}$.

   (c) Multiply $A = \begin{bmatrix} 1 & 1 \\ 0 & 2 \end{bmatrix}$ and $B = \begin{bmatrix} 2 & 0 \\ 1 & 2 \end{bmatrix}$ as $A \cdot B$ and as $B \cdot A$, calculate the difference of the two products $C = A \cdot B - B \cdot A$. This is called the *commutator* by mathematicians.

   (d) Add $A = \begin{bmatrix} 1 & 0 & -1 \\ 2 & 0 & 2 \\ 1 & 1 & 1 \end{bmatrix}$ and $B = \begin{bmatrix} 1 & 0 & 0 \\ 0 & 1 & 1 \\ 1 & -1 & -1 \end{bmatrix}$.

8. $A = \begin{bmatrix} 2 & 1 \\ 0 & -2 \end{bmatrix}$, $B = \begin{bmatrix} 1 & 2 & 3 \\ 3 & 2 & 1 \end{bmatrix}$. Find (a) $A + A^T$ and (b) $B + B^T$. (c) If A is square, then is $A + A^T$ symmetric? If so, why?

9. Show that, given the matrix A, then λA, where λ is any scalar, is equivalent to $\lambda I A$, which is equivalent to $A \lambda I = A \lambda$.

10. $\mathbf{x} = [1 \ \ 2]$, $\mathbf{y} = \begin{pmatrix} 2 \\ 3 \end{pmatrix}$. Find $\mathbf{x} \cdot \mathbf{y}$ and $\mathbf{y} \cdot \mathbf{x}$.

11. $\mathbf{x} = [1 \ \ 2 \ \ 3]$. (a) Find $\mathbf{x}\mathbf{x}^T$ and (b) $\mathbf{x}^T\mathbf{x}$.

12. Show that the x axis is a vector space over the field of real numbers.

13. Show that

$$\det \begin{bmatrix} a & tb \\ c & td \end{bmatrix} = t \det \begin{bmatrix} a & b \\ c & d \end{bmatrix}$$

t a scalar.

14. Show that

$$\det \begin{bmatrix} a + tb, & b \\ c + td, & d \end{bmatrix} = \det \begin{bmatrix} a & b \\ c & d \end{bmatrix}.$$

15. (a) Compute

$$\det \begin{bmatrix} 1 & 2 \\ 3 & 4 \end{bmatrix} \quad \text{and} \quad \det \begin{bmatrix} 2 & 1 \\ 4 & 3 \end{bmatrix}.$$

(b) Compute

$$\det \begin{bmatrix} a & b \\ c & d \end{bmatrix} \quad \text{and} \quad \det \begin{bmatrix} b & a \\ d & c \end{bmatrix}$$

and interpret the result.

16. The determinant of any matrix may be evaluated by the method of minors along any row or column. It is best to choose the row judiciously. Try for that column or row with the most zeros; among columns or rows with equal numbers of zeros, pick that one with the most 1's. For determinants of 2×2 and 3×3 matrices, the "method of arrows" works.

(a) Evaluate the following determinants by the method of arrows.

$$\begin{vmatrix} -1 & 1 \\ 1 & -1 \end{vmatrix}, \quad \begin{vmatrix} 1 & -7 \\ 4 & 2 \end{vmatrix}, \quad \begin{vmatrix} 1 & 0 & 1 \\ 0 & 1 & 0 \\ 1 & 0 & 1 \end{vmatrix}, \quad \begin{vmatrix} 1 & 0 & -1 \\ 0 & -1 & 0 \\ -1 & 0 & 1 \end{vmatrix}$$

(b) Evaluate

$$\begin{vmatrix} 7 & 16 & 27 \\ 4 & 0 & -9 \\ -2 & 0 & 24 \end{vmatrix}, \quad \begin{vmatrix} 2 & 3 & 3 \\ 3 & 2 & 2 \\ 3 & 3 & 2 \end{vmatrix}, \quad \text{and} \quad \begin{vmatrix} 1 & -2 & 0 \\ 1 & 1 & 0 \\ 0 & 0 & 1 \end{vmatrix}$$

both by arrows and along a column.

(c) The determinant of

$$\begin{bmatrix} 1 & 2 & 2 \\ 1 & 0 & 0 \\ 1 & -1 & -1 \end{bmatrix}$$

is zero, why?

(d) That of

$$\begin{bmatrix} 1 & 3 & 2 \\ 1 & 0 & 0 \\ 1 & -\frac{3}{2} & -1 \end{bmatrix}$$

is also zero, why?

(e) Evaluate the determinant of

$$\begin{bmatrix} 1 & 3 & 6 & 4 \\ 2 & 9 & 1 & -101 \\ 0 & -4 & 0 & 2 \\ 0 & 2 & 13 & 0 \end{bmatrix}$$

along the first column. Why is this the best choice?

(f) *More Advanced Part:* Try using the arrow method. Your answer will be wrong. Why?

(*Hint:* For an $n \times n$ matrix the determinant is defined as follows: The *determinant* is the *sum* of *all possible products* of elements of A in each of which there is *one and only one* element from each row and column, with each such product multiplied by $+1$ or -1 according to the following rule.

Join all pairs of elements in adjacent columns by line segments, count the number of such line segments with positive slope. If this number is *zero* or *even*, multiply the product by $+1$; if it is *odd*, multiply the product by -1.)

Two by two case as example:

$$\det \begin{bmatrix} a_{11} & a_{12} \\ a_{21} & a_{22} \end{bmatrix} = (\pm 1)a_{11} \cdot a_{22} + (\pm 1)a_{12}a_{21}.$$

In $a_{11}a_{22}$

$$\begin{matrix} a_{11} & \\ & \searrow \\ & a_{22} \end{matrix},$$

thus no element with positive slope. We assign $+1$ as a multiplier. In $a_{12}a_{21}$

$$\begin{matrix} & a_{12} \\ & \nearrow \\ a_{21} & \end{matrix},$$

thus one element with positive slope. We assign -1 as a multiplier. Hence,

$$\det \begin{bmatrix} a_{11} & a_{12} \\ a_{21} & a_{22} \end{bmatrix} = a_{11}a_{22} - a_{12}a_{21}.$$

17. (a) Show that

$$\begin{bmatrix} 1 & 0 & 0 \\ -1 & 27 & 1 \\ 2 & 3 & -32 \end{bmatrix} \quad \text{and} \quad \begin{bmatrix} 1 & -1 & 2 \\ 0 & 27 & 3 \\ 0 & 1 & -32 \end{bmatrix}$$

have the same determinant explicitly. Can you see why this could be anticipated?

(b) Show that

$$\begin{vmatrix} 1 & 2 & 1 \\ 1 & 3 & 2 \\ 1 & -1 & 0 \end{vmatrix} = \begin{vmatrix} 1 & 2+0 & 1 \\ 1 & 2+1 & 2 \\ 1 & -2+1 & 0 \end{vmatrix} = \begin{vmatrix} 1 & 2 & 1 \\ 1 & 2 & 2 \\ 1 & -2 & 0 \end{vmatrix} + \begin{vmatrix} 1 & 0 & 1 \\ 1 & 1 & 2 \\ 1 & 1 & 0 \end{vmatrix}$$

explicitly. Can you deduce anything general about

$$\begin{vmatrix} a_{11} & (\alpha_{12}+\beta_{12}) & a_{13} \\ a_{21} & (\alpha_{22}+\beta_{22}) & a_{23} \\ a_{31} & (\alpha_{32}+\beta_{32}) & a_{33} \end{vmatrix}$$

from this example? Can you prove your deduction?

(c) Why does

$$\det \begin{bmatrix} 1 & 0 \\ 0 & 1 \end{bmatrix} = -\det \begin{bmatrix} 0 & 1 \\ 1 & 0 \end{bmatrix}?$$

And why does

$$\det \begin{bmatrix} 1 & 0 & 1 \\ 0 & 1 & 1 \\ 0 & 0 & 1 \end{bmatrix} = -\det \begin{bmatrix} 1 & 1 & 0 \\ 0 & 1 & 1 \\ 0 & 1 & 0 \end{bmatrix}?$$

Why does

$$\det \begin{bmatrix} 1 & 0 & 1 \\ 0 & 1 & 1 \\ 0 & 0 & 1 \end{bmatrix} = \det \begin{bmatrix} 1 & 1 & 0 \\ 1 & 0 & 1 \\ 1 & 0 & 0 \end{bmatrix}?$$

Evaluate all these determinants to see that the claims in the questions are true.

18. Given

$$A = \begin{bmatrix} 1 & 3 & 5 \\ 0 & 4 & 6 \\ 2 & 1 & 7 \end{bmatrix},$$

find the cofactors of each element.

(a) Construct the matrix A' gotten from A by replacing each element with its cofactor.

(b) Calculate $(A')^T$, this is A^{adj}.

(c) Calculate A^T and the cofactors of each of its elements, construct $(A^T)'$, the matrix gotten by replacing each element of A^T by its cofactor. Verify that this is also A^{adj}.

(d) If $\det [A] \neq 0$, A^{-1} exists and $A^{-1} = A^{\text{adj}}/\det [A]$. Calculate A^{-1}.

(e) Use A^{-1} to solve the set of linear equations

$$\begin{aligned} x + 3y + 5z &= 1 \\ 0 + 4y + 6z &= 0 \\ 2x + y + 7z &= 0 \end{aligned}$$

by taking
$$\begin{bmatrix} x \\ y \\ z \end{bmatrix} = [A^{-1}] \begin{bmatrix} 1 \\ 0 \\ 0 \end{bmatrix}.$$

19. (a) $A = \begin{bmatrix} 1 & -1 \\ 1 & 0 \end{bmatrix}$ $B = \begin{bmatrix} 1 & 3 \\ 2 & 1 \end{bmatrix}$. If $C = A \cdot B$, calculate C and det $[C]$ directly. Also calculate det $[A]$ and det $[B]$, verify that det $[C]$ = det $[A] \cdot$ det $[B]$.

(b) If $C = A \cdot B$ has elements $c_{ij} = \sum\limits_{k=1}^{n} a_{ik}b_{kj}$, then use this fact to prove that $C^T = B^T \cdot A^T$.

(c) Calculate the adjoints and inverses of the matrices in part (a). Thus, verify that if $C = A \cdot B$
$$C^{\mathrm{adj}} = B^{\mathrm{adj}} \cdot A^{\mathrm{adj}}$$
$$C^{-1} = B^{-1} \cdot A^{-1}$$
by multiplication and direct calculation of C^{adj} and C^{-1}. This is a general property if both det $[A]$ and det $[B] \neq 0$.

20. Given the set of equations
$$x + 3y + 5z = 1$$
$$7x + 2y + 4z = 0$$
$$3x + 9y + z = -1$$
determine that these equations have a unique solution (x, y, z) and apply Cramer's rule to solve them. You should use the "method of arrows" to evaluate the determinants involved. Otherwise the method becomes tedious. For systems of more than three equations, the method of arrows fails, and this is the major reason that other methods are preferred for computer solution of systems of equations in many unknowns.

As an illustration, the student may calculate the determinants along the first row after doing the above.

21. Given the set of equations
$$\begin{bmatrix} 1 & -1 & 0 \\ 1 & -3 & 2 \\ 4 & 3 & 2 \end{bmatrix} \begin{bmatrix} x \\ y \\ z \end{bmatrix} = \begin{bmatrix} 1 \\ 1 \\ -1 \end{bmatrix}, \quad A\mathbf{x} = \mathbf{y}$$
verify their solvability. Then use the row and column operations method to obtain the solution. Then verify the value of $\mathbf{x}$ by Cramer's rule. (Note that when solving for $\mathbf{x}$ the determinant of the numerator differs from det A only in a_{31}.)

22. Show that
$$\begin{bmatrix} 1 & 3 & 0 \\ 1 & 2 & 1 \\ 1 & 4 & -1 \end{bmatrix} \begin{bmatrix} x \\ y \\ z \end{bmatrix} = \begin{bmatrix} -\frac{1}{2} \\ -1 \\ 0 \end{bmatrix}$$
cannot be solved fully. What is the rank of the matrix?

23. Let $A = \begin{bmatrix} 0 & 1 \\ 1 & 0 \end{bmatrix}$, $B = \begin{bmatrix} 1 & 1 \\ 1 & 0 \end{bmatrix}$, and $C = \begin{bmatrix} 0 & 1 \\ 1 & 1 \end{bmatrix}$.

(a) Find AB and $(AB)^{-1}$.
(b) Find A^{-1} and B^{-1}.
(c) Show that $(AB)^{-1} = B^{-1}A^{-1}$.
(d) Find ABC and $(ABC)^{-1}$.
(e) Find C^{-1}.
(f) Show that $(ABC)^{-1} = C^{-1}B^{-1}A^{-1}$.

Remark: The results of (c) and (f) are general.

24. Find a 2×2 matrix $A \neq 0$ such that $A^2 = 0$.

25. $A = \begin{bmatrix} 1 & 2 & 3 \\ 4 & 5 & 6 \end{bmatrix}, \quad B = \begin{bmatrix} 7 \\ 8 \end{bmatrix}$

(a) Find A^T and B^T.
(b) Find $A^T B$.
(c) Find $B^T A$.
(d) Find $A^T B^T$.
(e) Find AB.
(f) Find AA^T.
(g) Find BB^T and $B^T B$.

26. Let $\mathbf{x} = (a, b)$, and $\mathbf{y} = (c, d)$.

(a) Show that $\mathbf{x}$ and $\mathbf{y}$ are linearly dependent if $ad - bc = 0$.
(b) Show that they are linearly independent if $ad - bc \neq 0$. (This is an important result for solving systems of linear equations and for determinants.)

27. $A = \begin{bmatrix} 1 & 1 \\ 1 & 1 \end{bmatrix}, \quad B = \begin{bmatrix} 0 & 1 \\ 1 & 0 \end{bmatrix}.$

(a) Find A^2, A^3, A^4, and A^6.
(b) What is the generalized expression for A^n?
(c) Find B^2, B^3, B^4, B^5, and B^6.
(d) What is the generalized expression for B^n?
(e) Find $(B^n)^{-1}$.

28. Show that if $A \neq 0$, $B \neq 0$ and $A \cdot B = 0$, then neither A nor B has an inverse.

INDEX

Addition
   of matrices, 314
   of vectors, 307
Adjoint and cofactors of vectors, 335–39
   minor, 336–37
   signed minor, 337
Algebraic solution to linear programming
   basic solution of, 268–69
   feasible solution of, 268–69
   optimal solution of, 268, 270
   slack variables of, 267–68, 271
Annuity, present value of, 252
Approximation, numerical, integration by
   Simpson's rule, 242
   Taylor series, 242
   trapezoidal rule, 242–44
Argument of function, 17
Average rate of change, 61, 63–64, 221

Basic solution in algebraic solution to linear programming, 268–69
Basis of vector spaces, 311–12
Boundary points, 19
Bounded sets, 21–22

Calculus, fundamental theorem of, 221–29
Carrying costs, 100–101
Cartesian product set, 13–16
Chain rule, 76–77
Change, rate of, 60–61, 63–64, 221
Closed half-space, 19–22
Coefficients, matrices of, 313
Cofactors and the adjoint of vectors, 335–39
   minor, 336–37
   signed minor, 337
Collinear vectors, 306

Column and row operations with vectors, 321–27
   elementary, 321–23
   equivalent matrices and, 326
   linear programming in, 323–26
Column vectors, 305
Complement set, 10
Computation, logarithmic functions in, 206–7
Concave function, 82–83
   downward, 83
Constant
   derivative of, 70–71
   limit of, 49–50
Constant times value of function
   derivative of, 71–72
   limit of, 51
Constrained maxima and minima
   multiple constraints, 139–44
   one constraint, 133–39
Continuity
   of functions, limits and, 52–54
   See also Limits
Convergence sequences, $42n$
Convergent integral, 246
Convex-point-set, 22–23, 82
Convex sets, 22–23
Cost policy, minimum, 101–2
   objective function in, 102
Costs
   carrying, 100–101
   marginal, 108–11
   order, 101
Counting
   by one-to-one correspondence, 11–13
   by pairs, 13–15
Cramer's rule, 329

347

Critical points, 79–81
  extreme points as, 81
  relative maximum (minimum) as, 80
Cross partial derivatives, 120
Curve
  slope of, 82
  tangent to, 62–65, 82–84

Decision making, management, and linear programming, 259–88
  algebraic solution in, 267–71
    basic solution, 268–69
    feasible solution, 268–69
    optimal solution, 268, 270
    slack variables, 267–68, 271
  dual program in, 275–81
    dual to the dual, 277
    rules, 280–81
    shadow prices, 279
  feasibility variables in, 288
  graphic solution, feasible regions and extreme points in, 263–68
  parametric programming in, 286–87
  sensitivity analysis in, 260–61, 281–86
    degenerate solution, 284–86
  set-up of, 262–63
  simplex method in, 268–75
    pivot element, 272–75
    simplex, 270–71
    simplex tableau, 271–75
Definite integral, 222
Degenerate solution in sensitivity analysis, 284–86
Derivative functions, 59–66
  applications of, in inventory control, 100–111
    cost policy, minimum, 101–2
    marginal costs and revenues, 108–11
    model, 102–5
    profit maximizing, 105–8
    reorder quantity, optimal, 100–101
  critical points of, 79–81
    extreme points, 81
    relative maximum (minimum), 80
  differentiation, 66–69
    differential, 67–69
    See also Integration
  of exponential functions, 176–78
  general rules for, 70–78, 92–95
    chain rule, 76–77
    constant, 70–71
    constant times function, 71–72
    examples, 73–75, 77–78
    polynomial, 72–73
    proofs, 92–95
    sum of two (or more) functions, 72
  higher order, 78–79
  inflection points of, 87–89
  of inverse functions, 169–70
  of logarithmic functions, rules for, 183–90
  maxima and minima of, 82–87
  partial differentiation of, 115–53
    implicit differentiation, 131–32
    Lagrange multipliers, 144–46
    maxima and minima, 123–30, 133–44
    problems, 160–64
    procedure, 119–23
    summary, 153–55
  problems on, 95–98, 112–14
  summary of, 89–91, 111–12
Determinants, 327–35
  Cramer's rule, 329
  inversion through, 332–34
Diagonal of matrix, 317
Differentiable function, 65
Differential, 67–69
Differentiation, 66–69
  differential, 67–69
  implicit, 131–32
  integration and, 215–21
    fundamental theorem, 221–29
  partial, 115–53
    implicit, 131–32
    Lagrange multipliers, 144–46
    maxima and minima, 123–30, 133–44
    problems, 160–64
    procedure, 119–23
    summary, 153–55
Dimension of vector spaces, 312–13
Disjoint sets, 7–8
Distinct, definition of, 3–4, 9
Divergent integral, 246
Domain, 15–17
Downward concave function, 83
Dual to the dual in dual program of linear programming, 277
Dual program of linear programming, 275–81
  dual to the dual in, 277
  rules of, 280–81
  shadow prices in, 279

INDEX

e (irrational number), 173–76
  calculation of, 208
  table of values for, 253
Elasticity, logarithmic, 190, 194–201
  logarithmic derivative, 197–98
Elementary row and column operations with vectors, 321–23
Elements of sets, 2, 4–5
Empty (null) set, 5–6
Equality sets, 5
Equations and functions, 26–28
Equivalent matrices, 326
Exponential functions, 165–208
  derivative functions of, 176–78
  inverse functions of, 166–70
  irrational number e and, 173–76
    calculation, 208
  logarithmic functions and, 165–66, 178–83
    applications, 190–201
    computation, 206–7
    derivatives, rules for, 183–90
    problems on, 208–11
    summary of, 201–4
Extreme points, 22–23, 81
  graphic solutions to linear programming and, 263–68

Feasibility variables, 288
Feasible regions and graphic solutions to linear programming, 263–68
Feasible solution in algebraic solution to linear programming, 268–69
Field, vector, 304–9
Finite sets, 12–13
"Flop-over points," 88
Functions
  argument of, 17
  concave, 83
  constant times value of derivative, 71–72
    limit, 51
  continuity of, limits and, 52–54
  derivative, 59–66
    applications, in inventory control, 100–111
    critical points, 79–81
    differentiation, 66–69
    exponential, 176–78
  equations and, 26–28
  exponential, 165–208
    derivative functions, 176–78
    inverse functions, 166–70
    irrational number e, 173–76, 208
    logarithmic functions, 165–66, 178–83, 183–204, 206–11
    problems, 208–11
    summary, 201–4
  general rules, 70–78, 92–95
    higher order, 78–79
    inflection points, 87–89
    inverse functions, 169–70
    logarithmic, rules for, 183–90
    maxima and minima, 82–97
    partial differentiation, 115–55, 160–64
    problems, 95–98, 112–14
    summary, 89–91, 111–12
  integral, 223–24
  inverse, 166–70
  limits of, 43–54
    continuity, 52–54
    examples, 46–49
  logarithmic, 165–208
    applications, 190–201
    computation, 206–7
    derivatives, rules for, 183–90
    exponential, 170–73, 176–78, 204–5
    inverse, 166–70
    irrational number e, 173–76, 208
    problems, 208–11
    summary, 201–4
  primitive, 224–25, 227
  relations and, 18–19
  set, 15–19
  sum of, 50, 72
  value of, 17
Fundamental theorem of calculus, 221–29
Future value of sum, 251

Graphic solution to linear programming, feasible regions and extreme points in, 263–68

Half space, 19–22
Higher order derivatives, 78–79

Identity matrix, 316–17
Image in set functions, 17
Implicit differentiation, 131–32
Improper integrals, 244–46
Infinite sets, 12–13
Inflection points, 87–89

Instantaneous rate of change, 60–61,
    63–64, 221
Integrals
    convergent, 246
    definite, 222
    divergent, 246
    improper, 244–46
    mean for, first law of, 225–26
    negative of, 224
Integration, 213–52
    applications of, 238–42, 251–52
        future value of sum, 251
        present value of annuity, 252
        present value of future sum, 251–52
    convergent integral, 246
    definite integral, 222
    differentiation and, 215–21
        fundamental theorem, 221–29
    improper integrals, 244–46
    integral function, 223–24
    mean for integrals, first law of, 225–26
    multiple, 247–48
    negative of integral, 224
    by numerical approximation, 242–44
        Simpson's rule, 242
        Taylor series, 242
        trapezoidal rule, 242–44
    by parts, 236–38
    primitive function, 224–25
        value of, at any given point, 227
    problems on, 253–58
    rules for, 229–34, 248
    by substitution, 234–36
    summary of, 249–51
Interior points, 19
Intersection, 9–10
Inventory control
    cost policy, minimum, in, 101–2
        objective function, 102
    marginal costs and revenue in, 108–11
    model of, 102–5
    profit maximizing in, 105–8
    reorder quantity, optimal, in, 100–101
        carrying costs, 100–101
        order costs, 101
Inverse functions, 166–70
Inversion through determinants, 332–34
Inversion of vectors, 318–30
    nonsingular, 319
    rank and, 322

Irrational number e, 173–76
    calculation of, 208
    table of values of, 253

Kuhn-Tucker conditions, 152

Lagrange multipliers, 144–46
Left-hand limit, 45
Limits, 35–54
    of constant, 49–50
    of functions (general), 43–46, 50
        continuity, 52–54
        examples, 46–49
        problems on, 55–58
    of sequences, 37–40
        examples, 40–43
        summary of, 54–55
        theorems on, 49–51
Line segments, 21–22
Linear dependence (independence) in
    vector spaces, 310–11
Linear programming
    algebraic solution in, 267–71
        basic solution, 268–69
        feasible solution, 268–69
        optimal solution, 268, 270
        slack variables, 267–71
    dual program in, 275–81
        dual to the dual, 277
        rules, 280–81
        shadow prices, 279
    feasibility variables in, 288
    graphic solution, feasible regions and
        extreme points in, 263–68
    management decision making and,
        259–88
    parametric programming in, 286–87
    problems in, 290–302
    sensitivity analysis in, 260–61, 281–86
        degenerate solution, 284–86
    set-up of, 262–63
    simplex method in, 268–75
        pivot element, 272–75
        simplex, 270–71
        simplex tableau, 271–75
    summary of, 289–90
    vectors and, 323–26
Linear regression, 158–60, 190–94
Logarithmic functions, 165–208
    applications of, 190–201

INDEX

elasticity, 190, 194–201
  linear regression, 190–94
in computation, 206–7
derivatives of, rules for, 183–90
exponential functions and, 170–73, 204–5
  derivative functions, 176–78
inverse functions and, 166–70
irrational number *e* and, 173–76
  calculation, 208
problems on, 208–11
summary of, 201–4

Management decision making, and linear programming, 259–88
  algebraic solution in, 267–71
    basic solution, 268–69
    feasible solution, 268–69
    optimal solution, 268, 270
    slack variables, 267–68, 271
  dual program in, 275–81
    dual to the dual, 277
    rules, 280–81
    shadow prices, 279
  feasibility variables in, 288
  graphic solution, feasible regions and extreme points in, 263–68
  parametric programming in, 286–87
  sensitivity analysis in, 260–61, 281–86
    degenerate solution, 284–86
  set-up of, 262–63
  simplex method in, 268–75
    pivot element, 272–75
    simplex, 270–71
    simplex tableau, 271–75
Mapping, 15–16
Marginal costs and revenue, 108–11
Matrices, 313–16
  addition (subtraction) of, 314
  of coefficients, 313
  multiplication of, 314–16
  scalar multiplication and, 314
  special, 316–18
    diagonal, 317
    identity, 316–17
    symmetric, 317
    transpose, 318
    zero, 316–17
Maxima and minima, 80, 82–87
  partial differentiation and, 123–25
    constrained, 133–44

sufficient conditions, 125–30
See also Derivative functions
Mean for integrals, first law of, 225–26
Minima, *see* Maxima and minima
Minimum cost policy, 101–2
  objective function, 102
Minor, 336–37
Multiple integration, 247–48
Multiplication
  Lagrange multipliers, 144–46
  of matrices, 314–16
  scalar
    matrices, 314
    vectors, 305–6
Mutually exclusive sets, 7–8, 9

Negative of integral, 224
Neighborhood of point, 19
Noncountably infinite sets, 12–13
Nonnegativities in linear programming, 262–63
Nonsingular inversion of vectors, 319
Null (empty) set, 5–6
Number sets, 26
Numerical approximation, integration by
  Simpson's rule, 242
  Taylor, series, 242
  trapezoidal rule, 242–44

Objective function
  in linear programming, 262, 268, 277–78
  in minimum cost policy, 101–2
One-to-one correspondence, 11–15
  in inverse functions, 168–69
Open half space, 19–22
Open point set, 19–22
Operations, set, 9–10
  complement, 10
  difference, 10
  intersection, 9–10
  union, 9
Optimal solution in algebraic solution to linear programming, 268, 270
Order costs, 101
Ordered sets, 3–4
Orthogonal unit spaces, 309–10

Pairs, 13–15
Parametric programming, 286–87

Partial differentiation, 115–53
  implicit differentiation and, 131–32
  Lagrange multipliers and, 144–46
  maxima and minima and, 123–30, 133–44
    constrained, 133–44
    sufficient conditions, 125–30
  problems in, 160–64
  procedure for, 119–23
  summary of, 153–55
Partitions, 7–8
Parts, integration by, 236–38
Pivot element in simplex method of linear programming, 272–75
Point sets, 19–22
  convex, 22–23, 82
Points
  boundary, 19
  "flop-over," 88
  inflection, 87–89
  interior, 19
  neighborhood of, 19
  saddle, 126–28
Polynomials, derivative of, 72–73
Population (universal) set, 6
Power sets, 8–9
Present value of a future sum, 251–52
Primal program, 261
Primitive function, 224–25
  value of, at any point, 227
Problems
  on derivatives, maxima and minima, 95–98
    application, 112–14
  on exponential and logarithmic functions, 208–11
  on integration, 253–58
  on limits and continuity, 55–58
  on linear programming, 290–302
  on partial differentiation, 160–64
  on sets, 28–34
  on vectors and matrices, 342–46
Product of two (or more) functions, limit of, 50–51
Production function, 155–58
Program, primal, 261
Programming, linear
  algebraic solution in, 267–71
    basic solution, 268–69
    feasible solution, 268–69
    optimal solution, 268, 270
    slack variables, 267–71
  dual program in, 275–81
    dual to the dual, 277
    rules, 280–81
    shadow prices, 279
  feasibility variables in, 288
  graphic solution, feasible regions and extreme points in, 263–68
  management decision making and, 259–88
  parametric programming in, 286–87
  problems in, 290–302
  sensitivity analysis in, 260–61, 281–86
    degenerate solution, 284–86
  set-up of, 262–63
  simplex method in, 268–75
    pivot element, 272–75
    simplex, 270–71
    simplex tableau, 271–75
  summary of, 289–90
  vectors and, 323–26
Programming, parametric, 286–87
Proofs for derivative rules, 92–95
Proper subsets, 7

Quotient of two functions, limit of, 51

Range, 15–17
Rank and inversion of vectors, 322
Rate of change, instantaneous and average, 60–61, 63–64, 221
Regression, linear, 158–60, 190–94
Relations, 13–14, 16
  functions and, 18–19
Relative maxima and minima, 80
Reorder quantity, optimal
  carrying costs, 100–101
  order costs, 101
Resource constraints in linear programming, 262
Revenue, marginal, 108–11
Right-hand limit, 44
Row and column operations of vectors, 321–27
  elementary, 321–23
  equivalent matrices and, 326
  linear programming in, 323–26
Row vectors, 305

Saddle point, 126–28
Scalar multiplication
  of matrices, 314
  of vectors, 305–6
Secant, 62–63
  slope of, 63
Segments, line, 21–22
Sensitivity analysis, 260–61, 281–86
  degenerate solution, 284–86
Sequences, limits of, 37–40
  examples of, 40–43
Set functions, 15–19
Sets, 1
  concept of, 2–5
    definition, 2–3
    elements, 2, 4–5
    equality, 5
    ordered, 3–4
  counting, 10–15
    one-to-one correspondence, 11–13
    pairs, 13–15
  functions, 15–19
    relations and, 18–19
    set, 17–18
  operations, 9–10
    complement, 10
    difference, 10
    intersection, 9–10
    union, 9
  point sets, 19–22
    convex, 22–23, 82
  problems on, 28–34
  special, 5–9
    empty (null), 5–6
    partitions, 7–8
    power, 8–9
    subsets, 6–7
    universe (population), 6
  summary of, 23–25
Shadow prices, 279, 261
Signed minor, 337
Simplex method in linear programming, 268–75
  pivot element and, 272–75
  simplex and, 270–71
  simplex tableau and, 271–75
Simplex in simplex method of linear programming, 270–71
Simplex tableau in simplex method of linear programming, 271–75

Simpson's rule, 242
Slack variables in algebraic solution to linear programming, 267–68, 271
Slope
  of curve, rate of change of, 82
  of secant, 63
Spaces, vector, 309–13
  basis of, 311–12
  dimension of, 312–13
  linear dependence (independence), 310–11
  orthogonal unit spaces, 309–10
Special sets
  empty (null), 5–6
  partitions, 7–8
  power, 8–9
  subsets, 6–7
  universe (population), 6
Straight partial derivatives, 120
Subsets, 6–7
  proper, 7
Substitution, integration by, 234–42
Subtraction
  of matrices, 314
  of vectors, 307–8
Sufficient conditions, 125–30
Sum
  future value of, 251
  present value of future, 251–52
Sum of two (or more) functions
  derivative of, 72
  limit of, 50
Summary(ies)
  of derivative, maxima and minima, 89–91
    applications, 111–12
  of exponential and logarithmic functions, 201–4
  of integration, 249–51
  of limits and continuity, 54–55
  of linear programming, 289–90
  of partial differentiation, 153–55
  of sets, 23–25
  of vectors and matrices, 339–41
Symmetric matrix, 317

Tangent to curve, 62–65, 82–84
Taylor series, 242
Test limit value, 44–46

Theorems
   fundamental, 221–29
   on limits, 49–51
Transpose of matrix, 318
Trapezoidal rule, 242–44

Unbounded sets, 21–22
Union, 9
Unit spaces, orthogonal, 209–10
Universe (population) set, 6

Value of function, 17
Variables, feasibility, 288
Vectors
   addition of, 307
   cofactors and adjoint, 335–39
      minor, 336–37
      signed minor, 337
   collinear, 306
   column, 305
   determinants of, 327–35
      Cramer's rule, 329
      inversion through, 332–34
   field, 304–9
   inversion of, 318–30
      nonsingular, 319
      rank, 322
   matrices and, 313–18
      addition (subtraction), 314
      coefficients, 313
      diagonal, 317
      identity, 316–17
      multiplication, 314–16
      scalar multiplication, 314
      symmetric, 317
      transpose, 318
      zero, 316–17
   problems on, 342–46
   row and column operations of, 321–27
      elementary, 321–23
      equivalent matrices, 326
      linear programming, 323–26
   row vector, 305
   scalar multiplication and, 305–6
   spaces, 309–13
      basis, 311–12
      dimension, 312–13
      linear dependence (independence), 310–11
      orthogonal unit spaces, 309–10
   subtraction, 307–8
   summary of, 339–41

Well defined, definition of, 2–3

X-element technique, 25

Zero matrix, 316–17